Selected Titles in This Subseries

46 **Vladimir Arnold, Maxim Kontsevich, and Anton Zorich, Editors,** Pseudoperiodic topology (TRANS2/197)

45 **Ya. Eliashberg, D. Fuchs, T. Ratiu, and A. Weinstein, Editors,** Northern California symplectic geometry seminar (TRANS2/196)

44 **Alexander Astashkevich and Serge Tabachnikov, Editors,** Differential Topology, Infinite-Dimensional Lie Algebras, and Applications (D. B. Fuchs' 60th Anniversary Collection) (TRANS2/194)

43 **A. Yu. Morozov and M. A. Olshanetsky, Editors,** Moscow Seminar in Mathematical Physics (TRANS2/191)

42 **S. Tabachnikov, Editor,** Differential and Symplectic Topology of Knots and Curves (TRANS2/190)

41 **V. Buslaev, M. Solomyak, and D. Yafaev, Editors,** Differential Operators and Spectral Theory (M. Sh. Birman's 70th anniversary collection) (TRANS2/189)

40 **M. V. Karasev, Editor,** Coherent Transform, Quantization, and Poisson Geometry (TRANS2/187)

39 **A. Khovanskiĭ, A. Varchenko, and V. Vassiliev, Editors,** Geometry of Differential Equations (TRANS2/186)

38 **B. Feigin and V. Vassiliev, Editors,** Topics in Quantum Groups and Finite-Type Invariants (Mathematics at the Independent University of Moscow) (TRANS2/185)

37 **Peter Kuchment and Vladimir Lin, Editors,** Voronezh Winter Mathematical Schools (Dedicated to Selim Krein) (TRANS2/184)

36 **V. E. Zakharov, Editor,** Nonlinear Waves and Weak Turbulence (TRANS2/182)

35 **G. I. Olshanski, Editor,** Kirillov's Seminar on Representation Theory (TRANS2/181)

34 **A. Khovanskiĭ, A. Varchenko, and V. Vassiliev, Editors,** Topics in Singularity Theory (TRANS2/180)

33 **V. M. Buchstaber and S. P. Novikov, Editors,** Solitons, Geometry, and Topology: On the Crossroad (TRANS2/179)

32 **R. L. Dobrushin, R. A. Minlos, M. A. Shubin, and A. M. Vershik, Editors,** Topics in Statistical and Theoretical Physics (F. A. Berezin Memorial Volume) (TRANS2/177)

31 **R. L. Dobrushin, R. A. Minlos, M. A. Shubin, and A. M. Vershik, Editors,** Contemporary Mathematical Physics (F. A. Berezin Memorial Volume) (TRANS2/175)

30 **A. A. Bolibruch, A. S. Merkur′ev, and N. Yu. Netsvetaev, Editors,** Mathematics in St. Petersburg (TRANS2/174)

29 **V. Kharlamov, A. Korchagin, G. Polotovskiĭ, and O. Viro, Editors,** Topology of Real Algebraic Varieties and Related Topics (TRANS2/173)

28 **L. A. Bunimovich, B. M. Gurevich, and Ya. B. Pesin, Editors,** Sinai's Moscow Seminar on Dynamical Systems (TRANS2/171)

27 **S. P. Novikov, Editor,** Topics in Topology and Mathematical Physics (TRANS2/170)

26 **S. G. Gindikin and E. B. Vinberg, Editors,** Lie Groups and Lie Algebras: E. B. Dynkin's Seminar (TRANS2/169)

25 **V. V. Kozlov, Editor,** Dynamical Systems in Classical Mechanics (TRANS2/168)

24 **V. V. Lychagin, Editor,** The Interplay between Differential Geometry and Differential Equations (TRANS2/167)

23 **Yu. Ilyashenko and S. Yakovenko, Editors,** Concerning the Hilbert 16th Problem (TRANS2/165)

22 **N. N. Uraltseva, Editor,** Nonlinear Evolution Equations (TRANS2/164)

(*Continued in the back of this publication*)

Pseudoperiodic Topology

American Mathematical Society

TRANSLATIONS

Series 2 • Volume 197

Advances in the Mathematical Sciences — 46

(Formerly Advances in Soviet Mathematics)

Pseudoperiodic Topology

Vladimir Arnold
Maxim Kontsevich
Anton Zorich
Editors

American Mathematical Society
Providence, Rhode Island

2000 *Mathematics Subject Classification.* Primary 57Rxx, 58Exx, 37Axx, 37Cxx.

ABSTRACT. This book is an account of the present state of the art in pseudoperiodic topology, which is a young branch of mathematics, born at the boundary between the ergodic theory of dynamical systems, topology, and number theory, and related also to the theory of algorithms, convex integer polyhedra, Morse inequalities, real algebraic geometry, statistical physics, and algebraic number theory.

The book will be useful for researchers and graduate students working in dynamical systems, topology, and number theory.

Library of Congress Card Number 91-640741
ISBN 0-8218-2094-X
ISSN 0065-9290

∞ The paper used in this book is acid-free and falls within the guidelines
established to ensure permanence and durability.
Visit the AMS home page at URL: http://www.ams.org/

10 9 8 7 6 5 4 3 2 1 04 03 02 01 00 99

Contents

Preface

A *pseudoperiodic function* is the sum of a linear function and a periodic function. A *pseudoperiodic variety* is the set of common zeros of several pseudoperiodic functions.

An example of a pseudoperiodic curve is provided by the intersection of a periodic surface in three-space with a plane (which is virtually incommensurable with the periods lattice). Such curves occur in solid state physics (where the role of the periodic surface is played by the Fermi surface and the direction of the plane is defined by the magnetic field).

The present book is an account of the present state of the art in pseudoperiodic topology, which is a young branch of mathematics, born at the boundary between the ergodic theory of dynamical systems, topology, and number theory, and related also to algorithm theory, convex integer polyhedra, Morse inequalities, real algebraic geometry, statistical physics, and algebraic number theory. The authors represent the Moscow mathematical school.

The beginning of pseudoperiodic topology goes back to the classical works of Henri Poincaré on vector fields on tori (continued by A. Denjoy, A. Weil, C. L. Siegel, T. Cherry, and others). Following the ideas of Poincaré, A. N. Kolmogorov [**1**] proved in 1953 that the level lines of a closed one-form on the 2-torus, having no critical points, can be straightened by a suitable diffeomorphism of the torus (such that the form on the covering plane becomes the differential of a linear function).

This was the starting point of his celebrated works on invariant tori in Hamiltonian systems, now included in KAM-theory.

The study of the topology of level varieties of closed (but not exact) 1-forms is evidently related to pseudoperiodic topology.

From another point of view, such a form defines a *pseudo-Hamiltonian dynamical system* on a surface (the Hamilton function being replaced by a closed 1-form).

Studying the ergodic properties of such pseudo-Hamiltonian flows on surfaces of genus greater than one, I was led in 1963 to the model of *interval exchange* with its slow mixing. (This simple model was introduced in [**2**, Chapter 6, Section 2 ("Unsolved problems")]; see Russian Math. Surveys **18** (1963), p. 178, lines 15–18. Strangely, it seems that it had not been considered in earlier works on dynamical systems).

Studying the asymptotics of ergodic mean stabilization in this model in computer experiments (as well as in the original problem on pseudo-Hamiltonian flows), A. Zorich has discovered some striking phenomena.

Experiments have shown that some asymptotic characteristics of the memory loss in this system are universal: they are the same for almost all lengths of intervals, depending only on their number and on the order in which the intervals are permuted. This is one of few cases where an important new mathematical phenomenon has been discovered by computer experimentation.

The corresponding conjectures and theorems (due to A. Zorich and M. Kontsevich, and based on the study of the geodesic flows on Teichmüller spaces) are described below in Zorich's article, "How do the leaves of a closed 1-form wind around a surface?"

The problem on the intersection curve of a plane with a periodic surface in 3-space led S. P. Novikov (around 1981) to the remarkable conjecture that every infinite connected component of such a curve lives in a strip of finite width around a straight line of the plane.

The astonishing feature of this conjecture is the appearance of this line, which has no simple description in terms of the data of the problem and which is rather sensitive to a change in these data. Long attempts by Novikov and his students (especially S. Tsarev and I. Dynnikov) to prove this conjecture have led to the following striking answer: the conjecture is true "almost always", but in some "exceptional" cases there are counterexamples.

This situation is typical for pseudoperiodic topology: the Diophantine properties of the plane with respect to the period lattice are crucial, and the answer to this problem (as well as in the ergodic theory of continued fractions and the metric theory of Diophantine approximations) is universal not for all pseudoperiodic objects, but for almost all of them (in the Lebesgue measure sense).

This is also true in the problem of the ergodic properties of pseudo-Hamiltonian flows with singular points (which has been studied by Ya. G. Sinai and K. M. Khanin)—see the survey "Pseudoperiodic mappings" by D. A. Panov in the present book and my 1992 survey [**3**]. It seems that the situation is similar in many other problems of pseudoperiodic topology.

Perhaps this is the case for the other important conjecture by S. P. Novikov (discussed in A. Pajitnov's article "C^0-generic properties of boundary operators in the Novikov complex"), the conjecture on the exponential growth of the number of instantons (of the trajectories of the gradient of a pseudoperiodic function, on the manifold where the cyclic group acts, connecting one critical point to far-translated versions of the other).

This conjecture (which is still unproved, even for "generic" functions) has triggered a series of papers by several authors, including my theory on the asymptotics of intersections in dynamical systems (described in three papers, dedicated to the 60th birthdays of J. Moser, of S. Smale and of J. Milnor; see [**4, 5**] and the paper quoted by Pajitnov as [**2**]).

However, in this theory also, the crucial questions are open. This is, in particular, true for the conjecture on the typically exponential grows rate of the number of periodic orbits with the period. The problem on the possibility of superexponential growth in analytic systems is still open (despite the spectacular result by O. Kozlovski on the intersections growth rate).

The Gauss–Wiman–Kuzmin statistics of the elements of the continuous fraction expansion of a random irrational number (controlling the pseudoperiodic topology of the systems on a 2-torus) led me in 1988–1993 to a series of conjectures on "higher-dimensional continued fractions". Such a fraction is neither an algorithm nor a set of Diophantine approximations, but rather a certain infinite polyhedral hypersurface with integer vertices defined by a simplicial cones. It is the boundary of the convex hull of the semigroup of interior integer points of the cone.

The problems that I had then formulated included, among others, the generalization of the Lagrange theorem on periodic continued tractions (proved later by E. Korkina) as well as the computation of the statistics of different asymptotic characteristics of these polyhedra for generic simplicial cones. For instance, I suggested studying the mean number of vertices on a face, the distribution of the number of integer points on the edges and on the faces, and so on.

In formulating all these problems (published later in [**6**]), I had not realized that the very *existence* of the required universal statistics needs to be proved, not just the form of the statistics.

The existence proof is given in the remarkable article by M. Kontsevich and Yu. Sukhov in this book. It is notable that their article leaves open the most interesting problems on the character of the universal statistics whose existence they prove. We still do not know, say, whether the integer length of an edge of a random simplicial cone is shifted toward the longer or toward the shorter intervals as compared to the integer length of a random interval, whether there are more triangular than quadrangular faces, nor whether there are more faces of small integer area than of those with larger area (which, I hope, is true).

There is no proof of the coincidence of these universal statistics with the asymptotic averages obtained from simplicial cones defined by integral planes with bounded coefficients or defined by the eigenplanes of bounded integer matrices. There is no information on the asymptotic statistics of the triangulations of tori, corresponding to the "periodic" higher-dimensional continued fractions (even in the case of the 2-torus associated to cubical algebraic numbers fields).

One might regard the article of S. Gusein-Zade in this book as the first step in the pseudoperiodic version of Newton polyhedra theory and in real algebraic geometry. This domain (providing the "Morse inequality per volume unit" and "pseudoperiodic Harnak inequalities") will undoubtedly have important further development (relating it also to the statistics of quasicrystalls, as explained in [**7**]).

The theory of pseudoperiodic functions and varieties appears unexpectedly in many domains of mathematics. For instance, Hilbert [**8**] motivated his celebrated problem on smoothness in Lie group theory by the existence of analytic pseudoperiodic equations whose solutions are not smooth. Today we would call his equation the homological equation in the pseudoperiodic theory of circle rotation:

$$f(x+\alpha) - f(x) = g(x),$$

where g is the given and f the unknown 2π-periodic function (the mean value of g being equal to zero).

This equation has an analytic solution for almost every α. However, for some rotation angles α (incommensurable with 2π) the solutions of this equation (with an analytic right hand side g) are only finitely smooth (and sometimes even discontinuous).

This difficulty is typical for all the problems of pseudoperiodic topology. For almost all values of the parameters (in the sense of Lebesgue measure) the problems have natural and nice solutions. However, for some exceptional values of the parameter (forming everywhere dense sets) many kinds of pathology occur.

The authors of this book have done much to show how modern mathematics begets, from this sea of pathological counterexamples, remarkable general and universal laws, whose discovery would be unthinkable and whose formulation would be impossible in the naive set-theoretical setting.

V. I. Arnold

References

[1] A. N. Kolmogorov, *On dynamical systems with integral invariant on a torus*, Dokl. Akad. Nauk SSSR **93** (1953), 763–766; English transl. in his *Selected papers. Vol. 1: Mathematics and mechanics*, Kluwer, Dordrecht, 1991, pp. 344–348.

[2] V. I. Arnold, *Small denominators and problems of stability of motion in classical and celestial mechanics*, Uspekhi Mat. Nauk **18** (1963), no. 6, 91–192; English transl. in Russian Math. Surveys **18** (1963), no. 6.

[3] ———, *On some problems of pseudoperiodic topology*, Mat. Prosv. (3) **1997**, no. 1, 10–23. (Russian)

[4] ——— *Dynamics of complexity of intersections*, Bol. Soc. Brasil. Mat. (N.S.) **21** (1990), no. 1, 1–10.

[5] ——— *Bounds for Milnor numbers of intersections in holomorphic dynamical systems*, Topological Methods in Modern Mathematics (J. Milnor 60th Birthday Synpos., Stony Brook, NY, 1991; L. R. Goldberg and A. V. Phillips, editors), Publish or Perish, Houston, TX, 1993, pp. 379–390.

[6] ——— *Higher dimensional continued fractions*, Regular and Chaotic Dynamics **3** (1998), no. 3, 10–17.

[7] ——— *Remarks on quasicrystallic symmetries*, Phys. D **33** (1988), 21–25.

[8] D. Hilbert, *Mathematical problems*, Mathematical Developments Arising from Hilbert Problems, Proc. Sympos. Pure Math., vol. 28, Part 1, Amer. Math. Soc., Providence, RI, 1976, pp. 1–34.

Amer. Math. Soc. Transl.
(2) Vol. **197**, 1999

On the Topology of Quasiperiodic Functions

S. M. Gusein-Zade

ABSTRACT. We discuss the topology of level sets $V_c = \{f = c\}$ and of sets of lower values $M_c = \{f \leqslant c\}$ of a quasiperiodic function $f : \mathbb{R}^k \to \mathbb{R}$. Topological invariants of these sets are usually infinite. One can consider the density of a topological invariant, i.e., the limit of the value of the invariant for the intersection of M_c or V_c with a ball of large radius divided by the volume of the ball. We describe properties of some invariants as functions on c, and state some conjectures.

We discuss some properties which hold for a *generic* or for *almost any* function. Here "generic" means a function from a dense open subset. "Almost any function" means any function except one from a set of infinite codimension in the space of all functions. For a smooth (i.e., C^∞) function f on a manifold W, let V_c be its level set $\{x \in W : f(x) = c\}$ and let M_c be its set of lower values $\{x \in W : f(x) \leqslant c\}$ $(c \in \mathbb{R})$.

A generic smooth function f on a compact manifold W is a Morse function, i.e., it has only nondegenerate critical points with different critical values. The topology of the level set V_c and of the set of lower values M_c changes (in a well-known way) only when c passes through a critical value of f, and thus topological invariants of the spaces V_c and M_c (say, Betti numbers or the Euler characteristic) are piecewise constant functions on c. The number of critical points of f (say, with fixed Morse index) that belong to the space M_c has the same behavior.

It is not hard to see that the same property holds for almost any smooth function f on a compact manifold W: the topology of the sets V_c and M_c changes only when c passes through one of a finite number of critical values of f. In particular, their Betti numbers and the Euler characteristic are piecewise constant functions on c (with finite sets of discontinuity). Critical points of almost any smooth function are not, in general, nondegenerate. However, they are isolated, and both the total number of critical points of f in M_c and the number of such nondegenerate points with fixed Morse index possess this behavior.

If the manifold W is real analytic, the same property holds for any analytic function f on it.

A periodic (with k independent periods) smooth function f on the k-dimensional affine space $\mathbb{R}^k$ can be regarded as a smooth function $\widetilde{f}$ on the corresponding

1991 *Mathematics Subject Classification.* Primary 58C05, 58F05.
Partially supported by grants RFBR-98-01-00612 and INTAS-96-0713.

k-dimensional torus $\mathbb{T}^k$. Thus, again, for almost any smooth periodic function f, the topology of the space V_c or M_c is a piecewise constant function on c and changes only when c passes through a critical value of f (and there are only finitely many such critical values).

Topological invariants of the spaces V_c and M_c (such as Betti numbers or the Euler characteristic) and the number of critical points of f which belong to M_c are, generally speaking, infinite. In order to have a finite characteristic, one must consider the density of distribution of critical points of the function f. Let $B_r = \{x \in \mathbb{R}^k : \|x\| \leqslant r\}$ be the ball of radius r with center at the origin in the space $\mathbb{R}^k$. Let $\#_{mr}(c)$ be the number of nondegenerate critical points of the function f that have Morse index m and belong to the intersection $M_c \cap B_r$. Then the density of critical points (with Morse index m and critical values $\leqslant c$) is

$$\#_m(c) = \lim_{r\to\infty} \frac{\#_{mr}(c)}{V(B_r)},$$

where $V(B_r)$ is the volume of the ball B_r. Of course, this is just the number of corresponding critical points of the function $\widetilde{f}$ on the torus $\mathbb{T}^k$ divided by its volume (in the natural sense).

Let $b_{mr}(c)$ (respectively $\widetilde{b}_{mr}(c)$) be the mth Betti number of the intersection $M_c \cap B_r$ (respectively of $V_c \cap B_r$). Let

$$b_m(c) = \lim_{r\to\infty} \frac{b_{mr}(c)}{V(B_r)}, \qquad \widetilde{b}_m(c) = \lim_{r\to\infty} \frac{\widetilde{b}_{mr}(c)}{V(B_r)}.$$

One can prove that, for almost any smooth (and for any real analytic) periodic function f, these limits exist, and the functions $b_m(c)$ and $\widetilde{b}_m(c)$ are piecewise constant and change only when c passes through a critical value of f.

Now let f be a smooth quasiperiodic function on the affine space $\mathbb{R}^k$. This means that $f = \widetilde{f} \circ \pi \circ \varphi$, where φ is a nondegenerate linear map $\mathbb{R}^k \to \mathbb{R}^n$ $(n > k)$ such that the image of its composition $\pi \circ \varphi \colon \mathbb{R}^k \to \mathbb{T}^n$ with the factorization $\pi \colon \mathbb{R}^n \to \mathbb{T}^n = \mathbb{R}^n/\mathbb{Z}^n$ is dense in the torus $\mathbb{T}^n$, and where $\widetilde{f}$ is a smooth function on $\mathbb{T}^n$. We shall say that a quasiperiodic function $f = \widetilde{f} \circ \pi \circ \varphi$ is analytic if the corresponding function $\widetilde{f}$ on the torus $\mathbb{T}^n$ is analytic. The words "for almost any" or "for a generic" quasiperiodic function f mean "for almost any" or "for a generic" function $\widetilde{f} \colon \mathbb{T}^n \to \mathbb{R}$ defining the function f.

One can consider the density of critical points of the function f with critical values $\leqslant c$. However, in this case the set of critical values of f is not discrete. It is "spread" all over the real line, and for almost any quasiperiodic function its closure is the union of a finite number of segments. Thus the density of critical points belonging to the subspace M_c (if well defined) cannot be piecewise constant. One of the aims of the paper is to describe its behavior as a function on c.

As above, let B_r be the closed ball of radius r with center at the origin in the space $\mathbb{R}^k$, $\#_{mr}(c)$ the number of critical points of the function f that have Morse index m and belong to the intersection $M_c \cap B_r$, and

$$\begin{aligned} b_{mr}(c) &= \operatorname{rk} H_m(M_c \cap B_r; \mathbb{R}), \\ \widetilde{b}_{mr}(c) &= \operatorname{rk} H_m(V_c \cap B_r; \mathbb{R}), \\ \chi_r(c) &= \chi(M_c \cap B_r), \\ \widetilde{\chi}_r(c) &= \chi(V_c \cap B_r). \end{aligned}$$

For each of these values, let the same notation without the subscript r denote its density, defined as

$$\bullet(c) = \lim_{r\to\infty} \frac{\bullet_r(c)}{V(B_r)}$$

(if this limit exists; $\bullet = \#_m, b_m, \widetilde{b}_m, \chi$, or $\widetilde{\chi}$; e.g., $\chi(c) = \lim_{r\to\infty}(\chi_r(c)/V(B_r))$).

THEOREM 1. *For any analytic (and for almost any smooth) quasiperiodic function f and for any $c \in \mathbb{R}$, the density $\#_m(c)$ of the number of critical points of the function f that have Morse index m and belong to the subspace M_c (i.e., the corresponding limit) exists.*

PROOF. The statement follows from an explicit formula for $\#_m(c)$ (Lemma 5 below). Let L be the ($(n-k)$-dimensional) quotient space $\mathbb{R}^n/\operatorname{Im}\varphi$ of the linear space $\mathbb{R}^n$ by the image of the linear map $\varphi\colon \mathbb{R}^k \to \mathbb{R}^n$, and let $p\colon \mathbb{R}^n \to L$ be the corresponding projection. Let dV_k and dV_n be the standard volume forms on the spaces $\mathbb{R}^k$ and $\mathbb{R}^n$ ($dV_\ell = dx^1 \wedge \cdots \wedge dx^\ell$ for the standard coordinates $x^1, \dots, x^\ell$ in $\mathbb{R}^\ell$). The natural volume form dV_{n-k} on the space L is defined by the equation $dV_k = \varphi^*(dV_n/p^*dV_{n-k})$. This definition means the following. A (linear) section $s\colon L \to \mathbb{R}^n$ of the projection $p\colon \mathbb{R}^n \to L$ ($p \circ s = \mathrm{id}_L$) defines a splitting of the linear space $\mathbb{R}^n$ into the direct sum of the spaces $\varphi(\mathbb{R}^k)$ and L. Then, for any such splitting, the volume (in $\mathbb{R}^n$) of the direct product of two (measurable) subsets in the spaces $\mathbb{R}^k$ and L is equal to the product of their volumes. The volume form dV_{n-k} defines an orientation of the linear space L. The projection $p\colon \mathbb{R}^n \to L$ defines the $(n-k)$-form p^*dV_{n-k} on the torus $\mathbb{T}^n$ (since, for each point $x \in \mathbb{T}^n$, there is a natural isomorphism $T_x\mathbb{T}^n \cong \mathbb{R}^n$).

In what follows submanifolds are usually nonclosed. A subset M of a manifold W is a nonclosed submanifold in W if, for each point $x \in M$, M is an (ordinary, i.e., closed) submanifold of a neighborhood of x. For example, the set of nonsingular points of an analytic subset in $\mathbb{R}^n$ is a nonclosed submanifold of $\mathbb{R}^n$. Whenever we discuss submanifolds in the ordinary sense (which means "for each point $x \in W$" instead of "for each point $x \in M$" above), this will be explicitly emphasized.

Let M be an $(n-k)$-dimensional submanifold of the torus $\mathbb{T}^n$ such that, at each point $x \in M$, the (natural) projection of the tangent space $T_xM \subset T_x\mathbb{T}^n = \mathbb{R}^n$ to the space L is regular (i.e., $T_xM \cap \varphi(\mathbb{R}^k) = \{0\}$). The projection to L defines an orientation of M.

DEFINITION 2. The *transversal volume* $\widetilde{V}(M)$ of the submanifold M is the integral over M of the form p^*dV_{n-k}.

REMARK 3. The transversal volume of a submanifold is always positive (it may be equal to $+\infty$).

Let N_m be the set of points x of the torus $\mathbb{T}^n$ such that the restriction of the differential

$$df = \sum \frac{\partial f}{\partial x^i}\, dx^i$$

to the subspace $\varphi(\mathbb{R}^k) \subset T_x\mathbb{T}^n \cong \mathbb{R}^n$ is equal to zero and the restriction of the second differential

$$d^2 f = \sum \frac{\partial^2 f}{\partial x^i \partial x^j}\, dx^i\, dx^j$$

to the subspace $\varphi(\mathbb{R}^k)$ is a nondegenerate quadratic form with index (the number of negative squares in its normal form) equal to m.

LEMMA 4. *The subset N_m is a (generally speaking, nonclosed) smooth submanifold in the torus $\mathbb{T}^n$.*

PROOF. Let x be a point of N_m, and choose a splitting of the linear space $\mathbb{R}^n$ into the direct sum of the spaces $\varphi(\mathbb{R}^k)$ and L. With some (say, linear) coordinates on L this splitting defines local coordinates $y^1, \ldots, y^k, y^{k+1}, \ldots, y^n$ on a neighborhood U of x in $\mathbb{T}^n$ such that, for any point $x_0 = (y_0^1, \ldots, y_0^n) \in U$, we have

$$x_0 + \varphi(\mathbb{R}^k) = \{(y^1, \ldots, y^n) \in U : y^i = y_0^i \text{ for } i = k+1, \ldots, n\}.$$

Consider the map $d_k f\colon U \to \mathbb{R}^k$ which sends a point $x_0 \in U$ to

$$\left(\frac{\partial f}{\partial y^1}(x_0), \ldots, \frac{\partial f}{\partial y^k}(x_0)\right) \in \mathbb{R}^k.$$

Since the restriction of the second differential $d^2 f$ to the subspace $\varphi(\mathbb{R}^k) \subset T_x\mathbb{T}^n$ is nondegenerate, we have

$$\det\left(\frac{\partial^2 f}{\partial x^i \partial x^j}(x)\right)_{i,j=1,\ldots,k} \neq 0.$$

This means that the map $d_k f$ is regular at the point x (i.e., its differential is surjective), and thus $(d_k f)^{-1}(0) = N_m \cap U$ is a submanifold in a neighborhood of x. The quadratic form

$$\left(\frac{\partial^2 f}{\partial x^i \partial x^j}(x)\right)_{i,j=1,\ldots,k}$$

is nondegenerate in a neighborhood of x, and its index is locally constant. This implies the lemma. □

LEMMA 5. *For any analytic (and for almost any C^∞) quasiperiodic function f, one has*

$$\#_m(c) = \widetilde{V}(N_m \cap \{x \in \mathbb{T}^n : \widetilde{f}(x) \leqslant c\}). \tag{1}$$

PROOF. This follows directly from the following fact. Let $\mathbb{T}$ be an $(n-k)$-dimensional subtorus of $\mathbb{T}^k$ (e.g., a coordinate one) transversal to the subspace $\varphi(\mathbb{R}^k)$. Let us consider intersection points of the torus $\mathbb{T}$ with the image $\pi \circ \varphi(B_r)$ of the ball B_r. Then, as $r \to \infty$, the distribution of these points in the torus $\mathbb{T}$ tends to the uniform one. □

This completes the proof of Theorem 1. □

It is easy to see that for almost any quasiperiodic function f (smooth or analytic), the density of all critical points coincides with the density of nondegenerate ones.

The proof of Theorem 1 allows us to make a stronger statement.

THEOREM 6. *For any analytic (and for almost any C^∞) quasiperiodic function $f = \widetilde{f} \circ \pi \circ \varphi$, $\varphi\colon \mathbb{R}^k \to \mathbb{R}^n$, $\widetilde{f}\colon \mathbb{T}^n \to \mathbb{R}$, the function $\#_m(c)$ is piecewise analytic (respectively piecewise smooth). At a singular point c_0 (i.e., at a point where it is* *neither analytic nor smooth), from any side (i.e., for $c \to c_0+0$ or $c \to c_0-0$) the function $\#_m(c)$ has an asymptotic expansion of the form*

$$\sum_{\alpha}\sum_{q=0}^{n-k-1} a_{q,\alpha}|c-c_0|^{\alpha}(\ln|c-c_0|)^q, \tag{2}$$

where α belongs to a finite set of arithmetic progressions of rational numbers with difference equal to one. For almost any function f (analytic or C^∞), $\#_m(c)$ is a continuous function of c.

PROOF. For an analytic function f, the closure $\overline{N}_m$ of the set N_m is semianalytic and thus has a semianalytic Whitney stratification. For all $c \in \mathbb{R}$ except a finite number of values, the level sets $\{\widetilde{f}=c\}$ of the function $\widetilde{f}$ are transversal to all its strata [**3**]. If the level set $\{\widetilde{f}=c_0\}$ is transversal to all strata of the set $\overline{N}_m$ (from a Whitney regular stratification of it), then the function $\#_m(c)$ (given by the formula (1)) is analytic in a neighborhood of c_0.

The statement about the asymptotic expansion of the function $\#_m(c)$ follows from general properties of the Gelfand–Leray function (see [**1**, Chapter 7]) applied to a resolution of the variety $N=\{x\in\mathbb{T}^n : d\widetilde{f}|_{\varphi(\mathbb{R}^k)}=0\}$ with the differential form p^*dV_{n-k} on it.

The function $\#_m(c)$ is monotone increasing. It has a nonzero jump at a point c_0 if and only if there is a component of the submanifold N_m on which $f\equiv c_0$. This is a situation of infinite codimension.

This implies the statement of Theorem 1 for analytic functions. The statement for smooth functions follows from the fact that almost any smooth function at each point is locally R-equivalent to an analytic one (even to a polynomial). □

THEOREM 7. *For almost any smooth and for any analytic quasiperiodic function f one has*

$$\chi(c)=\sum_{m\geqslant 0}(-1)^m\#_m(c).$$

PROOF. For a closed (i.e., compact without boundary) manifold W, its Euler characteristic is

$$\chi(W)=\sum_{m\geqslant 0}(-1)^m\#_m,$$

where $\#_m$ is the number of critical points with Morse index m for a Morse function on W.

For a smooth function f on a compact manifold W with boundary ∂W, a critical point is either an ordinary critical point inside $W\setminus\partial W$, or a critical point of its restriction $f_{|\partial W}$ to the boundary ∂W. A critical point x of the latter type (i.e., $x\in\partial W$) is nondegenerate if it is nondegenerate as a critical point of the restriction $f_{|\partial W}$ and the gradient $df(x)$ of f at this point is not zero. A generic function f on a manifold W with boundary is Morse, i.e., it has only nondegenerate critical points with different critical values. In this case the Euler characteristic $\chi(W)$ of the manifold W equals

$$\chi(W)=\sum_{m\geqslant 0}(-1)^m[\#_m+\partial\#_{m-}],$$

where $\partial\#_{m-}$ is the number of critical points x of the function $f_{|\partial W}$ with Morse index equal to m, for which the value of the gradient $df(x)$ of f on the outpointing normal vector at the point x is negative.

This means that for a quasiperiodic function $f\colon \mathbb{R}^k \to \mathbb{R}$ the difference between

$$\chi_r(c) = \chi(M_c \cap B_r) \quad \text{and} \quad \sum_{m\geqslant 0}(-1)^m \#_{mr}(c)$$

is bounded by the number $\partial\#_r$ of critical points of the restriction of f to the sphere $S_r = \partial B_r$. One can show that for almost any smooth or for any analytic quasiperiodic function f,

$$\lim_{r\to\infty} \frac{\partial\#_r}{r^k} = 0.$$

This implies the statement. □

COROLLARY 1. *The "density of the Euler characteristic" $\chi(c)$ as a function on c has the same behavior as that described for the density of critical points $\#_m(c)$ in Theorems 1 and 6.*

It is easy to see that $\widetilde{\chi}(c)$ is also well defined, and moreover $\widetilde{\chi}(c) \equiv 0$ for k even and $\widetilde{\chi}(c) = 2\chi(c)$ for k odd. Thus it also has the same behavior.

Since

$$\chi_r(c) = \sum_{m\geqslant 0}(-1)^m b_{mr}(c),$$

one has

$$\chi(c) = \sum_{m\geqslant 0}(-1)^m b_m(c)$$

if the summands at the right-hand side are well defined.

CONJECTURE 1. *For almost any smooth and for any analytic quasiperiodic function $f\colon \mathbb{R}^k \to \mathbb{R}$, the functions $b_m(c)$ are well defined and have the behavior described for the functions $\#_m(c)$ in Theorems 1 and 6.*

Theorem 7 and Lemma 5 give an obvious formula for $\chi(c)$. For a generic quasiperiodic function $f\colon \mathbb{R}^k \to \mathbb{R}$, this formula can be rewritten in a more attractive way. For a generic smooth function $\widetilde{f}$ on the torus $\mathbb{T}^n$, the set

$$N = \{x \in \mathbb{T}^n : d\widetilde{f}_{\varphi(\mathbb{R}^k)} = 0\}$$

is a closed (!) smooth $(n-k)$-dimensional submanifold of $\mathbb{T}^n$. We can orient N in the following way. Let $x \in N$ be a point of N at which the projection $p\colon T_xN \to L$ is nondegenerate (this is true of a generic point $x \in N$). The restriction of the second differential $d^2\widetilde{f}(x)$ to the space $\varphi(\mathbb{R}^k)$ is a nondegenerate quadratic form. Let $i = i(x)$ be its index. The orientation of the space T_xN (and thus of N in a neighborhood of x) is defined as one which differs by the sign $(-1)^i$ from the orientation induced (by the projection p) from the orientation of the space L. It is not difficult to see that the described orientation is well defined: orientations of parts of the manifold N separated by the fold hypersurface in N are compatible. Theorem 7 and Lemma 5 give the following formula for the "density of the Euler characteristic" $\chi(c)$.

LEMMA 8. *For a generic quasiperiodic function*

$$f = \widetilde{f} \circ \pi \circ \varphi \qquad (\varphi\colon \mathbb{R}^k \to \mathbb{R}^n,\ \widetilde{f}\colon \mathbb{T}^n \to \mathbb{R}),$$

we have

$$\chi(c) = \int_{N \cap \{\widetilde{f} \leqslant c\}} \pi^* dV_{n-k}.$$

References

[1] V. I. Arnold, S. M. Gusein-Zade, and A. N. Varchenko, *Singularities of differentiable maps. Vol. II*, Birkhäuser, Boston, 1988.

[2] S. M. Gusein-Zade, *The number of critical points of a quasiperiodic potential*, Funktsional. Anal. i Prilozhen. **23** (1989), no. 2, 55–56; English transl., Functional Anal. Appl. **23** (1989), 129–130.

[3] A. N. Varchenko, *Theorems on the topological equisingularity of families of algebraic varieties and families of polynomials mappings*, Izv. Akad. Nauk SSSR Ser. Mat. **36** (1972), 957–1019; English transl., Math. USSR–Izv. **6** (1972), 949–1008.

MOSCOW STATE UNIVERSITY, DEPARTMENT OF MATHEMATICS AND MECHANICS, MOSCOW, 119899, RUSSIA.

E-mail address: `sabir@mccme.ru`

Amer. Math. Soc. Transl.
(2) Vol. **197**, 1999

Statistics of Klein Polyhedra and Multidimensional Continued Fractions

M. L. Kontsevich and Yu. M. Suhov

ABSTRACT. A (generalized) Klein polyhedron (KP) is defined as the convex hull of the set of lattice points inside a (convex) domain $\mathbb{D} \subset \mathbb{R}^n$. We focus on two cases, in which $\mathbb{D}$ is the interior (i) of a simplicial cone or (ii) of a paraboloid. We introduce a dynamical system with a group action and an invariant measure, which is naturally associated with the KP, and discuss statistical properties of the boundary of a (random) KP with respect to the invariant measure. The boundary of a KP is called a *veil*; it is formed by "faces" that are convex polyhedra of dimension $\leqslant n-1$. In the case of a simplicial cone, the dynamical systems that we construct are multidimensional generalizations of the well-known Gauss automorphism (related to the continued fraction representation of a number $x \in (0,1)$). The case of a paraboloid is important since it emerges as the limiting case for the domains in $\mathbb{R}^n$ of the form $\lambda\mathbb{D}$, as $\lambda \to +\infty$, where $\mathbb{D}$ is a given convex domain with smooth boundary. In particular, we are interested in the distribution of various parameters characterizing local combinatorial and geometric properties of the veil, such as the number of lattice points lying on a single face, the number of vertices of a face, etc. We find that the two cases under consideration (simplicial and parabolic) are quite different from this point of view. A by-product of our approach is that in the simplicial case it allows us to propose algorithms for searching for a "good" rational approximation for the components of a given vector θ; a particular algorithm of this kind was proposed earlier by J. Lagarias.

1. Introduction: the Gauss automorphism

This paper was conceived as an attempt to generalize certain aspects of the theory of continued fractions to the multidimensional case; the initial motivation was to provide an answer to a question proposed by V. I. Arnold at his seminar at the École Normale Supérieure, Paris, in the Spring of 1997. However, in the process of writing it, we decided to extend our considerations in other directions offered by the emerging approach. This approach may in particular be applied to the problem of asymptotic distribution of the number of integer points along the boundary $\partial(\lambda\mathbb{D})$ of a domain as $\lambda \to \infty$.

1991 *Mathematics Subject Classification.* Primary 11Hxx, 11J70; Secondary 58F11, 22D40.
This research was supported in part by I.H.E.S., Bures-sur-Yvette, EC Grant 'Training Mobility and Research' (Contracts CHRX-CT 93-0411 and ERBMRXT-CT 960075A), and INTAS Grant 'Mathematical Methods for Stochastic Discrete Event Systems' (INTAS 93-820).

Continued fraction theory, which was our original incentive, is indeed a popular subject, both theoretically and with regard to numerous applications. Attempts to develop a multidimensional analog were inspired in particular by the need to design algorithms that could find rational approximations for vectors and integer approximations for linear manifolds in $\mathbb{R}^n$. See, e.g., [**7, 16**] and references therein.

The continued fraction representation of a number $y \in (0,1)$:

$$y = (a_0 + (a_1 + (a_2 + (a_3 + \cdots)^{-1})^{-1})^{-1})^{-1}, \qquad a_i \in \mathbb{Z}_+ := \{1, 2, \dots\}, \tag{1.1}$$

associates with y a sequence of positive integers $(a_0, a_1, \dots)$ (accordingly, we shall write $y \sim (a_0, a_1, \dots)$). An important role in the "one-dimensional" theory of continued fractions is played by the Gauss automorphism

$$\gamma\colon (x_1, x_2) \in \mathbb{Q} \mapsto \left(\left\{\frac{1}{x_1}\right\}, \frac{1}{[x_1^{-1}] + x_2}\right), \tag{1.2}$$

with the invariant probability measure

$$\mu(\mathrm{d}x_1 \times \mathrm{d}x_2) = \frac{\mathrm{d}x_1\,\mathrm{d}x_2}{(1 + x_1 x_2)^2} \frac{1}{\ln 2} \tag{1.3}$$

on the unit square $\mathbb{Q} = (0,1) \times (0,1)$; see, e.g., [**11**]. Here, and below, $[x]$ and $\{x\} = x - [x]$ denote, respectively, the integer and fractional parts of a positive real number x. More precisely, if $x_1 \sim (a_0, a_1, \dots)$, and $x_2 \sim (a_{-1}, a_{-2}, \dots)$, then $\gamma(x_1, x_2) = (x_1', x_2')$, where $x_1' \sim (a_1, a_2, \dots)$ and $x_2' \sim (a_0, a_{-1}, \dots)$. In particular, the distribution of an entry a_i in the continued fraction representation of numbers $x_j \in (0,1)$, $j = 1, 2$, is given by

$$\rho_s^0 = \mu(a_i = s) = \log_2 \frac{(s+1)^2}{s(s+2)}, \qquad s \in \mathbb{Z}_+, \tag{1.4}$$

whereas the expected value of a_i is

$$E^0 = \sum_s s\rho_s^0 = \int_{\mathbb{Q}} \mu(\mathrm{d}x_1 \times \mathrm{d}x_2)[x_1^{-1}] = \int_{\mathbb{Q}} \mu(\mathrm{d}x_1 \times \mathrm{d}x_2)[x_2^{-1}] = \infty \tag{1.5}$$

(the meaning of the superscript ^{0} will become clear later).

Recently, following [**4, 5**], beautiful results were obtained in a series of papers starting with [**14**] (see also [**15**] and references therein), based on a geometric approach to multidimensional continued fraction theory. The basic notion here is that of a Klein polyhedron (KP) inside a simplicial cone in $\mathbb{R}^n$, $n \geqslant 2$, with the vertex at the origin. A KP is defined as the convex hull of the set of lattice points inside the cone (excluding the origin). The boundary of such a polyhedron may be regarded as an $(n-1)$-dimensional continued fraction; for $d = 2$ the standard theory is recovered. Following popular terminology, we call this boundary a *simplicial veil* (or simply a *veil*[1]). So far, the main attention in this approach has been focused on the analysis of an individual veil, rather than that of its "statistical" properties. In contrast, this paper (together with subsequent ones) aims to establish basic facts about the statistical behavior of the veil; we believe that such an approach may be of use in several applications, including the above-mentioned continued fraction algorithms and the popular problem of evaluating the number of integer points in a geometric locus (see, e.g., [**9, 10, 12**] and references therein).

[1] "Veil", besides its English meaning, is pronounced almost the same as the French word "voile" ("sail"), used in several French papers in this area

More precisely, the space of veils is endowed with a natural measure invariant under a group action. This leads us to a dynamical system (a veil flow) where the role of time is played by an appropriate subgroup of $\mathrm{SL}(2,\mathbb{R})$. One of the results of this paper, Theorem 3.1, asserts that with respect to the invariant probability measure, the expected number of lattice points on a "typical", or "average", face of a simplicial veil is infinite. (A face of a veil is itself a compact convex polyhedron of dimension $n-1$.) This is a direct generalization of the statement that the expected value of a "typical" entry of a continued fraction is infinite. In contrast, the expected number of lattice points lying on the boundary of a face (which is a union of compact convex polyhedra of dimension $n-2$) is finite. In general, the distribution of the number of lattice points on the lower-dimensional parts of the boundary of a face has several finite moments. Moreover, the distribution of the number of vertices of a face has a rapidly decaying tail: the probability that a face has s vertices is $\sim \exp(-s^{1/(n-2)})$; see Theorem 3.4. The next problem is to identify the distribution of these random variables. So far we have not been able to find concise formulas, but this does not mean that they do not exist.

In principle, the same approach may be used to study any local combinatorial parameter of a veil: degrees of vertices of a face, the integer lengths of the edges, etc.

A similar dynamical system may be defined when the simplicial cone is replaced by a paraboloid in $\mathbb{R}^n$. Here, as before, a notion of a veil can be introduced (we use the term parabolic veil, to distinguish it from a simplicial one). Properties of the dynamical system related to parabolic veils differ from those of the simplicial veil flow; in particular, the expected number of integer points on a face of a parabolic veil is finite; see Theorem 4.2. The specific role of parabolic veils is that they emerge as limits for veils associated with a general convex domain $\mathbb{D}$, in the course of "inflating", or scaling, $\mathbb{D}$ by the factor λ and letting $\lambda \to \infty$. See Theorems 4.6 and 4.7.

The rest of the paper is organized as follows. We begin with a revision of the Gauss automorphism (Section 2) and continue with results about simplicial veils (Section 3). Section 4 contains results on parabolic veils. In Section 5 we discuss the main constructions of the proof. It should be said that technical details of the proofs are systematically omitted, especially when they are not essentially new.

2. The Gauss automorphism: a geometric representation

As is well known, the Gauss automorphism $(\mathbb{Q},\gamma,\mu)$ (see (1.1)–(1.3)) is an Anosov map isomorphic to a Bernoulli shift with a finite entropy. A natural geometric representation of the Gauss automorphism is given in Proposition 2.1 and obtained on the basis of the following construction. Associated with a point $(x_1,x_2)\in\mathbb{Q}$ is a broken line $\mathcal{V}$ defined as the boundary $\partial\mathcal{C}$ of the convex hull $\mathcal{C}$ of the set of two-dimensional integer vectors inside a right angle Θ $(=\Theta(x_1,x_2))$, on the half-plane $\{(X_1,X_2): X_1>0\}$ (X_1 and X_2 are the plane coordinates). The angle Θ is formed by the half-lines

$$\mathbb{L}_+ = ((X_1,X_2): X_1>0, X_2/X_1 = 1/x),$$
$$\mathbb{L}_- = ((X_1,X_2): X_1>0, X_2/X_1 = -x).$$

Here $x_1\sim(a_0,a_1,\dots)$, $x_2\sim(a_{-1},a_{-2},\dots)$, and $x\sim(a_0,a_{-1},a_1,a_{-2},\dots)$. More precisely, $\mathcal{V}$ (which, bearing in mind the high-dimensional generalization, is called a *veil*) is formed by two broken "half-lines", $\mathcal{V}_\pm$, with the breaking points in $\mathbb{Z}^2$, which approach one of the above straight lines (say, $\mathcal{V}_+$ lies in the quarter-plane $X_1, X_2 > 0$ and approaches $\mathbb{L}_+$, and $\mathcal{V}_-$ lies in the quarter-plane $X_1 > 0$, $X_2 < 0$ and approaches $\mathbb{L}_-$). Further, a_i, $i \geqslant 0$, give the integer lengths of the sequence of the segments forming $\mathcal{V}_+$, and a_i, $i \leqslant -1$, the integer lengths of the sequence of the segments forming $\mathcal{V}_-$. (The integer length of a segment (or a vector) with endpoints in $\mathbb{Z}^2$ is defined as the number of points from $\mathbb{Z}^2$ lying in the interior of the segment, plus one.) The convex polygon $\mathcal{C}$ is called a *Klein polygon*.

A standard way to represent the Gauss automorphism is to construct a "natural" suspension flow whose base automorphism coincides with $(\mathbb{Q}, \gamma, \mu)$. This question was discussed in detail in [**1, 2, 3**] (see also [**19**]). Our construction seems simpler than the ones in those papers, and is carried out as follows. The previous construction of the veil $\mathcal{V}$ as the boundary of the convex hull of a part of $\mathbb{Z}^2$ can be repeated for any cone formed by a pair of lines on $\mathbb{R}^2$ through the origin. That is, the veil is geometrically determined by a unit frame $\mathbf{e} = (e_1, e_2)$. Here, e_1, e_2 are 2-vectors with $\det(e_1, e_2) = 1$. An important remark is that the integer lengths of the segments constituting the veil do not change when this frame (or pair of lines) is transformed by an element from the discrete group $\Gamma = \mathrm{SL}(2, \mathbb{Z})$, i.e., by a 2×2 integer unimodular matrix. Since the unit frames are associated with the group $\mathrm{SL}(2, \mathbb{R})$ of 2×2 real unimodular matrices, we conclude that the phase space $\mathcal{E}$ of the flow is the three-dimensional manifold G/Γ, where $G = \mathrm{SL}(2, \mathbb{R})$. That is, an element of $\mathcal{E}$ is an equivalence class, $\mathbf{Y}$, composed of unit frames $\mathbf{e}'$ of the form $\mathbf{e}z$, where $z \in \Gamma$ and $\mathbf{e}$ is a fixed unit frame representing $\mathbf{Y}$. It is convenient to write $\mathbf{e}(\mathbf{Y})$ and $e_j(\mathbf{Y})$, $j = 1, 2$. By $\mathcal{V}(\mathbf{Y})$ we denote the veil associated with the coset $\mathbf{Y}$.

The flow transformations are defined via the left action of the subgroup $\mathrm{D}^+(2, \mathbb{R}) \subset G$ consisting of diagonal 2×2 unimodular matrices with nonnegative entries, i.e., matrices of the form

$$\Lambda(\lambda) = \begin{pmatrix} \lambda & 0 \\ 0 & \lambda^{-1} \end{pmatrix}, \qquad \lambda > 0.$$

Namely, if $g \in \mathrm{D}^+(2, \mathbb{R})$, then the corresponding transformation $\sigma_g \colon \mathcal{E} \to \mathcal{E}$ is given by $\sigma_g \mathbf{Y} = g\mathbf{Y}$. The invariant measure ν is the Haar measure of G restricted to $\mathcal{E}$. Since $\nu(\mathcal{E}) < \infty$ (although $\mathcal{E}$ is not compact), we can assume that ν is a probability distribution (i.e., $\nu(\mathcal{E}) = 1$).

The dimension of the manifold $\mathrm{D}^+(2, \mathbb{R})$ equals one; in other words, $\{\sigma_g\}$ is a one-parameter family of maps $\mathcal{E} \to \mathcal{E}$. It is convenient to fix an isomorphism $\iota \colon \mathrm{D}^+(2, \mathbb{R}) \to \mathbb{R}$ by passing to logarithmic coordinates. Here

$$\iota\Lambda(\lambda) = \ln \lambda. \tag{2.1}$$

With such an isomorphism, we can consider the one-parameter group T of transformations $\tau_t = \sigma_{\iota^{-1}t}$, $t \in \mathbb{R}$, of $\mathcal{E}$, with the invariant measure ν. The function $\Phi \colon \mathbf{Y} \to \mathbb{R}_+$ establishing the suspension flow structure of $\{\sigma_g,\ g \in \mathrm{D}^+(2, \mathbb{R})\}$ may be defined by

$$\begin{aligned} \Phi(\mathbf{Y}) = \inf\{t \geqslant 0 : \text{the vector } e_1(\mathbf{Y}) \exp t - e_2(\mathbf{Y}) \exp(-t) \text{ is parallel} \\ \text{to a segment of the veil determined by the frame } \mathbf{e}(\mathbf{Y})\}. \end{aligned} \tag{2.2}$$

Note that the vector $e_1(\mathbf{Y})(\exp t) - e_2(\mathbf{Y})(\exp(-t))$ is parallel to the line passing through the endpoints of the frame $\Lambda(e^t)\mathbf{e}(\mathbf{Y})$.

Thus, define: (i) $\mathcal{M} \subset \mathcal{E}$ as the zero set of function Φ, (ii) the value $\varphi(\mathbf{Y})$, $\mathbf{Y} \in \mathcal{M}$, by

$$\varphi(\mathbf{Y}) = \min\,[t : t > 0,\ \tau_t \mathbf{Y} \in \mathcal{M}], \tag{2.3a}$$

and (iii) the map $\tau^0 \colon \mathcal{M} \to \mathcal{M}$ by

$$\mathbf{Y} \mapsto \tau_{\varphi(\mathbf{Y})}(\mathbf{Y}). \tag{2.3b}$$

The set $\mathcal{M}$ is the countable union of manifolds of dimension $\leqslant 2$. Also, define the probability measure ν^0 on $\mathcal{M}$ by using the fact that $\mathrm{D}^+(2,\mathbb{R})$ acts in a locally free fashion in a neighborhood of $\mathcal{M}$. Formally,

(a) we write $\nu(d\mathbf{Y})$, where $\mathbf{Y} \in \mathcal{E}$ is of the form $\tau_t \mathbf{Y}^0$, with $\mathbf{Y}^0 \in \mathcal{M}$ and $t \in \mathbb{R}$ close to 0, as $d\tilde{\nu}^0(d\mathbf{Y}) \times dt$, obtaining a measure $\tilde{\nu}^0$ on $\mathcal{M}$, and then
(b) we normalize $\tilde{\nu}^0$ to get a probability distribution ν^0. From the construction it follows that ν^0 is invariant under τ^0.

Pictorially, a point $\mathbf{Y} \in \mathcal{M}$ is associated with a veil whose segments are labelled by the integers, and the label 0 is attached to the segment parallel to the vector $e_1(\exp t) - e_2(\exp(-t))$. We call it the *ground segment* and denote it by $\mathbb{P}(\mathbf{Y})$. The map τ^0 is represented as the shift of all labels by one. The result of the above construction is the following.

Proposition 2.1. *The flow $(\mathcal{E}, T, \nu)$ is a suspension flow, with base automorphism $(\mathcal{M}, \tau^0, \nu^0)$. Furthermore, the latter automorphism is isomorphic to the Gauss automorphism $(\mathbb{Q}, \gamma, \mu)$.*

The logarithmic coordinates allow us to define the logarithmic projection Π of the veil $\mathcal{V}(\mathbf{Y})$, $\mathbf{Y} \in \mathcal{E}$, to the time-axis $\mathbb{R}^1$. Geometrically,

(i) we perform the central projection Π_{ce} from the origin of the segments of $\mathcal{V}(\mathbf{Y})$ (with marked lattice points on them) to the hyperbola $\mathbb{U}(\mathbf{Y})$ given by the equation $X_{1,\mathbf{Y}} X_{2,\mathbf{Y}} = 1$, $X_{1,\mathbf{Y}}, X_{2,\mathbf{Y}} > 0$, in the coordinates $X_{i,\mathbf{Y}}$, $i = 1, 2$, related to the frame $\mathbf{e}(\mathbf{Y})$, and then
(ii) we pass to the coordinate $t = \ln x_{1,\mathbf{Y}}$ on $\mathbb{U}(\mathbf{Y})$.

As a result, we obtain a partition of $\mathbb{R}^1$ into disjoint intervals (the endpoints of these intervals are the images under the projection Π_{ce} of the vertices of $\mathcal{V}(\mathbf{Y})$), with marked points on them (which are the images of the lattice points lying inside the segments of $\mathcal{V}(\mathbf{Y})$). The intervals constituting the partition are called *cells* (again in view of future multidimensional generalizations). The image of a vertex of $\mathcal{V}(\mathbf{Y})$ is also called a *vertex* (of a cell partition).

It is convenient to label the cells forming a partition by integers. If $\mathbf{Y} \in \mathcal{M}$, there exists a distinguished cell, which is the projection, under Π, of the ground segment $\mathbb{P}(\mathbf{Y})$ of $\mathcal{V}$; this cell is labelled by 0. Otherwise (i.e., when $\mathbf{Y} \notin \mathcal{M}$) there is a distinguished pair of adjacent cells (or, if one prefers, a vertex of the partition). This vertex (also called *ground*) is the image, under Π, of the vertex of $\mathcal{V}(\mathbf{Y})$ which is the first one to be reached by the translations, towards the veil, of the line $\mathbb{G}(\mathbf{Y})$ given by the equation $\langle x, e_1(\mathbf{Y})\rangle + \langle x, e_2(\mathbf{Y})\rangle = 0$. Note that $\mathbb{G}(\mathbf{Y})$ is a line through the origin, parallel to the vector $e_1(\mathbf{Y}) - e_2(\mathbf{Y})$. One of two cells adjacent to the ground vertex, say the furthest to the right (in the sense of the direction of the coordinate t), is labelled by 0.

Once the 0th label is attached, we assign the numbers $j \in \mathbb{Z}^1$ observing the following principle: two adjacent cells are assigned labels that differ by one, and of an adjacent pair the cell furthest to the right is assigned the greater label.

As a result, a point $\mathbf{Y} \in \mathcal{M}$ is identified with the sequence $\mathbf{a} = (a_i,\, i \in \mathbb{Z}^1)$, where a_i is the number of marked points inside the ith cell, plus one. We also can define a function ϕ on $\mathcal{M}$ whose value $\phi(\mathbf{a}) > 0$ is determined by the condition that the vector $e_1(\mathbf{Y})\exp(\phi(\mathbf{a})) + e_2(\mathbf{Y})\exp(-\phi(\mathbf{a}))$ is parallel to the segment of the veil closest to the ground, namely the one projected to cell one. On the other hand, a point $\mathbf{Y} \in \mathcal{E}\setminus\mathcal{M}$ is identified with the pair $(\mathbf{a}, t^0)$, where $\mathbf{a}$ is as above, and $0 < t^0 < \phi(\mathbf{a})$ is determined by the condition that the vector $e_1(\mathbf{Y})(\exp t^0) + e_2(\mathbf{Y})(\exp(-t^0))$ is again parallel to the segment of the veil which is projected to cell one.

PROPOSITION 2.2. *Given the above identification, the flow $(\mathcal{E}, T, \nu)$ is isomorphic to the suspension flow with base automorphism $(\mathbb{Q}, \gamma, \mu)$ and function ϕ.*

The proof of Proposition 2.2 follows the arguments developed in [**1, 2, 3**], and is omitted.

A straightforward calculation with probability distribution (1.3), (1.4) leads to the following asymptotical formulas. Let $N_s(\mathbf{Y}, R)$, $s \in \mathbb{Z}_+$, denote the number of edges of the veil $\mathbf{Y}$ containing $s+1$ lattice points and lying within a circle in $\mathbb{R}^2$ of radius $R > 0$ centered at the origin, and $N(\mathbf{Y}, R)$ the total number of edges within this circle. Then for ν-a.a. $\mathbf{Y} \in \mathcal{E}$,

$$N_s(\mathbf{Y}, R) \simeq \frac{2\ln R}{\zeta(2)} \ln \frac{(s+1)^2}{s(s+2)}, \tag{2.4}$$

$$N(\mathbf{Y}, R) \simeq \ln 2\, \frac{2\ln R}{\zeta(2)}. \tag{2.5}$$

Our approach may be extended to the case of "larger" groups, where $G = \mathrm{SL}(2,\mathbb{R}) \ltimes \mathbb{R}^2$ (the group of the affine transformations of $\mathbb{R}^2$) and $\Gamma = \mathrm{SL}(2,\mathbb{Z}) \ltimes \mathbb{Z}^2$, with the same group T as before. Pictorially, this means that the corner point of the angle is not necessarily at zero. Similar calculations show that if the corner is distributed with a smooth density on the unit square $\mathbb{Q}$, then

$$N_s(\mathbf{Y}, R) \simeq \frac{\ln R}{\zeta(2)} \ln \frac{(s-1)^{s-1}(s+1)^{3(s+1)}}{s^{3s}(s+2)^{s+2}}, \tag{2.6}$$

whereas $N(\mathbf{Y}, R)$ has the same asymptotics (2.5). If the corner is fixed at a rational point, the limit of $N_s(\mathbf{Y}, R)/\ln R$ may be calculated explicitly. The resulting formula is quite cumbersome, but the total number $N_s(\mathbf{Y}, R)$ again behaves as in (2.5). We conjecture that $N_s(\mathbf{Y}, R)$ has the same asymptotics as before for any position of the corner on $\mathbb{R}^2$.

3. Simplicial veils in higher dimensions

We begin with a dynamical system that is a natural generalization of the flow $(\mathcal{E}, T, \nu)$ (it is convenient to use the system of notation from the preceding section). A veil in n-dimensional space $\mathbb{R}^n$ is constructed in a similar fashion. Given a simplicial cone, we take the convex hull $\mathcal{C}$ of the set of integer points lying inside the cone; the *veil* $\mathcal{V}$ is the boundary of $\mathcal{C}$. It is an $(n-1)$-dimensional polyhedral surface "approaching" the boundary of the simplicial cone. The plane pieces of this surface are called *faces*; each face is in turn a convex polyhedron; its boundary is decomposed into plane faces of dimension $n-2$, etc. As before, we consider veils modulo the action of the group $\Gamma = \mathrm{SL}(n,\mathbb{Z})$. For simplicity, the corresponding equivalence classes are again called *veils*; for such a class we still use the previous notation $\mathcal{V}$. The simplicial cone is determined by the unit frame $\mathbf{e} = (e_1, \ldots, e_n)$

(with $\det(e_1, \dots, e_n) = 1$); thus the phase space of our dynamical system is the $(n^2 - 1)$-dimensional fundamental domain $\mathcal{E} = G/\Gamma$, where $G = \mathrm{SL}(n, \mathbb{R})$. A point of $\mathcal{E}$ is denoted by $\mathbf{Y}$, and is the coset $\mathbf{e}z$, $z \in \Gamma$, where $\mathbf{e} = (e_1, \dots, e_n)$ is a unit frame representing $\mathbf{e}$. As before, we use the notation $\mathbf{Y} \sim (\mathbf{e})$, $\mathbf{e}(\mathbf{Y})$ and $e_j(\mathbf{Y})$, $j = 1, \dots, n$. Furthermore, we again denote by $\mathcal{V}(\mathbf{Y})$ the veil associated with the coset $\mathbf{Y}$.

The flow transformations on $\mathcal{E}$ are defined, as before, via the left action of the group $\mathrm{D}^+(n, \mathbb{R})$ of real $n \times n$ unimodular diagonal matrices with nonnegative entries, i.e., matrices of the form

$$\Lambda(\underline{\lambda}) = \begin{pmatrix} \lambda_1 & 0 & \dots & 0 \\ 0 & \lambda_2 & \dots & 0 \\ \vdots & \vdots & \ddots & \vdots \\ 0 & 0 & \dots & \lambda_n \end{pmatrix}, \qquad \underline{\lambda} = (\lambda_1, \dots, \lambda_n),\ \lambda_1, \dots, \lambda_n > 0,\ \lambda_1 \cdots \lambda_n = 1, \tag{3.1}$$

namely, $\sigma_{\Lambda(\underline{\lambda})} \colon \mathbf{Y} \in \mathcal{E} \mapsto \Lambda(\underline{\lambda})\mathbf{Y}$. The invariant measure ν is again the Haar measure of G restricted to $\mathcal{E}$. As before, $\nu(\mathcal{E}) < \infty$ (see [**6**]); we again assume that ν is a probability distribution.

The dimension of the manifold $\mathrm{D}^+(n, \mathbb{R})$ equals $n - 1$. The logarithmic coordinates on $\mathrm{D}^+(n, \mathbb{R})$ may be introduced via the isomorphism $\iota \colon \mathrm{D}^+(2, \mathbb{R}) \to \mathbb{R}^{n-1}$, where

$$\iota\Lambda(\underline{\lambda}) = (\ln \lambda_1, \dots, \ln \lambda_{n-1}). \tag{3.2}$$

With such an isomorphism, we can talk about the group of transformations $\tau_{\underline{t}} = \sigma_{\iota^{-1}\underline{t}}$, $\underline{t} = (t_1, \dots, t_{n-1}) \in \mathbb{R}^{n-1}$, on $\mathcal{E}$, with the invariant measure ν. It is well known (see [**17, 21**]) that the flow $(\mathcal{E}, T, \nu)$ is ergodic. However, it is not difficult to check that its $(n-1)$-dimensional entropy equals zero. The same is true of the k-dimensional entropy (i.e., the entropy of the action of a k-parameter subgroup $\tau_{\underline{t}}$, $\underline{t} \in \mathbb{H}^k$, where $\mathbb{H}^k$ is a k-dimensional subspace in $\mathbb{R}^{n-1}$, $1 < k < n-1$). On the other hand, the one-dimensional entropy (i.e., the entropy of the action of a one-parameter subgroup $\gamma_{\underline{t}}$, $\underline{t} \in \mathbb{H}^1$, where $\mathbb{H}^1$ is a straight line in $\mathbb{R}^{n-1}$) is positive.

The flow $(\mathcal{E}, T, \nu)$ admits a kind of a suspension flow representation. Namely, let $\mathbb{G}(\mathbf{Y})$ be the $(n-1)$-dimensional hyperplane consisting of vectors of the form $\sum \mu_i e_i(\mathbf{Y})$, where $\sum \mu_i = 0$. Denote by $\mathcal{M}$ the following subset of $\mathcal{E}$:

$$\begin{aligned} \mathcal{M} = \{Y \in \mathcal{E} : &\text{ one of the } (n-1)\text{-dimensional faces} \\ &\text{of the veil } \mathcal{V}(\mathbf{Y}) \text{ is parallel to the hyperplane } \mathbb{G}(\mathbf{Y})\}. \end{aligned} \tag{3.3}$$

Let $\mathbb{P}(\mathbf{Y})$, $\mathbf{Y} \in \mathcal{M}$, denote the convex $(n-1)$-dimensional polyhedron that is the part of the veil $\mathcal{V}(\mathbf{Y})$ lying on the face figuring in the definition of $\mathcal{M}$ (more precisely, the whole class of the polyhedra obtained by the action of $\mathrm{SL}(n-1, \mathbb{Z}) \ltimes \mathbb{Z}^{n-1}$, the group of the affine integer transforms of $\mathbb{R}^{n-1}$). The set $\mathbb{P}(\mathbf{Y})$ is called the *ground polyhedron*, or *ground face* (of the veil $\mathcal{V}(\mathbf{Y})$, $\mathbf{Y} \in \mathcal{M}$). $\mathcal{M}$ is the countable union of submanifolds of $\mathcal{E}$ of dimension $n(n-1)$. The probability measure ν^0 on $\mathcal{M}$ is again constructed from the measure ν by using the fact that $\mathcal{M}$ is transversal to the orbits $\{\gamma_{\underline{t}}, \underline{t} \in \mathbb{R}^{n-1}\}$.

However, instead of a single basic map $\gamma^0 \colon \mathcal{M} \to \mathcal{M}$, we now have, for ν^0-almost all $\mathbf{Y} \in \mathcal{M}$, a collection of maps

$$\gamma^0_{1,\mathbf{Y}}, \dots, \gamma^0_{b(\mathbf{Y}),\mathbf{Y}} \colon \mathcal{M} \to \mathcal{M}, \tag{3.4}$$

preserving the measure ν^0. Here, the numbers $1, \ldots, b(\mathbf{Y})$ label the $(n-2)$-dimensional faces forming the boundary $\partial\mathbb{P}$. Pictorially speaking, each map $\gamma^0_{j,\mathbf{Y}}$, $j = 1, \ldots, b(\mathbf{Y})$, takes the veil $\mathcal{V}(\mathbf{Y})$ to a veil for which the ground polyhedron is the one adjacent to $\mathbb{P}(\mathbf{Y})$ via the jth face of $\partial\mathbb{P}(\mathbf{Y})$, and the "combinatorial structure" of $\mathcal{V}(\mathbf{Y})$ is preserved.

The combinatorial structure of the veil may be described, as before, in terms of a logarithmic projection Π. Namely, we again (i) take the central projection Π_{ce}, from the origin, of the $(n-1)$-dimensional faces of $\mathcal{V}(\mathbf{Y})$ (with marked lattice points on them) to the hyperboloid $\mathbb{U}(\mathbf{Y})$ given by the equation $X_{1,\mathbf{Y}} \cdots X_{n,\mathbf{Y}} = 1$, $X_{1,\mathbf{Y}}, \ldots, X_{n,\mathbf{Y}} > 0$, in the coordinates $X_{i,\mathbf{Y}}$, $i = 1, \ldots, n$, related to the frame $\mathbf{e}(\mathbf{Y})$, and then (ii) pass to the coordinates $\underline{t} = (\ln X_{1,\mathbf{Y}}, \ldots, \ln X_{n-1,\mathbf{Y}})$ on $\mathbb{U}(\mathbf{Y})$. As in Section 2, we obtain a partition of the linear time-manifold $\mathbb{R}^{n-1}$ into disjoint domains, with marked points on them (which are the images of the lattice points lying on the faces of $\mathcal{V}(\mathbf{Y})$). The boundaries of these domains are the images, under Π, of the faces of $\mathcal{V}(\mathbf{Y})$ of lower dimension (down to the vertices). The domains constituting the partition (together with the collections of lattice points in them) are called *cells*, and the images of the vertices of $\mathcal{V}(\mathbf{Y})$ are again called *vertices*.

Unlike the two-dimensional case, for $n \geqslant 3$ there is in general no lattice-like labelling of the cells of the given partition. However, for a point $\mathbf{Y} \in \mathcal{M}$ we can pick a particular cell, and for a generic point $\mathbf{Y} \in \mathcal{E} \setminus \mathcal{M}$ a particular vertex. Namely, if $\mathbf{Y} \in \mathcal{M}$, then the distinguished cell is the image, under Π, of the ground polyhedron $\mathbb{P}(\mathbf{Y})$ (we call it the *ground cell*). On the other hand, if $\mathbf{Y} \notin \mathcal{M}$, then the distinguished vertex (also called *ground*) is the one we first meet by translating the $(n-1)$-dimensional hyperplane $\mathbb{G}(\mathbf{Y})$ in the direction of $\mathcal{V}(\mathbf{Y})$. Natural parameters here are the area of the ground cell for $\mathbf{Y} \in \mathcal{M}$ and the total area of the cells incident to the ground vertex for a generic $\mathbf{Y} \in \mathcal{E} \setminus \mathcal{M}$.

One of the first questions about the flow $(\mathcal{E}, T, \nu)$ that one can try to address is about the "statistical" properties of the ground polyhedron $\mathbb{P}(\mathbf{Y})$, $\mathbf{Y} \in \mathcal{M}$, with respect to the measure ν^0. Denote by $\mathcal{M}_s$ the subset of $\mathcal{M}$ where $\mathbb{P}(\mathbf{Y})$ contains s lattice points, and set $\rho^0_s = \nu^0(\mathcal{M}_s)$, $s \in \mathbb{Z}_+$. In this paper we establish the following result.

THEOREM 3.1. *The expected value is*

$$E^0 = \sum_{s \in \mathbb{Z}_+} s\rho^0_s = \infty. \tag{3.5}$$

Pictorially, the assertion of Theorem 3.1 may be interpreted as the statement that "on average" the number of lattice points on a face (or in a cell) of a ν^0-"typical" veil $\mathcal{V}(\mathbf{Y})$, $\mathbf{Y} \in \mathcal{M}$, is infinite. The precise form of such a statement may vary with the choice of the concept of averaging. We comment here on two (from our point of view, most natural) concepts: the sample average and the log-volume-weighted average. Here, one takes a "large" domain in $\mathbb{R}^{n-1}$ (say a ball of a large radius R centered at the origin), and counts the numbers of integer points within the cells of the veil that are entirely inside the domain. In the case of the sample average, the total number of such integer points is divided by the number of cells. In the case of the log-volume-weighted average, it is divided by the volume of the domain. One then passes to the limit as the domain is inflated to eventually cover all of $\mathbb{R}^{n-1}$.

THEOREM 3.2. *Suppose that*

$$N_R^1(\mathbf{Y}) = \frac{M_R(\mathbf{Y})}{m_R(\mathbf{Y})}, \qquad N_R^2(\mathbf{Y}) = \frac{M_R(\mathbf{Y})}{v_R},$$

where $M_R(\mathbf{Y})$ *is the total number of lattice points in cells that are entirely inside the ball* $\mathbb{B}_R \subset \mathbb{R}^{n-1}$ *of radius* R *centered at the origin,* $m_R(\mathbf{Y})$ *is the number of such cells, and* v_R *is the (Euclidean) volume of* $\mathbb{B}_R$. *Then, with* ν*-probability one,*

$$E^1 := \lim_{R\to\infty} N_R^1(\mathbf{Y}) = \infty, \qquad E^2 := \lim_{R\to\infty} N_R^2(\mathbf{Y}) = \infty. \tag{3.6}$$

For the proof of Theorems 3.1 and 3.2, see Section 5. Formulas (3.5) and (3.6) extend relation (1.3) to the multidimensional case. It is a challenging problem to determine the distribution of the random variable N on $\mathcal{M}$ (cf. (1.2)): at present this seems beyond our reach. Numerical results approximating this distribution will be published separately.

A similar problem arises with other parameters of the (random) polygon $\mathbb{P}(\mathbf{Y})$: the total number of lattice points on the boundary $\partial\mathbb{P}(\mathbf{Y})$, the number of lattice points lying on the faces of $\partial\mathbb{P}(\mathbf{Y})$ of dimension $n-3$, etc., and finally the number of vertices in $\mathbb{P}(\mathbf{Y})$. Similarly, one can consider the distribution of the number of vertices of $\mathbb{P}(\mathbf{Y})$ of a given degree r (including or excluding the edges of faces adjacent to $\mathbb{P}(\mathbf{Y})$), the number of edges with vertices of given degrees (r, r'), etc. In general, one may speak of the distribution of any "local" combinatorial structure of the veil $\mathcal{V}(\mathbf{Y})$. In this paper, we shall comment on the distribution of the number of vertices of $\mathbb{P}(\mathbf{Y})$. The following assertion is proved in Section 5.

LEMMA 3.3. *There exists a constant* δ $(=\delta(d)) > 0$ *such that for any two distinct vertices* v, v' *of the veil* $\mathcal{V}(\mathbf{Y})$ *the log-distance satisfies*

$$\operatorname{dist}_{\mathbb{R}^{n-1}}(\Pi v, \Pi v') := \|\Pi v - \Pi v'\| \geqslant \delta.$$

Lemma 3.3, combined with a thorough case-by-case analysis of a number of geometric situations, leads to

THEOREM 3.4. *Let* V $(= V(\mathbf{Y}))$ *denote the number of vertices in* $\mathbb{P}(\mathbf{Y})$, $\mathbf{Y} \in \mathcal{M}$. *Then, for positive constants* c *and* $c^{\pm}$, *we have*

$$c^- \leqslant \nu^0(V(\mathbf{Y}) = s)\exp(cs^{1/(n-2)}) \leqslant c^+. \tag{3.7}$$

We conclude this section with a brief discussion of possible algorithms for finding rational (or integer) approximations that can be designed on the basis of the approach proposed in this paper. Let $\mathbb{K} \subset \mathbb{R}^n$ be a linear subspace of dimension $l < n$; we first consider the problem of finding a point $m \in \mathbb{Z}^n$ close to $\mathbb{K}$. Fix a "nice" set, e.g., a d-dimensional ball $\mathbb{B}$ centered at the origin and of sufficiently large radius. The basic idea is simple: given $\varepsilon > 0$, we apply a linear transformation $A_{\varepsilon,\mathbb{K}} \in G$ to $\mathbb{B}$ and get a "tube-like" domain (e.g., an ellipsoid) $\mathbb{T} = A_{\varepsilon,\mathbb{K}}\mathbb{B} \subset \mathbb{R}^n$ "stretched" along $\mathbb{K}$ so that the width of $\mathbb{T}$ in the directions orthogonal to $\mathbb{K}$ is $< \varepsilon$. The transformation $A_{\varepsilon,\mathbb{K}}$ can be specified to have $\mathbb{K}$ as the eigenspace with eigenvalue of size $\sim \varepsilon^{-(n-l)/l}$, and the orthogonal complement $\mathbb{K}^{\perp}$ as the eigenspace with the eigenvalue of size $\sim \varepsilon$. If the volume of $\mathbb{B}$ (which is equal to the volume of $\mathbb{T}$) is sufficiently large, then by Minkowski's convex body theorem (see, e.g., [**8**] and [**13**]), $\mathbb{T}$ will contain a nonzero integer point. This is the point that we are looking for. One can also formulate the question the other way around: find a point in the unimodular lattice $A_{\varepsilon,\mathbb{K}}^{-1}(\mathbb{Z}^n)$ inside the body $\mathbb{B}$.

Observe that the transformations $A_{\varepsilon,\mathbb{K}}$ form a one-parameter subgroup of G. Thus, our problem is reduced to that of calculating the action of a "large" element of such a subgroup at a point of $\mathcal{E}$. This can be implemented numerically by using a partition of $\mathcal{E}$ into disjoint semialgebraic sets.

If we want to find a rational approximation for a given vector $\theta = (\theta_1, \dots, \theta_n)$, we form an $(n+1)$-dimensional vector $\bar{\theta} = (\theta_1, \dots, \theta_n, 1)$ and repeat the above construction for the one-dimensional space $\mathbb{K} \subset \mathbb{R}^{n+1}$ along $\bar{\theta}$. Then an integer $\varepsilon/2$-approximation for $\mathbb{K}$ will straightforwardly lead to an ε-approximation for θ. It is not hard to assess the accuracy of such an algorithm: its speed will be similar to that of the algorithm proposed in [**16**].

4. Parabolic veils

We now take $G = \mathrm{SL}(n, \mathbb{R}) \ltimes \mathbb{R}^n$, $\Gamma = \mathrm{SL}(n, \mathbb{R}) \ltimes \mathbb{Z}^n$. The group T will be of the form $\mathrm{SO}(n-1, \mathbb{R}) \ltimes \mathbb{R}^{n-1}$ and will be identified with the subgroup of G generated by the following transformations $\pi_\bullet : \mathbb{R}^n \to \mathbb{R}$:

$$\pi_{\underline{t}} \colon (x_1, \dots, x_{n-1}, x_n) \mapsto \Big(x_1 + t_1, \dots, x_{n-1} + t_{n-1}, x_n + \sum_{1 \leqslant i \leqslant n-1} (2t_i x_i + t_i^2)\Big),$$
$$\underline{t} = (t_1, \dots, t_{n-1}) \in \mathbb{R}^{n-1}, \tag{4.1a}$$

$$\pi_A \colon (x_1, \dots, x_{n-1}, x_n) \mapsto \Big(\sum_{1 \leqslant j \leqslant n-1} a_{1,j} x_j, \dots, \sum_{1 \leqslant j \leqslant n-1} a_{n-1,j} x_j, x_n\Big),$$
$$A = (a_{ij}) \in \mathrm{SO}(n-1, \mathbb{R}). \tag{4.1b}$$

These transformations preserve the paraboloid $\{x_n = \sum_{1 \leqslant j \leqslant n-1} x_j^2\}$. As before, $\mathcal{E} = G/\Gamma$. Pictorially, a point $\mathbf{Y} \in \mathcal{E}$ is associated with a lattice of unit volume in $\mathbb{R}^n$, not necessarily containing the origin.

As before, let ν be the probability measure on $\mathcal{E}$ generated by the Haar measure on G. We are interested in the flow $(\mathcal{E}, T, \nu)$ (with the left action of the group T). It follows from Theorem 4.1 below that this flow is ergodic (see also [**18**], where a more general assertion is established). However, for $n \geqslant 2$ its restriction to any 1-parameter subgroup $\mathbb{R} \subset \mathbb{R}^{n-1} \subset T$ is not Bernoulli (it has zero entropy).

THEOREM 4.1. *Any noncompact closed subgroup $T' \subseteq T$, including T itself, acts ergodically on $(G/\Gamma, \nu)$.*

As before, we consider the lattice points in the domain $X_n \geqslant \sum_{1 \leqslant j \leqslant n-1} X_j^2$ (i.e., inside the paraboloid) and denote by $\mathcal{C}(\mathbf{Y})$ their convex hull; the boundary $\partial\mathcal{C}(\mathbf{Y})$ is again called a *(parabolic) veil* and denoted $\mathcal{V}(\mathbf{Y})$. Set

$$\mathcal{M} = \{\mathbf{Y} \in \mathcal{E} : \text{veil } \mathcal{V}(\mathbf{Y}) \text{ contains a horizontal face on which } x_n \text{ is constant}\}. \tag{4.2}$$

Such a face is denoted $\mathbb{P}(\mathbf{Y})$ and again called the *ground polyhedron*, or *ground face* (of the veil $\mathcal{V}(\mathbf{Y})$, $\mathbf{Y} \in \mathcal{M}$). Further, it is convenient to consider the subset $\mathcal{M}_s \subset \mathcal{M}$ where the ground face contains s lattice points.

In the same fashion as before, we introduce a probability measure ν^0 on $\mathcal{M}$. The probabilities

$$\rho_s^0 = \nu^0(\mathcal{M}_s), \qquad s = 1, 2, \dots, \tag{4.3a}$$

may be considered as the "parabolic" analog of the "simplicial" probabilities (1.4), and the expected value

$$E^0 = \sum_{s \in \mathbb{Z}_+} s\rho_s^0 \tag{4.3b}$$

as the analog of (1.5). However, in contrast to the simplicial case, the expected values are finite.

THEOREM 4.2. *For parabolic veils, in any dimension $n \geqslant 2$,*

$$E^0 < \infty. \tag{4.4}$$

The proof of Theorem 4.2 is based on a version of Minkowski's convex body theorem: apart from some "degenerate" cases, the number of lattice points on a face is $\leqslant c_0$ times the $(n-1)$-dimensional volume of the face, where c_0 $(= c_0(n))$ is a constant. By direct calculation, both the expected value of the $(n-1)$-volume of $\mathbb{P}(\mathbf{Y})$ and the contribution from the degenerate cases are finite, which leads to (4.4).

THEOREM 4.3. *Consider the projection $\mathbb{R}^n \to \mathbb{R}^{n-1}$ given by*

$$(X_1, \ldots, X_{n-1}, X_n) \mapsto (X_1, \ldots, X_{n-1}).$$

As in the simplicial case, set

$$N_R^1(\mathbf{Y}) = \frac{M_R(\mathbf{Y})}{m_R(\mathbf{Y})}, \qquad N_R^2(\mathbf{Y}) = \frac{M_R(\mathbf{Y})}{v_R},$$

where $M_R(\mathbf{Y})$ is the total number of the integer points in the cells that are projected entirely inside the ball $\mathbb{B}_R \subset \mathbb{R}^{n-1}$ of radius R centered at the origin, $m_R(\mathbf{Y})$ is the number of such cells, and v_R is the Euclidean volume of $\mathbb{B}_R$. Then, for parabolic veils, in any dimension $n \geqslant 2$, with ν-probability one, the limits

$$E^1 = \lim_{R\to\infty} N_R^1(\mathbf{Y}), \qquad E^2 = \lim_{R\to\infty} N_R^2(\mathbf{Y}) \tag{4.5}$$

are finite constants. The ratio $r = r(n) := E^2/E^1$ is the average number of cells per unit volume in the orthogonal projection of the veil to the horizontal hyperplane $\mathbb{R}^{n-1}$.

The proof of Theorem 4.3 combines arguments from the proofs of Theorem 4.2, Theorem 4.1, and Theorem 3.2 (see Section 5).

In the case $n = 2$, the flow $(\mathcal{E}, T, \nu)$ is a suspension flow, with base automorphism $(\mathcal{M}, \phi^0, \nu^0)$, where ϕ^0 is the Poincaré return flow map $\mathcal{M} \to \mathcal{M}$. Following the analogy with the simplicial case, we call $(\mathcal{M}, \phi^0, \nu^0)$ the *parabolic Gauss automorphism*. It would be interesting to find a "natural" realization of this automorphism on a "nice" four-dimensional manifold. In this paper we give "more explicit" formulas for parabolic probabilities ρ_s:

THEOREM 4.4. *For $n = 2$,*

$$\rho_s^0 \simeq cs^{-7/3}, \tag{4.6}$$

where c is a positive constant.

One can calculate the appropriate integrals numerically and obtain, for example, that the average density $r(2)$ of cells in Theorem 4.3 for $n = 2$ is $\simeq 0.69254\ldots$.

There are several alternative approaches to studying statistical properties of the parabolic veil. For example, given $\mathbf{Y} \in \mathcal{E}$, we can specify the face of $\mathcal{V}(\mathbf{Y})$

that is closest to the origin $O = (0, \dots, 0) \in \mathbb{R}^n$; apart from a set of ν-measure zero, this is a well-defined object, which we denote by $\mathbb{P}^{\rm ne}(Y)$. Further, denote by $N^{\rm ne}(\mathbf{Y})$ the number of the lattice points inside $\mathbb{P}^{\rm ne}(Y)$. Finally, denote by $\Sigma^{\rm ne}(\mathbf{Y})$ the $(n-1)$-dimensional Euclidean area of the set

$$\{\underline{t} = (t_1, \dots, t_{n-1}) \in \mathbb{R}^{n-1} : \mathbb{P}^{\rm ne}(Y) = \mathbb{P}^{\rm ne}(\pi_{\underline{t}}\mathbf{Y})\}.$$

THEOREM 4.5. *The expected values*

$$E^{\rm ne} = \int_{\mathcal{E}} \mathrm{d}\nu(\mathbf{Y})\, N^{\rm ne}(\mathbf{Y}), \tag{4.7}$$

$$S^{\rm ne} = \int_{\mathcal{E}} \mathrm{d}\nu(\mathbf{Y})\, \Sigma^{\rm ne}(\mathbf{Y}) \tag{4.8}$$

are both finite.

We conclude this section with a theorem emphasizing the significance of parabolic veils. First, note that the above construction can be repeated for any paraboloid $x_n = \mathbf{Q}(x_1, \dots, x_{n-1})$, where $\mathbf{Q}$ is a positive definite quadratic form. The corresponding expected values (4.7), (4.8) will be denoted by $E^{\rm ne}(\mathbf{Q})$ and $S^{\rm ne}(\mathbf{Q})$, respectively. Consider a convex domain $\mathbb{D} \subset \mathbb{R}^n$, with a C^2-smooth boundary $\partial\mathbb{D}$, such that the origin O is in $\mathbb{D}$. Denote by $\mathbf{Q}(\mathbf{x})$ the second quadratic form of $\partial\mathbb{D}$ at the point $\mathbf{x} \in \partial\mathbb{D}$ and by v the volume element on $\partial\mathbb{D}$ induced by the Riemannian metric of the ambient space $\mathbb{R}^n$. We assume that the form $\mathbf{Q}(\mathbf{x})$ is strictly positive at each point $x \in \partial\mathbb{D}$. Given $\lambda > 0$, consider the rescaled domain $\lambda\mathbb{D} = \{\mathbf{x} \in \mathbb{R}^n : \lambda^{-1}\mathbf{x} \in \mathbb{D}\}$. Now, given $\mathbf{Y} \in \mathcal{E}$, consider the boundary $\widetilde{\mathcal{V}}(\lambda, \mathbf{Y})$ of the convex hull $\widetilde{\mathcal{C}}(\lambda, \mathbf{Y})$ of the set of lattice points inside $\lambda\mathbb{D}$. Further, given $\mathbf{x} \in \partial\mathbb{D}$, let $N^{\rm ne}(\mathbf{x}, \lambda, \mathbf{Y})$ denote the number of lattice points lying on the face of $\widetilde{\mathcal{V}}(\lambda, \mathbf{Y})$ that is closest to the point $\lambda\mathbf{x} \in \partial(\lambda\mathbb{D})$ (if there are two or more such faces, we sum over them). Finally, $\bar{N}(\lambda, \mathbf{Y})$ will denote the total number of faces of $\widetilde{\mathcal{V}}(\lambda, \mathbf{Y})$.

THEOREM 4.6. (i) *For ν-a.a. $\mathbf{Y} \in \mathcal{E}$*

$$\lim_{\lambda\to\infty} \lambda^{-n(n-1)/(n+1)} \bar{N}(\lambda, \mathbf{Y}) = r(n) \int_{\partial\mathbb{D}} \mathrm{d}v(\mathbf{y}) \sqrt{\det \mathbf{Q}(\mathbf{y})}. \tag{4.9}$$

(ii) *There exists a positive constant c $(= c(n))$ such that for any $\mathbf{x} \in \partial\mathbb{D}$, and for ν-a.a. $\mathbf{Y} \in \mathcal{E}$,*

$$\lim_{\lambda\to\infty} \lambda^{-n(n-1)/(n+1)} N(\mathbf{x}, \lambda, \mathbf{Y}) = c\, \sqrt{\det \mathbf{Q}(\mathbf{y})}. \tag{4.10}$$

The proof of Theorem 4.6 is based on the ergodicity property stated in Theorem 4.1 and additional straightforward (although tedious) estimates. The main idea is that after applying affine transformations preserving the volume element, one can approximate large pieces of $\partial(\lambda\mathbb{D})$ by standard paraboloids.

A "dual" point of view is also possible: given $\lambda > 0$ and $\mathbf{Y} \in \mathcal{E}$, pick points $\mathbf{x} \in \partial\mathbb{D}$ such that the tangent hyperplane $\mathbb{T}(\lambda, \mathbf{x})$ through $\lambda\mathbf{x}$ to $\partial(\lambda\mathbb{D})$ is parallel to a face of $\widetilde{\mathcal{V}}(\lambda, \mathbf{Y})$. This gives a (finite) set $\mathbf{X} = \{\mathbf{x}\} \subset \partial\mathbb{D}$; the face associated with $\mathbf{x}$ is denoted by $\mathbb{F}_\lambda(\mathbf{x}, \mathbf{Y})$. We can also consider functions $\xi \colon \mathbf{x} \in \mathbf{X} \mapsto \xi(\mathbf{x}) \in \mathbb{Z}_+$ describing the number of lattice points on $\mathbb{F}_\lambda(\mathbf{x}, \mathbf{Y})$, the number of vertices in $\mathbb{F}_\lambda(\mathbf{x}, \mathbf{Y})$, etc.; in principle, any "local" combinatorial characteristic of the veil may be considered. To be definite, assume that $\xi(\mathbf{x})$ equals the number of lattice points in $\mathbb{F}_\lambda(\mathbf{x}, \mathbf{Y})$. As $\mathbf{Y}$ varies along the probability space $(\mathcal{E}, \nu)$, this defines a random "marked" point field Ξ_λ on $\partial\mathbb{D}$, with marks from $\mathbb{Z}_+$; a sample of such a random field is a pair $(\mathbf{X}, \psi)$. The distribution of Ξ_λ is determined by a sequence of its moment measures κ_m, $m \in \mathbb{Z}_+$. Here κ_m is the Borel measure on $(\partial\mathbb{D})^m \times \mathbb{Z}_+^m$ defined by

$$\kappa_m(B) = \int_{\mathcal{E}} \nu(\mathrm{d}\mathbf{Y}) N_B(\mathbf{Y}), \tag{4.11}$$

where $B \subseteq \partial\mathbb{D}^m \times \mathbb{Z}_+^m$, and $N_B(\mathbf{Y})$ is the number of pairs of "vectors"

$$((\mathbf{x}_1, \dots, \mathbf{x}_m), (n_1, \dots, n_m)) \in B$$

such that $\mathbf{x}_i \in \mathbf{X}$, $n_i = \psi(\mathbf{x}_i)$, $1 \leqslant i \leqslant m$.

It is not hard to check that the projection of κ_m to the first component $\partial\mathbb{D}^m$ of the Cartesian product $(\partial\mathbb{D})^m \times \mathbb{Z}_+^m$ is absolutely continuous with respect to $dv(\mathbf{x}_1) \times \cdots \times dv(\mathbf{x}_m)$, where v is the Riemannian volume on $\partial\mathbb{D}$. Taking the corresponding Radon–Nikodým derivative leads to a nonnegative function

$$((\mathbf{x}_1, \dots, \mathbf{x}_m), (n_1, \dots, n_m)) \in (\partial\mathbb{D})^m \times \mathbb{Z}_+^m \mapsto \mathfrak{k}_m((\mathbf{x}_1, n_1), \dots, (\mathbf{x}_m, n_m)).$$

"Physically speaking", the sum

$$\bar{\mathfrak{k}}_m(\mathbf{x}_1, \dots, \mathbf{x}_m) = \sum_{n_1, \dots, n_m \in \mathbb{Z}_+} \mathfrak{k}_m((\mathbf{x}_1, n_1), \dots, (\mathbf{x}_m, n_m))$$

describes the joint m-fold density of points $\mathbf{x} \in \mathbf{X}$ (at $(\mathbf{x}_1, \dots, \mathbf{x}_m)$), and the ratio

$$\frac{\mathfrak{k}_m((\mathbf{x}_1, n_1), \dots, (\mathbf{x}_m, n_m))}{\bar{\mathfrak{k}}_m(\mathbf{x}_1, \dots, \mathbf{x}_m)}$$

describes the "conditional probability" of the event that $\psi(\mathbf{x}_i) = n_i$, $i = 1, \dots, m$, "given" that $\mathbf{x}_1, \dots, \mathbf{x}_m \in \mathbf{X}$.

THEOREM 4.7. *As* $\lambda \to \infty$,

(i) *the measure* $\lambda^{-n(n-1)/(n+1)}\kappa_1$ *converges in variation to the measure*

$$\kappa^0(\mathrm{d}\mathbf{x}) = v(\mathrm{d}\mathbf{x})\sqrt{\mathbf{Q}(\mathbf{x})};$$

(ii) *for any* $m > 1$, *the measure* $\lambda^{-mn(n-1)/(n+1)}\kappa_m$ *converges in variation to* $(\kappa^0)^m$, *the Cartesian power of* κ^0.

5. About the proofs

This section contains an outline of the proofs of the results stated in the paper. The omitted technicalities are, as a rule, straightforward and do not contain any essential novelties.

PROOF OF THEOREMS 3.1 AND 3.2. The proofs of these theorems are very similar. To verify Theorem 3.1, it suffices to check that the integral over $\mathcal{M}^0$ is infinite. Here, $\mathcal{M}^0$ is the component of $\mathcal{M}$ for which the ground polyhedron $\mathbb{P}(\mathbf{Y})$ is an $(n-1)$-dimensional simplex (i.e., a convex polyhedron with n vertices), and the integer distance from the origin to the ground face of the veil $\mathcal{V}(\mathbf{Y})$ equals one. The measure $\nu^0(\mathrm{d}\mathbf{Y})$ on $\mathcal{M}^0$ is proportional to

$$\prod_{1 \leqslant j \leqslant n} \mathrm{d}y_j \Big/ (V(\mathbf{Y}))^n. \tag{5.1}$$

Here, $\mathbf{Y}_j \,(= y_j(\mathbf{Y}))$, $j = 1, \dots, n$, are coordinates of n vertices of the simplex $\mathbb{P}(\mathbf{Y})$; each $\mathbf{Y}_j$ is an $(n-1)$-dimensional vector $(y_j^1, \dots, y_j^{n-1})$. Further, $V(\mathbf{Y})$ is the $(n-1)$-dimensional volume of $\mathbb{P}(\mathbf{Y})$.

Pictorially, by changing $\mathbf{y} = (y_j,\ 1 \leqslant j \leqslant n)$, we vary the "location" of the simplex, its "shape" and a "homothetic" factor $\ell > 0$ (the latter is just a scaling factor). Both the location and the shape variables run over compact domains; the only unrestricted quantity is ℓ. It is convenient to perform a change of variables and separate the variable ℓ from the rest. After such a change, the left-hand side of (5.1) can be written as

$$\mathrm{d}\mathbf{y}_{\mathrm{loc}}\,\mathrm{d}\mathbf{y}_{\mathrm{sh}}\,\mathrm{d}\ell\, J/V^n. \tag{5.2}$$

Here $\mathbf{y}_{\mathrm{loc}}$ is the collection of variables responsible for the location of the simplex, and $\mathbf{y}_{\mathrm{sh}}$ is that determining its shape. Further, $J = J(\mathbf{y}_{\mathrm{loc}}, \mathbf{y}_{\mathrm{sh}}, \ell)$ is the Jacobian and $V = V(\mathbf{y}_{\mathrm{loc}}, \mathbf{y}_{\mathrm{sh}}, \ell)$, as before, the volume of the simplex.

The next key point of the proof of Theorem 3.1 is the following observation, based on Minkowski's convex body theorem: the number of integer points inside a convex polyhedron grows proportionally to its (Euclidean) volume, where the proportionality coefficient is bounded below and above provided that the ratio of the radii of the inscribed and circumscribed spheres of the polyhedron (which in our case is a simplex) are kept within an interval (κ_0, κ_1), where $\kappa_0,\ \kappa_1 > 0$ are fixed numbers. Since the last condition extracts a subset of positive measure, in the right-hand side of (2.5) we can replace $N(\mathbf{Y})$ by the volume $V(\mathbf{Y})$.

Thus we must analyze the integrability, in the variable $\ell > 0$, of the expression

$$\mathrm{d}\ell\,\frac{JV}{V^n} = \mathrm{d}\ell\,\frac{J}{V^{n-1}}. \tag{5.3}$$

Now observe that the determinant $J \sim \ell^{(n-1)^2-1}$ (in fact,

$$(n-1)^2 - 1 = n(n-1) - (n-1) - 1$$

is the number of scalar variables, among the $\mathbf{Y}_j,\ 1 \leqslant j \leqslant n$, tied to ℓ). On the other hand, $V^{n-1} \sim \ell^{(n-1)^2}$. Therefore (5.3) behaves like $\mathrm{d}\ell/\ell$, which yields the divergence.

REMARKS. (1) The expression $\mathrm{d}\ell\, J/V^n$ gives a convergent integral: this reflects the fact that ν^0 is a probability measure on $\mathcal{M}$.

(2) Observe that the order of divergence $\mathrm{d}\ell/\ell$ emerging from the above argument does not depend on the dimension n (and is the same as for $n = 2$).

Now let us turn to the proof of Theorem 3.2. The fact that

$$\lim_{R\to\infty} N_R^1(\mathbf{Y}) = \infty$$

is a direct consequence of Theorem 3.1 and the ergodicity of the flow $(\mathcal{E}, T, \nu)$. After this, to check that $\lim_{R\to\infty} N_R^2(\mathbf{Y}) = \infty$, it suffices to check that $m_R(\mathbf{Y})/v_R$ is separated from 0 with ν-probability one as $R \to \infty$. This essentially means that a majority of the cells lying inside the ball $\mathbb{B}_R$ are of "moderate" volume. A sufficient condition is that the integral

$$I(\mathbf{Y}) = \sqrt{V_\Delta}\int_{\mathbb{P}(\mathbf{Y})} \frac{\mathrm{d}y}{\prod_{1\leqslant k\leqslant n} \mathrm{dist}(y, \partial_k\Delta)} \tag{5.4}$$

is uniformly bounded from below as a function of $\mathbf{Y} \in \mathcal{M}^0$, provided that the radii of the inscribed and circumscribed spheres of the ground polyhedron $\mathbb{P}(\mathbf{Y})$ (which is a simplex for $\mathbf{Y} \in \mathcal{M}$) are confined to a fixed interval (R_0, R_1), with $R_0, R_1 > 0$. Here $\Delta\,(= \Delta(\mathbf{Y}))$ is the $(n-1)$-dimensional simplex obtained as the intersection of the translated hyperplane $\mathbb{G}(\mathbf{Y})$ when it reaches the ground face of $\mathcal{V}(\mathbf{Y})$ with the interior of the positive orthant (or the positive simplicial cone) generated by vectors $e_1(\mathbf{Y}), \dots, e_n(\mathbf{Y})$. Recall that

$$\mathbb{G}(\mathbf{Y}) = \Big\{x \in \mathbb{R}^n : x = \sum \mu_i e_i(\mathbf{Y}), \text{ where } \sum \mu_i = 0\Big\}.$$

Thus, $\Delta \supset \mathbb{P}(\mathbf{Y})$. As above, $V(\Delta)$ denotes the (Euclidean) volume of the simplex Δ. Furthermore, $\partial_k \Delta$ is an $(n-2)$-dimensional face of the boundary $\partial\Delta$ (the symbol k is simply a label for such a face; altogether there are n of them). Finally, the number $\operatorname{dist}(y, \partial_k\Delta)$ is the (Euclidean) distance from the point $y \in \mathbb{P}(\mathbf{Y})$ to $\partial_k\Delta$.

The factor $\sqrt{V(\Delta)}$ is bounded below by the square root of the volume of the $(n-1)$-dimensional ball of radius R_0. The volume of this ball also provides a lower bound for the volume of the simplex $\mathbb{P}(\mathbf{Y})$. Thus, it remains to produce an upper bound for the product

$$\prod_{1 \leqslant k \leqslant n} \operatorname{dist}(y, \partial_k\Delta).$$

We present the corresponding argument in the case $n = 3$, where the simplex Δ is a plane triangle (and so is $\mathbb{P}(\mathbf{Y})$). The modifications needed for the general case are purely technical and are left to the reader.

Recall that $\mathbb{P}(\mathbf{Y})$ is a triangle with integer vertices, say $z^{(1)}$, $z^{(2)}$, $z^{(3)}$, and that there must be no integer points in $\Delta \setminus \mathbb{P}(\mathbf{Y})$. Thus, for all i, i', with $1 \leqslant i < i' \leqslant 3$, there must be a side of $\mathbb{P}(\mathbf{Y})$ that crosses the segment with the endpoints

$$z^{(i)} - (z^{(i')} - z^{(i)}) = 2z^{(i)} - z^{(i')}.$$

From this fact it is easy to deduce that two sides of the triangle Δ cross a circle of radius $3R_1$ which contains $\mathbb{P}(\mathbf{Y})$, and thus the distance from $\mathbf{Y} \in \mathbb{P}(\mathbf{Y})$ to any of them is $\leqslant 3R_1$ (recall that R_1 is a bound for the radius of the circumscribed circle for $\mathbb{P}(\mathbf{Y})$).

Now observe that each of the sides $(z^{(i')} - z^{(i)})$ of the ground triangle $\mathbb{P}(\mathbf{Y})$ forms, with either of the two aforementioned sides of Δ, an angle $\geqslant \theta^0$, where $\theta^0 \,(= \theta^0(R_0, R_1)) > 0$. Take one of the sides, say $z^{(1)} - z^{(2)}$, and draw, on the translated plane $\mathbb{G}(\mathbf{Y}) + s$, lines parallel to $z^{(1)} - z^{(2)}$ through the integer points. The distance between two such subsequent lines is $\leqslant 2R_1$, and the spacing (i.e., the distance between two neighboring integer points) on each line is $\leqslant \|z^{(1)} - z^{(2)}\|$. It is then easy to see that the third side of Δ should be at distance $\leqslant R_2 = R_2(R_0, R_1)$ from $y \in \mathbb{P}(\mathbf{Y})$. This leads to the desired bound for $\prod_{1 \leqslant k \leqslant 3} \operatorname{dist}(y, \partial_k\Delta)$. □

Proof of Lemma 3.3. We shall check that if $\tilde{\delta} > 0$ is small enough, then for any two vertices v, v' of the veil the inequality $\operatorname{dist}_{\mathbb{R}^{n-1}}(\Pi v, \Pi v') \leqslant \delta$ is impossible. Indeed, assume the opposite. Without loss of generality, we can assume that one of the points Πv, $\Pi v'$ is at the origin. Then, for the inverse images in $\mathbb{R}^n$, we see that v is of the form $(\lambda, \dots, \lambda)$ and v' is of the form $(\lambda' + \vartheta_1 \dots, \lambda' + \vartheta_n)$, where

$$|\vartheta_i| \leqslant c\tilde{\delta} \quad \text{and} \quad \left|\frac{\lambda}{\lambda'} - 1\right| \leqslant c\tilde{\delta}, \tag{5.5}$$

and c is a positive constant (depending on the dimension n only). Then, for $\tilde{\delta}$ small enough, $w = 2v - v'$ is an integer point with positive coordinates, i.e., $w \in \mathcal{C}(\mathbf{Y})$, the Klein polyhedron. Thus, $v = (w + v')/2$ cannot be a vertex. □

PROOF OF THEOREM 4.1. The quotient projections

$$G \twoheadrightarrow G_0 = \mathrm{SL}(n, \mathbb{R}), \qquad \Gamma \twoheadrightarrow \Gamma_0 = \mathrm{SL}(n, \mathbb{Z})$$

determine a fiber bundle

$$\pi \colon G/\Gamma \to G_0/\Gamma_0,$$

with the natural action of the group G. The fiber $\pi^{-1}(x)$, $x \in G_0/\Gamma_0$, of π is a torus equivalent to $\mathbb{R}^n/\mathbb{Z}^n$. Thus, points $X \in G/\Gamma$ may be written as pairs $X = (x, \theta)$, $x \in G_0/\Gamma_0$, $\theta \in \pi^{-1}(x)$. On the base G_0/Γ_0, we have the probability measure ν_0 generated by the Haar measure on $\mathrm{SL}(n, \mathbb{R})$; the measure ν is generated by ν_0 and the normalized Lebesgue measures λ_x on $\pi^{-1}(x)$. Let $U \subset G/\Gamma$ be a T'-invariant set with $\nu(U) < 1$, and χ_U its indicator function. We want to prove that $\nu(U) = 0$. The action of G (and hence that of T) preserves the affine structure on the tori; this enables us to define a function $F_U \geqslant 0$ on the dual fiber bundle

$$\widetilde{\pi} \colon D \to G_0/\Gamma_0,$$

where the fiber $\widetilde{\pi}^{-1}(x)$, $x \in G_0/\Gamma_0$, is the first cohomology group

$$H^1(\pi^{-1}(x), \mathbb{Z}) \sim \mathbb{Z}^n.$$

The function F_U is simply the square of the modulus of the Fourier transform of χ_U in the fiber variable θ (unlike the Fourier transform itself, its modulus is correctly defined). By the Parseval identity, F_U obeys

$$\sum_{\vartheta \in \widetilde{\pi}^{-1}(x)} F_U(x, \vartheta) \leqslant 1.$$

Furthermore, there is a natural action of G on D, through the projection $G \twoheadrightarrow G_0 = \mathrm{SL}(n, \mathbb{R})$, and hence an action of T', through the induced projection $T' \twoheadrightarrow T'_0$, where $T'_0 \subseteq G_0$ is a noncompact subgroup and F_U is T'_0-invariant.

It suffices to prove that the support $\operatorname{supp} F_U$ is not concentrated in the zero section $D_{(0)} = \{(x, \vartheta^0_x) : x \in G_0/\Gamma_0\}$ of the bundle D, where ϑ^0_x is the nil-cohomology class from $H^1(\pi^{-1}(x), \mathbb{Z})$. Indeed, if $\operatorname{supp} F_U \subseteq D_{(0)}$, then χ_U is λ_x-a.s. constant on $\pi^{-1}(x)$ for ν-a.a. $x \in G_0/\Gamma_0$. Thus, U is the union of fibers, modulo a set of ν-measure 0. That is, $\nu(U \triangle (\pi^{-1}(U_0))) = 0$, where $U_0 \subseteq G_0/\Gamma_0$; the set U_0 must have $\nu_0(U_0) > 0$ and be invariant under the action of T'_0. Therefore, by Moore's ergodic theorem [**17, 21**], $U_0 = G_0/\Gamma_0$, which implies that $\nu(U) = 1$.

Let us write

$$D \setminus D_{(0)} = \bigcup_{1 \leqslant k < \infty} D_{(k)}, \tag{5.6}$$

where

$$D_{(k)} = \{(x, \vartheta) : x \in G_0/\Gamma_0,\ \vartheta \in H^1(\pi^{-1}(x), \mathbb{Z}),\ \vartheta = k\vartheta_0,$$
$$\vartheta_0\ (= \vartheta_0(x)) \text{ is a primitive element of } \mathbb{Z}^d\}. \tag{5.7}$$

Note that $D_{(0)}, D_{(1)}, \dots,$ are the connected components of D. Furthermore, it is easy to see that $D_{(0)} \sim G_0/\Gamma_0$. Finally, as we show below, $D_{(k)} \sim G_0/\Gamma_1$, $k = 1, 2, \dots,$ where Γ_1 is the subgroup of Γ_0 preserving the first coordinate in $\mathbb{Z}^n$.

Indeed, observe that D is equivalent to the set of pairs (Λ, γ), where Λ is a lattice in $\mathbb{R}^n$, with $0 \in \Lambda$ and $\mathrm{vol}(\mathbb{R}^n/\Lambda) = 1$, and γ is a homomorphism $\Lambda \to \mathbb{Z}$; in this language, $D_{(k)} \sim \{(\Lambda, \gamma) : \gamma(\Lambda) = k\mathbb{Z}\}$, and, clearly, $D_{(k)} \simeq D_{(1)}$, $k \geqslant 1$. So it is enough to check that $D_{(1)} \sim G_0/\Gamma_1$, which is clear from the fact that G_0/Γ_0 is equivalent to the set of lattices $\Lambda \subset \mathbb{R}^n$ with $0 \in \Lambda$ and $\mathrm{vol}(\mathbb{R}^n/\Lambda) = 1$, and hence G_0/Γ_1 itself is equivalent to the set of pairs (Λ, γ), where Λ is as before and γ is an epimorphism $\Lambda \to \mathbb{Z}$ (the set of such epimorphisms is denoted by $\mathrm{Ep}(\Lambda, \mathbb{Z})$).

So, we assume that $F_U \restriction_{D_{(k)}} \neq 0$ for some $k \geqslant 1$. Hence we obtain a function on G_0/Γ_1, which we again denote by F_U, such that

(i) $F_U \geqslant 0$, and $F_U > 0$ on a set of positive measure;
(ii) $\sum_{\gamma \in \mathrm{Ep}(\Lambda,\mathbb{Z})} F(\Lambda, \gamma) \leqslant 1$; and
(iii) F_U is invariant under the action of a noncompact subgroup $T_0' \subset G_0$.

It follows from (i) and (ii) that F_U is integrable on G_0/Γ_0.

We shall use the following geometric fact concerning the left quotient space $T_0' \setminus G_0$:

LEMMA 5.1. *The double coset $T_0' \setminus G_0/\Gamma_1$ contains a subset $\mathcal{V}$ such that the inverse image $\mathcal{W}$ of $\mathcal{V}$ under the projection $T_0' \setminus G_0 \to T_0' \setminus G_0/\Gamma_1$ has the following property. The set $\mathcal{W}$ is an open subset of full measure of the manifold $T_0' \setminus G_0$, it is invariant under the right action of Γ_1, and Γ_1 acts on $\mathcal{W}$ discontinuously and freely. That is, for any $v \in \mathcal{W}$ there exists an open neighborhood $\mathcal{W}(v) \subseteq \mathcal{W}$ such that the map $(\beta, u) \in \Gamma_1 \times \mathcal{W}(v) \mapsto \beta(u) \in \mathcal{W}$ is an injection.*

PROOF OF LEMMA 5.1. It suffices to check the property stated in the lemma for $T' = T$. Observe that the image T_0 of the group T under the projection $G \twoheadrightarrow G_0$ is the stabilizer of the cylinder

$$\mathbb{V}_0 = \left\{(x_1, \ldots, x_n) \in \mathbb{R}^n : \sum_{j=1}^{n-1} x_j^2 = 1\right\}.$$

The left quotient space $T_0 \setminus G_0$ is the set of such cylinders $\mathbb{V}$, i.e., the set of non-negative quadratic forms $q_{\mathbb{V}}$ in n variables of rank $n-1$. Together with the above description of the right quotient space G_0/Γ_1, this leads to the following description of the double quotient space $T_0 \setminus G_0/\Gamma_1$: it is equivalent to the set of equivalence classes, under the action of G_0, of triples $(\Lambda, \gamma, q_{\mathbb{V}})$. Here the pair (Λ, γ) is the same as in the case of G_0/Γ_1, and $q_{\mathbb{V}}$ is a quadratic form in n variables of rank $n-1$ (i.e., a cylinder $\mathbb{V}$ in $\mathbb{R}^n$).

Given a representative $(\Lambda, \gamma, \mathbb{V})$ of a double coset from $T_0 \setminus G_0/\Gamma_1$, consider the linear continuation $\widetilde{\gamma}$ $(= \widetilde{\gamma}(\gamma))$ of the map γ from Λ to the whole space $\mathbb{R}^n$. Call the triple $(\Lambda, \gamma, \mathbb{V})$ *good* if the linear form $\widetilde{\gamma}$ does not vanish along the axis of the cylinder $\mathbb{V}$. The set $\widetilde{\mathcal{V}}$ of good triples forms an $\mathrm{SO}(n-1, \mathbb{R})$-orbifold, and in particular is a Hausdorff space. Namely, the nondegenerate quadratic form $\widetilde{q} = \widetilde{\gamma}^2 + q_{\mathbb{V}}$ determines a Riemannian volume on $\mathbb{R}^n$; in some coordinates this volume has the form $\mathrm{d}x_1 \cdots \mathrm{d}x_n$, while

$$\widetilde{q}(x_1, \ldots, x_n) = \sum_{j=1}^{n-1} x_j^2 \quad \text{and} \quad \widetilde{\gamma}(x_1, \ldots, x_n) = \eta x_n,$$

where $\eta > 0$. Since such coordinates are defined only up to the action of the group $\mathrm{SO}(n-1, \mathbb{R})$, this results in an orbifold structure, rather than that of a manifold, on $\widetilde{\mathcal{V}}$. However, after removing a suitable set of smaller dimension from $\widetilde{\mathcal{V}}$, the remaining set $\mathcal{V}$ forms a manifold, and the action of T_0 on its inverse image $\mathcal{W}$ is discontinuous and free. □

We continue with the proof of Theorem 4.1. By Lemma 5.1, we can take the inverse image V_0 of $\mathcal{V}$ under the projection $G_0/\Gamma_0 \to T_0' \setminus G_0/\Gamma_1$. The set V_0 is again of full measure, and we can pass from F_U to its restriction to V_0, for which we retain the previous notation F_U; clearly properties (i)–(iii) above hold for the new F_U. Then F_U must be integrable on G_0/Γ_1. However, this function arises from a function F_U^1 on $\mathcal{V}$; by Fubini's theorem, the integral of F_U equals that of F_U^1 times the volume of the group T_0'. The latter is infinite because T_0' is noncompact. This contradiction completes the proof of Theorem 4.1. □

Acknowledgements. We are grateful to V. I. Arnold for his encouragement and constant interest. We thank D. Anosov, M. Gordin, B. Gurevich, I. Ibragimov, A. Katok, S. Kerov, J. Lagarias, J. Lebowitz, G. Margulis, J. Marklof, N. Moshchevitin, M. Ratner, D. Ruelle, Ya. Sinai, A. Starkov, and A. Vershik for their interest and useful remarks. MLK and YMS thank, respectively, DPMMS, University of Cambridge, and IHES, Bures-sur-Yvette, for their hospitality during visits in 1997/8 and 1998/9. Special thanks go to S. Shea-Simonds for checking the style of the paper.

References

[1] R. Adler and L. Flatto. *Cross section maps for geodesic flows.* I, Ergodic Theory and Dynamical Systems (Proc. Special Year, Maryland, 1979-1980; A. Katok, ed.). Progress in Math., Vol. 21. Birkhäuser, Boston, 1982, pp. 103–161.

[2] ———, *The backward continued fraction map and geodesic flow*, Ergodic Theor. Dyn. Systems **4** (1984), 487–492.

[3] ———, *Geodesic flows, interval maps and symbolic dynamics*, Bull. (N.S.) Amer. Math. Soc. **25** (1991), 229–334.

[4] V. I. Arnold, *Statistics of integral convex polygons*, Funktsional. Anal. i Prilozhen. **14** (1980), no. 2, 1–3; English transl. in Functional Anal. Appl. **14** (1980), no 2.

[5] ———, *A-graded algebras and continued fractions*, Comm. Pure Appl. Math. **42** (1989), 993–1000.

[6] A. Borel, *Linear algebraic groups*, Springer-Verlag, Berlin, 1991.

[7] A. J. Brentjes. *Multi-dimensional continued fraction algorithms*, Math. Centre Tracts, No. **145**, Math. Centrum, Amsterdam, 1981.

[8] J. W. A. Cassels. *An introduction to the geometry of numbers*, Springer-Verlag, Berlin, 1959.

[9] Y. Colin de Verdière, *Nombre de points entiers dans une famille homothétique de domaines de* $\mathbf{R}^n$, Ann. Sci. École Norm. Sup. (4) **10** (1977), 559–575.

[10] ———, *Nombre de points entiers dans une famille homothétique de domaines de* $\mathbf{R}^n$, Séminaire Goulaouic–Schwartz (1076/1977), Équations aux Dérivées Partielles et Analyse Fonctionnelle, Exposé 23, École Polytech., Palaiseau, 1977.

[11] I. P. Cornfeld, S. V. Fomin, and Ya. G. Sinai, *Ergodic theory*, Springer-Verlag, Berlin, 1982.

[12] W. Duke, Z. Rudnick, and P. Sarnak. *Density of integer points on affine homogeneous varieties*, Duke Math. J. **71** (1993), 143–180.

[13] P. M. Gruber and C. F. Lekkerkerker. *Geometry of numbers*, North-Holland, Amsterdam, 1987.

[14] E. Korkina, *Two-dimensional continued fractions. The simplest examples*, Trudy Mat. Inst. Steklov. **209** (1995), 143–166; English transl., Proc. Steklov Math. Inst. **209** (1995), 124–144.

[15] G. Lachaud, *Voiles et polyèdres de Klein*, 1, 2. Preprints No. 95-22. Laboratiore de Mathématiques Discrètes, CNRS, Marseille-Luminy, 1995.

[16] J. C. Lagarias, *Geodesic multidimensional continued fractions*, Proc. London Math. Soc. **(3) 69** (1994), 464–488.

[17] G. A. Margulis, *Discrete subgroups of semisimple Lie groups*, Springer–Verlag, Berlin, 1991.

[18] ———, *Dynamical and ergodic properties of subgroup actions on homogeneous spaces with applications to number theory*, Proc. Internat. Congr. Math. (Kyoto, 1990), Vol. 1, Math. Soc. Japan, Kyoto, 1991, pp. 193–215.

[19] C. Series, *The modular surface and continued fractions*, J. London Math. Soc. **31** (1985), 69–80.
[20] M. M. Skriganov, *Ergodic theory on* SL(n)*, Diophantine approximations and anomalities in the lattice point problem,* Invent. Math. **132** (1998), 1–72.
[21] R. Zimmer. *Ergodic theory and semisimple groups*, Birkhäuser, Boston, 1984.

I.H.E.S., Bur-sur-Yvette, F-91440, France; Institute for Problems of Information Transmission, Russian AS, Moscow 101447, Russia

Institute for Problems of Information Transmission, Russian AS, Moscow 101447, Russia; Statistical Laboratory, DPMMS, University of Cambridge, Cambridge CB2 1SB, and St John's College, Cambridge CB2 1TP, UK

Amer. Math. Soc. Transl.
(2) Vol. **197**, 1999

C^0-Generic Properties of Boundary Operators in the Novikov Complex

A. Pajitnov

1. Introduction

1.1. The Novikov complex and the exponential growth conjecture.

1.1.1. *Morse complex of a Morse function.* We start by recalling one of the basic constructions of Morse theory, the Morse complex. Let M be a closed manifold, and $g\colon M \to \mathbb{R}$ a Morse function. Let $S(g)$ be the set of critical points of g, and $S_k(g)$ the set of critical points of g of index k. Consider the free abelian group $C_k(g)$, freely generated by $S_k(g)$. We shall introduce the homomorphisms $\partial_k\colon C_k(g) \to C_{k-1}(g)$, which will endow $C_*(g)$ with the structure of a chain complex, called the *Morse complex*. The homomorphisms depend on some additional choices. Namely, one must choose:

(1) a gradient-like vector field[1] v for g satisfying the transversality condition (Definition 2.21 below), and
(2) orientations of the stable manifolds of the zeros of v.

Having made these choices, we proceed as follows. Let

$$p, q \in S(f), \qquad \operatorname{ind} p = \operatorname{ind} q + 1.$$

Let $L(p,q)$ be the set of orbits of v joining p to q. The transversality condition for v implies that $L(p,q)$ is finite. The choice of orientations of the stable manifolds allows us to associate to each $\gamma \in L(p,q)$ a sign $\varepsilon(\gamma) \in \{-1,+1\}$. Set

$$n(p,q) = \sum_{\gamma \in L(p,q)} \varepsilon(\gamma).$$

The integer $n(p,q)$ is called the *incidence coefficient* corresponding to p, q. Now define the homomorphism ∂_k on the generators $p \in S_k(f)$ by the formula

$$\partial_k(p) = \sum_{r \in S_{k-1}(f)} n(p,r)r.$$

One can prove that $\partial_k \circ \partial_{k+1} = 0$ and that the homology of the resulting complex is isomorphic to $H_*(M)$. (See for example [**25**, Appendix] for the details.)

1991 *Mathematics Subject Classification.* Primary 57R70, 58F25.

[1]See [**16**, Definition 3.1] for the definition of gradient-like vector fields.

1.1.2. *Novikov complex of a Morse map to the circle.* In the early eighties S. P. Novikov generalized the construction of the Morse complex to the case of Morse maps $f\colon M \to S^1$. We shall describe the construction in detail a bit later (see 5.1.2). Here we just summarize the properties of the resulting chain complex.

Let $f\colon M \to S^1$ be a Morse map, where M is a closed connected manifold. We assume that f is not homotopic to zero. As before, let $S(f)$ be the set of critical points of f and $S_k(f)$ the set of critical points of f of index k. Just as the classical Morse complex, the Novikov complex is a free complex generated in dimension k by the set $S_k(f)$, but the base ring here is much larger. Namely, consider the ring of all Laurent power series in one variable with integral coefficients and finite negative part; that is,

$$\mathbb{Z}((t)) = \Big\{\lambda = \sum a_k t^k \mid \forall k,\ a_k \in \mathbb{Z} \text{ and } \exists N = N(\lambda) \text{ such that } a_k = 0 \text{ if } k < N\Big\}.$$

Let $\mathcal{C}\colon \bar{M} \to M$ be the infinite connected cyclic cover such that $f \circ \mathcal{C}$ is homotopic to zero (note that $\mathcal{C}$ is uniquely determined by this last property). To define the Novikov complex, one must make some additional choices. Namely, we must choose:

(1) a gradient-like vector field v for f, satisfying the transversality condition (the definition of gradient-like vector field for the maps to the circle copies that for Morse functions);
(2) orientations of the stable manifolds of the critical points of f; and
(3) for every $p \in S(f)$, a lifting $\bar{p}$ of p to $\bar{M}$ (that is, a point $\bar{p} \in \bar{M}$ satisfying $\mathcal{C}(\bar{p}) = p$).

Having made these choices, one applies the procedure described in 1.1.1 (with the corresponding modifications, see 5.1.3 for details) and obtains, for every pair p, q of critical points of f satisfying $\operatorname{ind} p = \operatorname{ind} q + 1$, an element $n(\bar{p}, \bar{q}; v) \in \mathbb{Z}((t))$, called the *Novikov incidence coefficient* of p, q. (This coefficient depends on the choice of the liftings $\bar{p}, \bar{q} \in \bar{M}$, see formula (5.5).) We shall also write

$$n(\bar{p}, \bar{q}; v) = \sum_i n_k(\bar{p}, \bar{q}; v) t^k.$$

One can prove that $\partial_{k+1} \circ \partial_k = 0$ for every k. Thus we obtain a chain complex $C_*(f, v)$ such that $C_p(f, v)$ is a free $\mathbb{Z}((t))$-complex freely generated by $S_p(f)$. One can prove that

$$H_*(C_*(f, v)) \approx H_*(\bar{M}) \bigotimes_{\mathbb{Z}[t, t^{-1}]} \mathbb{Z}((t)).$$

(See [**25**] for details.)

REMARK 1.1. In this paper we introduce the notion of the f-gradient, which is a generalization of the notion of gradient-like vector field for f (Definition 2.3). The gradient of f with respect to an arbitrary Riemannian metric is an f-gradient, as well as a gradient-like vector field for f. One defines the Novikov incidence coefficient and the Novikov complex for f-gradients as described above (see 5.1.3). We shall use this in the following sections of the Introduction.

1.1.3. *Morse–Novikov theory.* The Novikov complex was introduced by Novikov in [**20**]. In that paper analogs of the classical Morse inequalities were obtained.

These inequalities (called *Novikov inequalities*) were then studied in a series of papers.[2]

Farber [**9**] proved that these inequalities are sharp in the case $\pi_1 M = \mathbb{Z}$, $\dim M \geqslant 6$. J.-C. Sikorav [**41**] applied these inequalities to the theory of Lagrangian intersections. In [**33, 34**] the author obtained some results concerning the sharpness of these inequalities for $\pi_1 M = \mathbb{Z}^m$, $\dim M \geqslant 6$. The study of the Novikov complex itself advanced slower. In [**40**] yet another version of the Novikov complex, defined over a completion of the group ring of $\pi_1(M)$, was given. In [**29, 25**] I proved that the simple homotopy type of this chain complex is equal to the simple homotopy type of the completed chain simplicial complex of the universal covering of M. In [**30, 26**] I proved a theorem on the realization of the abstract chain complexes as Novikov complexes of Morse maps $M \to S^1$. This theorem implies the results of [**9**] and of [**34**]. The algebraic result by Ranicki [**39**] shows that the main theorem of [**30, 26**] also implies the Farrell fibration theorem [**10**]. In [**13**] Latour has developed an approach to Morse–Novikov theory based on Cerf's method [**6**].

1.1.4. *Novikov exponential growth conjecture.* A priori the Novikov incidence coefficients $n(\bar{p}, \bar{q}; v)$ are elements of $\mathbb{Z}((t))$, that is, power series of the form

$$\sum_{i=-N}^{\infty} n_k(\bar{p}, \bar{q}; v) t^k, \qquad n_k(\bar{p}, \bar{q}; v) \in \mathbb{Z}.$$

From the outset Novikov conjectured that these power series had some nice analytic properties. Namely, he conjectured that the positive part of the power series $n(\bar{p}, \bar{q}; v)$ converges in a disc of nonzero radius with center 0 (equivalently: the coefficients $n_k(\bar{p}, \bar{q}; v)$ grow at most exponentially with k). For several years the conjecture has been discussed and adjusted. Now there are several versions of the conjecture. To discuss these versions it is convenient to introduce a definition.

DEFINITION 1.2. (1) Let $f\colon M \to S^1$ be a Morse map, and let v be an f-gradient, satisfying the transversality condition. The pair (f, v) will be called a *Morse–Novikov pair.*

(2) Let (f, v) be a Morse–Novikov pair. We say that the *exponential growth property* holds for the pair (f, v) if for every $p, q \in S(f)$ with $\operatorname{ind} p = \operatorname{ind} q + 1$ there are $C, D > 0$ such that for every $k \in \mathbb{Z}$ we have

$$|n_k(\bar{p}, \bar{q}; v)| \leqslant Ce^{kD}.$$

The initial version of Novikov exponential growth conjecture said that the exponential growth property holds for every Morse–Novikov pair (f, v).

As far as I know, the first published version of the conjecture appeared in V. I. Arnold's paper "Dynamics of Intersections", 1989. Arnold wrote in [**2**, p. 83]:

> The author is indebted to S. P. Novikov who has communicated the following conjecture which was the starting point of the present paper. Let $p\colon \widetilde{M} \to M$ be a covering of a compact manifold M with fiber $\mathbb{Z}^n$ and let α be a closed 1-form on M such that $p^*\alpha = df$ where $f\colon \widetilde{M} \to \mathbb{R}$ is a Morse function. The Novikov conjecture states that "generically" the number of trajectories of the vector field $-\operatorname{grad} f$ on $\widetilde{M}$ starting at a critical point x of the function

[2] We make no attempt to give a complete survey here.

> f of index k and connecting it with those critical points y having index $k-1$ and satisfying $f(y) \geqslant f(x)-n$, grows in n more slowly than some exponent, e^{an}.

Thus the version of the Novikov exponential growth conjecture cited in [**2**] says, in particular, that the exponential growth property holds for every generic pair (f, v), where v is a *Riemannian gradient* of f. Note that this version is stated in a somewhat more general setting than our present one: it deals with Morse 1-forms, and not only with Morse maps to the circle. Note also that Arnold has conjectured something a bit stronger that the original Novikov exponential growth conjecture. Namely, this version deals with the *absolute number* of trajectories joining p with q.

Here is the version of the conjecture published in 1993 by Novikov [**23**, p. 229]:

> CONJECTURE. For any closed quantized analytic 1-form ω the boundary operator ∂ in the Morse complex has all coefficients $a_{pq}^{(i)} \in K$ with positive part convergent in some region $|t| \leqslant r_1^*(\omega) \leqslant r_1$. (If we replace S by $(-S)$ and t by t^{-1} we have to replace r_1 by r_1^{-1}.) In particular, $r_1^* \neq 0$ and there will be coefficients $a_{pq}^{(i)}$ with radius of convergence not greater than $r_1(M, [\omega])$ if at least one jumping point really exists.[3]

Thus the version of [**23**] says that the exponential growth property holds for every Morse pair (f, v), where f is analytic and v is a Riemannian gradient of f.

We proceed to versions of the Novikov exponential growth conjecture that are in a sense stronger.

DEFINITION 1.3. We say that the *rationality property* holds for the Morse–Novikov pair (f, v), if for every $p, q \in S(f)$ with $\operatorname{ind} p = \operatorname{ind} q + 1$ the power series $n(\bar{p}, \bar{q}; v)$ is the Laurent series of a rational function.

It is obvious that the rationality property for a Morse–Novikov pair (f, v) implies the exponential growth property for (f, v).

M. Farber conjectured that the rationality property holds for every Morse–Novikov pair (f, v). P. Vogel conjectured that the rationality property holds for every Morse–Novikov pair (f, v). He motivated his conjecture by the fact that the Novikov incidence coefficients are results of the same procedure applied to the fundamental cobordism of f (see the definition in 1.2.2 below).

1.2. C^0-generic rationality of Novikov incidence coefficients. Contents of the paper.

1.2.1. *Contents of the paper.* In this subsection we give a brief outline of the contents of the present paper; the reader will find more information on the principal ideas in their relation to the main theorem in 1.2.4 and in the comments and introductions to different sections.

About 1995 I proved that the Novikov incidence coefficients are C^0-generically rational functions. The results first appeared in the e-print `dg-ga 9603006`

---

[3] For the reader's convenience, we explain the terminology in this citation. Novikov denotes by K the ring denoted by $\mathbb{Z}((t))$ in the present paper; p, q correspond to critical points with $\operatorname{ind} p = \operatorname{ind} q + 1$, and $a_{pq}^{(i)}$ is $n(p, q; v)$ in our notation. A closed 1-form w is called *quantized* if its cohomology class belongs to $H^1(M, \mathbb{Z})$.

(14 March 1996). They were announced in [**27**] and published in [**35**], which is a version of [**31**] revised for publication.

The aim of the present paper is to generalize this result and to provide a systematic treatment of the subject. For a Morse map $f\colon M \to S^1$, we define a new class of vector fields, which contains the gradient-like vector fields for f and the Riemannian gradients for f. These vector fields are called f-gradients (see Definition 2.3). We give (Section 2) a systematic and self-contained presentation of all the necessary results and constructions from Morse theory; so Section 2 can be considered as an introduction to the corresponding part of Morse theory.

Subsections 2.1.5, 2.1.6, 2.3.2, 2.3.3, 2.4.1, 2.4.2 contain the standard Morse-theoretical techniques, generalized to the case of f-gradients. In subsections 2.1.4 and 2.2, 2.5, we generalize the techniques of [**35**] to the case of f-gradients.

Section 3 contains a construction which is a Morse-theoretic analog of the triangulation of a manifold with simplices of arbitrary small diameter. This construction makes no use of triangulations of the manifold; we do not leave the framework of Morse theory, Morse functions, and their gradients. The content of this section does not differ much from the corresponding chapter of [**35**]; we have only revised the exposition on order to make it easier to read.

Section 4 is technically the central part of the work. It deals with gradients of Morse functions on cobordisms. We construct a small perturbation of a given gradient, so that the gradient descent map associated to the perturbed gradient respects a given handle decomposition of the components of the boundary of the cobordism.

Section 5 contains the proof of our main results. Let $\mathcal{GT}(f)$ be the set of all f-gradients satisfying the transversality condition. We introduce a subset $\mathcal{G}\mathfrak{C}\mathcal{CT}(f) \subset \mathcal{GT}(f)$ which is open and dense in the C^0 topology (see Proposition 5.4). The next theorem asserts that C^0-generically the Novikov incidence coefficients are rational functions. Denote by $\widetilde{L}$ the subring of the ring $\mathbb{Q}(t)$ formed by all the rational functions of the form $P(t)/t^m Q(t)$, where $m \in \mathbb{N}$ and P, Q are polynomials with integer coefficients and $Q(0) = 1$.

THEOREM 5.5. *Let $v \in \mathcal{G}\mathfrak{C}\mathcal{CT}(f)$ and $x, y \in S(f)$, $\operatorname{ind} x = \operatorname{ind} y + 1$. Then $n(x, y; v) \in \widetilde{L}$,*

The gradients belonging to $\mathcal{G}\mathfrak{C}\mathcal{CT}(f)$ have remarkable properties. In particular, the incidence coefficients $n(x, y; v)$ are invariant under C^0-small perturbations of $\mathcal{GCT}(f)$, if this perturbation has support outside $S(f)$. Here is the precise statement.

THEOREM 5.7. *Let $v \in \mathcal{G}\mathfrak{C}\mathcal{CT}(f)$ and let U be an open neighborhood of $S(f)$. Then there is a $\delta > 0$ such that for every $w \in \mathcal{G}\mathfrak{C}\mathcal{CT}(f)$ with $\|w - v\| < \delta$ and $v|U = w|U$ and for every $x, y \in S(f)$ with $\operatorname{ind} x = \operatorname{ind} y + 1$ we have*

$$n(x, y; v) = n(x, y; w).$$

There are some naturally arising versions of the Novikov complex defined over completions of group rings, for example the completion of the group ring $\mathbb{Z}[\pi_1(M)]$. (One may consider these versions as "equivariant analogs" of the ordinary Novikov complex.) There is a corresponding generalization of Theorem 5.5 for this situation. To cite this theorem here, let $f\colon M \to S^1$ be a Morse map, which we assume to be primitive, i.e., assume that the homomorphism $f_*\colon \pi_1(M) \to \pi_1(S^1) = \mathbb{Z}$

is epimorphic. To the homomorphism $\xi = f_*$ we associate a certain completion $\widehat{\Lambda}_\xi$ of the group ring $\Lambda = \mathbb{Z}[\pi_1(M)]$ (see 5.1.1 for the definition). There is a noncommutative localization procedure that allows us to define a certain subring of $\widehat{\Lambda}_\xi$, which is a natural analog of the ring $\widetilde{L}$. This subring is denoted by $\operatorname{Im}\ell$ (see 5.3.2 for the definition). The following statement is a part of Theorem 5.14.

THEOREM. *Let $v \in \mathcal{G}\mathfrak{CCT}(f)$. Then for every $x, y \in S(f)$ with* $\operatorname{ind} x = \operatorname{ind} y + 1$, *the Novikov incidence coefficient $\hat{n}(\hat{x}, \hat{y}; v)$ is in* $\operatorname{Im}\ell$.

1.2.2. *Preliminaries on Morse maps $M \to S^1$.* In this subsection we introduce some terminology that will also be used in Section 5.

Let M be a closed connected manifold and let $f\colon M \to S^1$ be a Morse map, not homotopic to zero. We shall assume that the homotopy class

$$[f] \in [M, S^1] = H^1(M, \mathbb{Z})$$

is indivisible (the general case is easily reduced to this one by considering the map $\tilde{f} = f/n$ for some $n \in \mathbb{N}$).

The map f lifts to a Morse function $F\colon \bar{M} \to \mathbb{R}$. Let t be the generator of the structure group ($\approx \mathbb{Z}$) of $\mathcal{C}$, so that for every x we have $F(xt) < F(x)$. (We invite the reader to look at Figure 1 to visualize these constructions.)

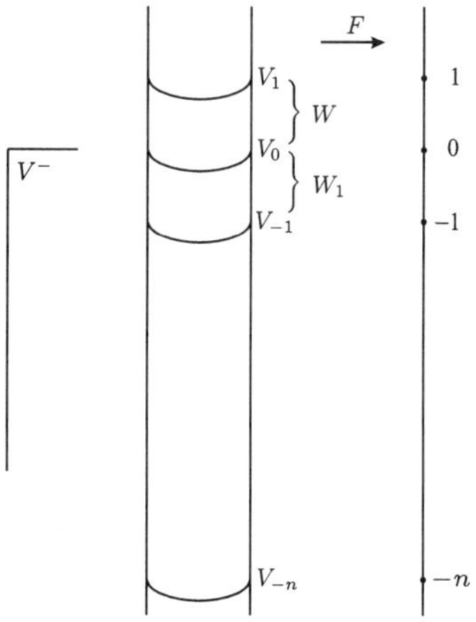

FIGURE 1

We shall assume that f is primitive; that is, $f_*\colon H_1(M) \to H_1(S^1)$ is epimorphic (one can always reduce the situation to this case by considering the map F/n for some $n \in \mathbb{N}$.) In this case we have $F(xt) = F(x) - 1$. To simplify the notation, we shall assume that $1 \in S^1$ is a regular value of f, and therefore every $n \in \mathbb{N}$ is a regular value of F. Denote $f^{-1}(1)$ by V. Set

$$V_\alpha = F^{-1}(\alpha), \quad W = F^{-1}([0,1]), \quad V^- = F^{-1}(]-\infty, 1]).$$

Note that $V_\alpha t = V_{\alpha-1}$.

The cobordism W can be thought of as the result of cutting M along V. We shall sometimes call W the *fundamental cobordism* of f. Denote Wt^s by W_s; then $\bar{M}$ is the union $\bigcup_{s\in\mathbb{Z}} W_s$, the neighboring copies W_s and W_{s+1} intersecting in V_{-s}. For any $n \in \mathbb{Z}$ the restriction of $\mathcal{C}$ to V_n is a diffeomorphism $V_n \to V$.

The $\mathcal{C}$-preimage of a subset $A \subset M$ will be denoted by $\bar{A}$. On the other hand, if $x \in M$, we shall reserve the symbol $\bar{x}$ for liftings of x to $\bar{M}$ (that is, $\bar{x} \in \bar{M}$ and $\mathcal{C}(\bar{x}) = x$). Let v be an f-gradient. Lift v to $\bar{M}$; we obtain a complete vector field on $\bar{M}$, which is obviously an F-gradient. We shall denote this F-gradient by the same symbol v, since there is no possibility of confusion. The restriction of this F-gradient to any cobordism $W_{[\lambda,\mu]} = F^{-1}([\lambda,\mu])$ (where λ, μ are regular values of F) is an $(F|W_{[\lambda,\mu]})$-gradient. Note that if v satisfies the transversality condition, so does the F-gradient $v|W_{[\lambda,\mu]}$.

The set of all f-gradients will be denoted by $\mathcal{G}(f)$, and the set of all f-gradients satisfying the transversality condition will be denoted by $\mathcal{GT}(f)$.

By definition the *Novikov incidence coefficient* $n_i(p,q;v)$ is equal to the incidence coefficient $n(p,qt^i;v)$ of the critical points p and qt^i of the Morse function F, restricted to the cobordism $F^{-1}([a,b])$, where the interval $[a,b]$ is chosen large enough to contain $F(p)$ and $F(qt^i)$.

1.2.3. *First step of the proof.* We continue to use the terminology of the preceding subsection.

Let $p,q \in W$ be critical points of F, $\operatorname{ind} p = \operatorname{ind} q + 1$. Set $n = \dim M$ and $l+1 = \operatorname{ind} p$. For simplicity, we assume that $F(p)$ is the lowest critical level of $F|W$, so that the interval $]0,F(p)[$ is regular. Similarly, we assume that the interval $]F(q),1[$ is regular and that $\bar{p} \in W_0$, $\bar{q} \in W_1$. Let $S(p)$ denote the intersection of the stable manifold of p with V_0; this is an oriented l-dimensional embedded sphere in V_0. The intersection $S(q)$ of the unstable manifold of q with V_1 is a cooriented $(n-l-1)$-dimensional sphere in V_1.

Let $v \in \mathcal{GT}(f)$. Let K_1 be the union of all unstable discs of critical points of $F|W$, and K_0 the union of all stable discs of critical points of $F|W$. Let $(-v)^{\leadsto}$ be the diffeomorphism of $V_1 \setminus K_1$ to $V_0 \setminus K_0$ which sends each point x to the point of intersection of the $(-v)$-trajectory starting at x with V_0. Using the presentation of $\bar{M}$ described above, we obtain the identity

$$n_k(p,q;v) = ((-v)^{\leadsto})^k\big(S(p)\big) \# S(q). \tag{1.1}$$

Here are some comments on this formula. 1) The symbol $\#$ stands for the algebraic intersection index of two manifolds; it can be checked that this number is well defined in our situation. 2) $(-v)^{\leadsto}$ is not defined on all of V_1; nevertheless, we denote by $(-v)^{\leadsto}(A)$ the set $(-v)^{\leadsto}(A \setminus K_1)$, where $A \subset V_1$. 3) We identify V_1 and V_0 with V, so that the iterations of $(-v)^{\leadsto}$ make sense and are applicable to $S(p) \subset V_0$.

Should $(-v)^{\leadsto}$ be an everywhere defined diffeomorphism, the intersection index in (1.1) would coincide with the intersection index of some homology classes, and our result would be a consequence of linear algebra. Indeed, let $[p]$ be the homology class of the sphere $S(p)$ in $H_l(V)$, let $]q[\in H^l(V)$ be the cohomology class dual to $S(q)$, and let $L\colon H_*(V) \to H_*(V)$ be the homomorphism induced by $(-v)^{\leadsto}$. Then

$$((-v)^{\leadsto})^k(S(p)) \# S(q) = \langle]q[, L^k([p])\rangle t^k$$

(here and in what follows we denote by $\langle \cdot, \cdot \rangle$ the canonical pairing between homology and cohomology). A simple computation shows that the power series

$$\sum_{k=0}^{\infty} \langle]q[, L^k([p])t^k\rangle$$

is a rational function with denominator $1 - \det L \cdot t$ (see Lemma 5.6 for details).

These considerations were well known to S. P. Novikov and V. I. Arnold at the end of the eighties. They constitute the motivation for the Novikov exponential growth conjecture. I learned the algebraic Lemma 5.6 from a lecture by Arnold in Moscow in 1990. The paper [**2**] gave rise to a new domain of investigation: the theory of topological complexity of intersection. The basic objects of study in this theory are intersections of the form $A^n X \cap Y$, where X and Y are submanifolds of a manifold M and A is a diffeomorphism $M \to M$. One tries to establish the asymptotic properties of this intersection as $n \to \infty$. See [**3**] and the bibliography there for the present state of this theory.

1.2.4. *Outline of the proof of the main theorem.* The exposition in this section is rather informal, but we think that it will be useful to the reader to go through it, since it explains the main ideas of what follows.

We continue the arguments of the previous subsection. If f has critical points, then the map $(-v)^{\leadsto}$ is no longer an everywhere defined diffeomorphism, and the indices $n_i(p, q; v)$ do not admit any simple homological interpretation. We shall show that for a C^0-generic f-gradient v we can still obtain a formula of the type

$$n(p, q; v) = \sum_{k=0}^{\infty} \langle]q[, L^k([p]) \rangle t^k,$$

where, however, the meaning of the objects $[p]$, $]q[$, L will be more complicated than in the model situation discussed in 1.2.3.

Namely, we shall show that for any f and any $p, q \in S(f)$ one can find an C^0-open-and-dense subset $G \subset \mathcal{GT}(f)$, disjoint compact sets $A, B \subset V$, and a homomorphism $L\colon H_l(V \setminus B, A) \to H_l(V \setminus B, A)$ such that the following condition is fulfilled for any $v \in G$:

$S(p)$ does not intersect B, $S(q)$ does not intersect A, the group $H_*(V \setminus B, A)$ is finitely generated, and

$$n(p, q; v) = \sum_{k=0}^{\infty} \langle]q[, L^k([p]) \rangle t^k, \tag{1.2}$$

where $[p]$ is the image in $H_l(V \setminus B, A)$ of the fundamental class $[S(p)]$, and $]q[$ is the element of $H^l(V \setminus B, A)$ dual to $S(q)$.

REMARK 1.4. The homomorphism L is *not* induced by any continuous map of the pair $(V \setminus B, A)$ to itself.

Now we describe the sets A, B, the homomorphism L, and the C^0-small perturbations of v satisfying (1.2). We use the terminology of 2.1.4, 2.1.5. We start with an auxiliary Morse function $\phi\colon V \to \mathbb{R}$ and a ϕ-gradient u satisfying the almost transversality condition. For $i \leqslant n-1$ consider the set $D(\text{ind}{\leqslant}i\,; u)$ (recall from Definition 2.16 that it is the union of all the stable manifolds of v of dimension $\leqslant i$). These sets form a kind of cellular decomposition of V, and we shall call $D(\text{ind}{\leqslant}i\,; u)$ the *i-skeleton* of V. (This is an abuse of terminology, but in this subsection we shall not use the term *skeleton* in its usual meaning.) The sets $D(\text{ind}{\leqslant}j\,; -u)$ form the dual decomposition of V. We also have "thickenings" of this skeleton: the sets $B_\delta(\text{ind}{\leqslant}i\,; u)$, $D_\delta(\text{ind}{\leqslant}i\,; u)$, $C_\delta(\text{ind}{\leqslant}i\,; u)$. Here $B_\delta(\text{ind}{\leqslant}i\,; u)$ is an open set, and $D_\delta(\text{ind}{\leqslant}i\,; u)$, $C_\delta(\text{ind}{\leqslant}i\,; u)$ are compact. One may think of $D_\delta(\text{ind}{\leqslant}i\,; u)$ as the union of all handles of V of thickness δ and indices $\leqslant i$.

For δ sufficiently small, the sets $D_\delta(\text{ind}{\leqslant}i\,; u)$ and $D_\delta(\text{ind}{\leqslant}n-i-2\,; -u)$ are disjoint.

Put $A = D_\delta(\text{ind}\leqslant l-1\,; u)$, $B = D_\delta(\text{ind}\leqslant n-l-2\,; -u)$.

Returning to the cobordism W with boundary $\partial W = V_0 \sqcup V_1$, we note that for every geometric object associated with V there are two copies of it that correspond to V_0 and to V_1, because both V_0 and V_1 are diffeomorphic to V. These objects will be denoted by the same letter but with indices 0 or 1, respectively. In this way, we obtain Morse functions ϕ_0, ϕ_1, compact sets A_0, A_1, etc.

The homomorphism L will be constructed for vector fields v satisfying an additional restriction, namely condition $(\mathfrak{C}\mathcal{Y})$ (see Definition 4.7). The principal part of this condition is given by the following two formulas:

$$(-v)^{\leadsto}(C_\delta(\text{ind}\leqslant j\,; u_1)) \cup (D_\delta(\text{ind}\leqslant j+1\,; v) \cap \partial_0 W) \subset B_\delta(\text{ind}\leqslant j\,, u_0) \text{ for every } j, \quad \text{(B1)}$$

$$\overset{\leadsto}{v}(C_\delta(\text{ind}\leqslant j\,; -u_0)) \cup (D_\delta(\text{ind}\leqslant j+1\,; -v) \cap \partial_1 W) \subset B_\delta(\text{ind}\leqslant j\,; -u_1) \text{ for every } j \quad \text{(B0)}$$

(see Definitions 4.7, 4.2 for the details).

The conditions (B0), (B1) may be regarded as analogs of the cellular approximation conditions; however, an additional restriction is imposed: the soles of the δ-thickened handles of v of dimension less than or equal to $j+1$ must belong to the δ-thickened j-skeleton of V.

If condition $(\mathfrak{C}\mathcal{Y})$ is satisfied, then we construct the homomorphism

$$L = \mathcal{H}_l(-v)\colon H_l(V \setminus B, A) \to H_l(V \setminus B, A).$$

To clarify the idea of this construction, we make a simplifying assumption: we assume that the homology class $x \in H_l(V \setminus B, A)$ is represented by an oriented closed submanifold N of $V \setminus B$ of dimension l. Set $N' = (-v)^{\leadsto}(N_1)$; then N' is a submanifold of V_0 (noncompact in general). The condition (B1) implies $N' \subset V_0 \setminus B_0$. The condition (B0) implies that of the discs $D(a, -v)$, the manifold N_1 can intersect only those for which $\operatorname{ind} a \leqslant l$. Therefore the set $N' \cup (D(\text{ind}\leqslant l\,; v) \cap V_0)$ is compact, and the set $N' \setminus B_\delta(\text{ind}\leqslant l-1\,; u_0)$ is also compact. Thus the fundamental class of N' modulo A_0 is well defined as an element of $H_l(V_0 \setminus B_0, A_0)$. Identifying V_0 and V, we obtain an element of $H_*(V \setminus B, A)$, which is by definition $\mathcal{H}_l(-v)(x)$.

The detailed construction of $\mathcal{H}_l(-v)$ is the subject of subsection 4.2.

Now we explain how to achieve properties (B0), (B1) by a C^0-small perturbation of the initial f-gradient v (that is, how to prove that the condition $(\mathfrak{C}\mathcal{Y})$ is C^0-dense).

As we have already noted, for a compact submanifold R in V_1, the submanifold $(-v)^{\leadsto}(R)$ in V_0 is not necessarily compact; to make it compact, we must add the soles of the descending disks of v (thus losing the manifold property). Therefore, it should be expected that the correspondence $\cdot \mapsto (-v)^{\leadsto}(\cdot)$ is well defined on an appropriate category of stratified manifolds. As will be shown in subsection 2.2, this is indeed so. The results of 2.2 imply that (under some mild transversality conditions) the set

$$\mathcal{R}_j = (-v)^{\leadsto}(D(\text{ind}\leqslant j\,; u_1)) \cup (D(\text{ind}\leqslant j+1\,; v) \cap V_0) \tag{1.3}$$

is compact for every j. Since this set is the union of submanifolds of V_0 of dimension not exceeding j, we can push $\mathcal{R}_j$ away from $D(\text{ind}\leqslant n-j-2\,; -u_0)$; then, applying a diffeomorphism $\Phi(T, -u_0)$ for T sufficiently large, we obtain an isotopy of V_0 that pushes $\mathcal{R}_j$ into $B_\delta(\text{ind}\leqslant j\,; u_0)$. After some work with thickenings, we obtain an isotopy of V_0 such that the set $(-v)^{\leadsto}(D_\delta(\text{ind}\leqslant j\,; u_1)) \cup (D_\delta(\text{ind}\leqslant j+1\,; v) \cap V_0)$ is also pushed inside $B_\delta(\text{ind}\leqslant j\,; u_0)$. This isotopy may be realized by modifying v; the new gradient will satisfy (B0) and (B1). This procedure is described in detail in subsection 4.1. The resulting gradient can be chosen so as to be C^0-close to the initial one. In other words, the pair (ϕ_0, u_0) can be chosen in such a way that $\Phi(\tau, -u_0)$ pushes $V_0 \setminus B_\delta(\mathrm{ind}\leqslant n-j-2\,; -u_0)$ inside $B_\delta(\mathrm{ind}\leqslant j\,; u_0)$ already for τ sufficiently small. Such flows will be called *quick flows.* The construction of quick flows is the subject of Section 3.

As we have already mentioned, our construction of quick flows differs from the ordinary constructions of a "small handle decomposition" in that it uses no triangulations of the manifold. A description of the basic idea of this construction can be found at the beginning of Section 3.

1.2.5. *Further remarks.* 1. Note that the C^0 topology is very important for us throughout the paper. One reason why the C^0 topology is relevant has to do with the following property of integral curves of differential equations: the integral curves (on a given finite interval) of a C^0*-small* perturbation v' of a C^1-vector field v are C^0-small perturbations of the integral curves of v (see, e.g., [**4**, Chapter 5, §4]).

See subsection 2.6 for a detailed exposition of C^0 continuity properties. We think in general that the C^0 topology is much more natural and efficient in Morse theory than is generally believed at present.

2. The subsection 5.3 of the paper is devoted to the generalization of the Novikov exponential growth conjecture to the case of incidence coefficients with values in group rings. These results (in a somewhat weaker form) were announced in [**27**]. Unfortunately, I have no time to include in the present paper the proofs of the other results announced in [**27**]; that will be the subject of the second part of this paper.

1.3. Terminology.

1.3.1. *Terminological conventions.* Here I have gathered some definitions and notational conventions that will be used throughout the paper.

The closed disc of radius r and center a in Euclidean space $\mathbb{R}^n$ is denoted by $D^n(a,r)$.

The open disc of radius r and center a in Euclidean space $\mathbb{R}^n$ is denoted by $B^n(a,r)$.

Let v be a vector field on a manifold W (possibly with boundary). We denote by $\gamma(x,t;v)$ the value at t of the integral curve of v starting at x. If $A \subset \mathbb{R}$ and $\gamma(x,\,\cdot\,;v)$ is defined on A, then $\gamma(x,A;v)$ denotes the set $\{\gamma(x,t;v) \mid t \in A\}$.

A subset $X \subset W$ is said to be v*-invariant* if $\gamma(x,t;v)$ is defined for every $t \geqslant 0$ and every $x \in X$, and $\gamma(x,[0,\infty[\,;v) \subset X$. A subset $X \subset W$ is $(\pm v)$*-invariant* if it is v-invariant and $(-v)$-invariant.

A subset $X \subset W$ is said to be *weakly* v*-invariant* if $\gamma(x,t;v) \in X$ whenever $x \in X$, $t \geqslant 0$, and $\gamma(x,t;v)$ is defined. The definition of the notion of weak $(\pm v)$-invariance is left to the reader.

For any (time dependent) vector field w on a closed manifold M and any $t \in \mathbb{R}$, we denote by $\Phi(w,t)$ the diffeomorphism of M given by $x \mapsto \gamma(x,t;w)$.

The closure of a set U is denoted by $\overline{U}$. The short bar is reserved for objects related to the infinite cyclic covering (e.g., $\mathcal{P}\colon \bar{M} \to M$).

Let $R\colon A \to B$ be a diffeomorphism of Riemannian manifolds. For $x \in A$ set $E(R,x) = \|R'(x)\|$. The number

$$\max\Big(\sup_{x\in A} E(R,x), \sup_{y\in B} E(R^{-1},y)\Big)$$

is called the *expansion constant* of R.

A *cobordism* is a compact manifold W together with a presentation $\partial W = \partial_0 W \sqcup \partial_1 W$, where $\partial_1 W$ and $\partial_0 W$ are compact manifolds of dimension $\dim W - 1$ without boundary (one or both of them can be empty). The manifold $W \setminus \partial W$ will be denoted by W°.

A *Riemannian cobordism* is a cobordism endowed with a Riemannian metric.

Let W be a Riemannian cobordism of dimension n, and let $q \in W^\circ$. Let $\delta_0 > 0$ be so small that the restriction of the exponential map $\exp_q : T_q W \to W$ to the disc $B^n(0, \delta_0)$ is a diffeomorphism onto its image, and this image belongs to W°. Let $0 < \delta < \delta_0$. Denote by $B_\delta(q)$, resp. $D_\delta(q)$, the $\exp_q$-images of the open Euclidean ball $B^n(0, \delta)$, resp. closed Euclidean ball $D^n(0, \delta)$, with center at 0. Then $B_\delta(q)$, resp. $D_\delta(q)$, are Riemannian δ-discs (open, resp. closed). In what follows we shall use the notation $B_\delta(q)$, $D_\delta(q)$ only when $0 < \delta_0 < \delta$.

1.3.2. *Some differences between the notation of this paper and that of* [**26, 31**]. The gradients satisfying the almost transversality condition were called "almost good" in [**31, 32, 26**].

The diffeomorphism denoted by $(-v)^{\leadsto}$ in the present paper was denoted by $\overset{\leadsto}{v}$ in [**31, 32, 26**]. Our present notation is the same as that of [**35**]; it seems more natural because this diffeomorphism is a shift in the *positive* direction along the integral curves of the field $-v$. In [**26**] the sets $B_\delta(\text{ind}{\leqslant}s\,; v)$, $D_\delta(\text{ind}{\leqslant}s\,; v)$, $D(\text{ind}{\leqslant}s\,; v)$ were denoted by $B_\delta({\leqslant}s; v)$, $D_\delta({\leqslant}s; v)$, $D({\leqslant}s; v)$. In [**26**] the set $D(\text{ind}{\leqslant}s\,, v)$ was denoted by $K(\text{ind}{\leqslant}s\,; v)$.

2. Morse functions and their gradients

2.1. Basic definitions and constructions.

2.1.1. *Basic definitions.* Recall that for us a *cobordism* is a compact manifold W together with a presentation $\partial W = \partial_0 W \sqcup \partial_1 W$, where $\partial_1 W$ and $\partial_0 W$ are compact manifolds of dimension $\dim W - 1$ without boundary (one or both of them can be empty). The manifold $W \setminus \partial W$ is denoted by W°.

DEFINITION 2.1. A *Morse function* $f : W \to [a, b]$ is a C^∞ map $f : W \to \mathbb{R}$ such that

$$f(W) \subset [a, b], \quad f^{-1}(b) = \partial_1 W, \quad f^{-1}(a) = \partial_0 W,$$

and such that all the critical points of f are nondegenerate and belong to W°.

We shall denote $f^{-1}(\lambda)$ by V_λ and $f^{-1}([\alpha, \beta])$ by $W_{[\alpha,\beta]}$, if there is no possibility of confusion.

Recall that for a critical point a of f there is a well defined bilinear form on $T_a M$, called the *Hessian* of f, and denoted by $d^2 f(a)$. For a nondegenerate critical point a, the dimension of the negative eigenspace of $d^2 f(a)$ is called the *index* of a. The set of critical points of f is denoted by $S(f)$, and the set of critical points of f of index k is denoted by $S_k(f)$.

To introduce the notion of gradient vector field for f, first recall some notions and results from the general theory of vector fields. Our reference for that theory is the book by R. Abraham and J. Marsden [**1**], but we modify their terminology a bit. Let M be a manifold without boundary. We say that a zero p of a vector field v on M is *elementary* if the matrix of the derivative $v'(p)$ has no eigenvalues with zero real part. Recall the following Hadamard–Perron local stable manifold theorem ([**1**, p. 85]).

THEOREM 2.2. *Let p be an elementary zero of a vector field v. Then there exist a neighborhood $U(p)$ and two submanifolds*

$$W^{\mathrm{un}}_{\mathrm{loc}}(p,v) \subset U(p), \qquad W^{\mathrm{st}}_{\mathrm{loc}}(p,v) \subset U(p),$$

having the following properties:

$$W^{\mathrm{un}}_{\mathrm{loc}}(p,v) = \{x \in U(p) \mid \gamma(t,x;v) \text{ is defined and belongs to } U(p) \text{ for all } t \leqslant 0, \text{ and } \gamma(t,x;v) \to p \text{ as } t \to -\infty\},$$

$$W^{\mathrm{st}}_{\mathrm{loc}}(p,v) = \{x \in U(p) \mid \gamma(t,x;v) \text{ is defined and belongs to } U(p) \text{ for all } t \geqslant 0, \text{ and } \gamma(t,x;v) \to p \text{ as } t \to \infty\},$$

$$W^{\mathrm{un}}_{\mathrm{loc}}(p,v) \pitchfork W^{\mathrm{st}}_{\mathrm{loc}}(p,v), \qquad W^{\mathrm{un}}_{\mathrm{loc}}(p,v) \cap W^{\mathrm{st}}_{\mathrm{loc}}(p,v) = \{p\}.$$

The tangent spaces to $W^{\mathrm{st}}_{\mathrm{loc}}(p,v)$, resp. $W^{\mathrm{un}}_{\mathrm{loc}}(p,v)$, will be denoted by $T_-(p;v)$, resp. $T_+(p;v)$.

Let W be a manifold with boundary. Let v be a vector field on W such that for all $x \in \partial W$ we have $v(x) \neq 0$. We can apply the above theorem to every elementary zero p of $v|W^\circ$. Thus we obtain a neighborhood $U(p)$, a local unstable manifold $W^{\mathrm{un}}_{\mathrm{loc}}(p,v)$, a local stable manifold $W^{\mathrm{st}}_{\mathrm{loc}}(p,v)$, and the corresponding subspaces $T_+(p;v)$, $T_-(p;v)$ of T_pW.

DEFINITION 2.3. Let $f\colon W \to [a,b]$ be a Morse function on a cobordism W. A vector field v on W is called *gradient for f*, or simply *f-gradient*, if

(1) for every $x \in W \setminus S(f)$ we have $df(x)(v(x)) > 0$;
(2) the set of zeros of v equals $S(f)$ and each zero of v is elementary;
(3) for every $p \in S(f)$ the vector subspaces $T_-(p,v)$ and $T_+(p,v)$ are orthogonal with respect to the bilinear form $d^2f(a)$.

REMARK 2.4. The same definition is of course valid for Morse functions on manifolds without boundary (the case of closed manifolds is already contained in the definition).

The basic example of an f-gradient is the Riemannian gradient. Namely, let G be a Riemannian metric on W; recall that the *Riemannian gradient* $\operatorname{grad} f$ is the vector field defined by the formula

$$\langle \operatorname{grad} f(x), h\rangle = df(x)(h) \quad \text{for } x \in M.$$

LEMMA 2.5. *The Riemannian gradient of a Morse function f is an f-gradient in the sense of the preceding definition.*

PROOF. Property 1) and the first half of 2) are obvious. Proceeding to the local properties in a neighborhood of a critical point $p \in S(f)$, consider the geodesic coordinate system centered at p. We can assume therefore that f and $v = \operatorname{grad} f$ are defined in a neighborhood U of $0 \in \mathbb{R}^n$, and that for every $y \in U$ the matrix $G(y)$ of our Riemannian metric in U satisfies $G(y) = I + O(|y|^2)$. Write the Taylor decomposition of f as

$$f(y) = f(0) + \tfrac{1}{2}\langle Ay, y\rangle + O(|y|^3),$$

where $\langle\,\cdot\,,\,\cdot\,\rangle$ stands for the Euclidean scalar product, $|\cdot|$ stands for the Euclidean norm, and A is a nondegenerate symmetric linear operator. Then we have

$$f'(y)h = \langle Ay, h\rangle + O(|y|^2)(h), \qquad v(y) = Ay + O(|y|^2).$$

Thus $v'(0) = A$ and all the eigenvalues of $v'(0)$ are real and nonzero, which proves the second half of 2).

In order to prove the last required property, we shall make some more coordinate changes. Choose a base in $\mathbb{R}^n$ formed by the eigenvectors of A. First we make a diagonal linear change of base so that in the new base all the eigenvalues of the Hessian of f have absolute value 2. Note that in the coordinates determined by this base the matrix of $v'(0)$ is still the same: $v'(0) = A$.

Now we use Lemma 2.8 to find a local coordinate system $x = (x_1, \ldots, x_n)$ around 0 such that in these coordinates

$$f \circ \varphi^{-1}(x_1, \ldots, x_n) = f(0) - \sum_{i=1}^{k} x_i^2 + \sum_{i=k+1}^{n} x_i^2.$$

We write $x_- = (x_1, \ldots, x_k)$, $x_+ = (x_{k+1}, \ldots, x_n)$, so that $x = (x_-, x_+)$. The diffeomorphism φ which performs this coordinate change sends our vector field v to a vector field, say w; we have $w(0) = 0$. Using item 2) of the cited lemma, we can assume that the matrix of $w'(0)$ is still the same: $w'(0) = A$. Thus

$$w(x) = (A_- x_-, A_+ x_+) + \theta(x),$$

where A_- (resp. A_+) is a linear symmetric operator $\mathbb{R}^k \to \mathbb{R}^k$ (resp. $\mathbb{R}^{n-k} \to \mathbb{R}^{n-k}$) having a base of eigenvectors with real negative (resp. real positive) eigenvalues, and $\theta(x)$ is $o(|x|)$.

For $1 \geqslant \varepsilon > 0$, define the cones

$$C_\varepsilon = \{x = (x_-, x_+) \mid |x_+| \leqslant |x_-|\}, \qquad R_\varepsilon = \{x = (x_-, x_+) \mid \varepsilon |x_-| \leqslant |x_+| \leqslant |x_-|\}.$$

Let

$$Q_r = \{x = (x_-, x_+) \mid |x_+| \leqslant r,\ |x_-| \leqslant r\}.$$

The boundary $\partial(R_\varepsilon \cap Q_r)$ is the union of three parts:

$$\partial_0 = \{x \in R_\varepsilon \cap Q_r \mid |x_+| = |x_-|\}, \qquad \partial_1 = \{x \in R_\varepsilon \cap Q_r \mid |x_+| = \varepsilon |x_-|\},$$
$$\partial_2 = \{x \in R_\varepsilon \cap Q_r \mid |x_-| = r\}.$$

To prove that $T_0 W_- = \mathbb{R}^k$, we shall show that for every $\varepsilon > 0$ there is a positive $r > 0$ such that every $x = (x_-, x_+)$ in $W_- \cap Q_r$ is in $C_\varepsilon \cap Q_r$.

Using the formula above for $w(x)$, it is not difficult to show that for r sufficiently small the vector field w points inward $R_\varepsilon \cap Q_r$ at the points of $\partial_1 \cup \partial_2$. Thus, if $x \in R_\varepsilon \cap Q_r$, then the trajectory $\gamma(\,\cdot\,) = \gamma(x, \cdot\,; w)$ cannot leave $R_\varepsilon \cap Q_r$, and, therefore, for every $t \geqslant 0$ we have $\varepsilon|\gamma_-(t)| \leqslant |\gamma_+(t)| \leqslant |\gamma_-(t)|$. Now we see that γ_+ satisfies the differential equation

$$\gamma_+'(t) = A_+ \gamma_+(t) + \theta_+(\gamma(t)),$$

where $\theta_+(\gamma(t)) = o(|\gamma(t)|) = o(|\gamma_+(t)|)$, which implies (for r sufficiently small) $(|\gamma_+|^2)'(t) \geqslant 0$, a contradiction. □

REMARK 2.6 (on the terminology). Our present terminology is different from that of [**31**] and from that of [**26**]. For the reader's convenience, we cite here the definitions from [**16, 26, 31**], and establish relations between them.

Here is the definition of gradient-like vector field ([**16**, Definition 3.1]):

Let f be a Morse function for the triad $(W^n; V, V')$. A vector field ξ on W^n is a *gradient-like vector field for* f if

(1) $\xi(f) > 0$ throughout the complement of the set of critical points of f, and

(2) given any critical point p of f, there are coordinates

$$(\vec{x}, \vec{y}) = (x_1, \ldots, x_\lambda, x_{\lambda+1}, \ldots, x_n)$$

in a neighborhood U of p such that $f = f(p) - |\vec{x}|^2 + |\vec{y}|^2$ and ξ has coordinates $(-x_1, \ldots, -x_\lambda, x_{\lambda+1}, \ldots, x_n)$ throughout U.

Here is the definition of f-gradient and related objects from [**26**, §2.2]:

Let $f\colon W \to [a, b]$ be a Morse function on W (for the definition see [**16**, Definition 2.3]). Denote by $S(f)$ the set of critical points of f and by $S_k(f)$ the set of critical points of f of index k. For each $p \in S(f)$ choose a chart $\Phi(p)\colon U(p) \to B^n(0, r)$ (where $U(p)$ is a neighborhood of p in W, and $n = \dim W$) so that

$$f \circ \Phi(p)^{-1}(x_1, \ldots, x_n) = f(p) - (x_1^2 + \cdots + x_k^2) + (x_{k+1}^2 + \cdots + x_n^2),$$

where $k = \operatorname{ind} p$. We call this chart the *standard chart* and we consider it as a part of the data of the Morse function f. The neighborhood $U(p)$ is called the *standard coordinate neighborhood.* We say (cf. [**16**, §3]) that a vector field v on W is a *gradient* for f, or an *f-gradient*, if 1) $df(v) > 0$ except at the critical points of f, and 2) for any $p \in S(f)$ the vector fields $\Phi(p)_*(v)$ and $(-x_1, \ldots, -x_k, x_{k+1}, \ldots, x_n)$ are equal in some neighborhood of 0 (where $k = \operatorname{ind} p$).

Here is the definition of f-gradient and related objects from [**31**, Definition 1.1].

Let $f\colon M \to \mathbb{R}$ be a Morse function on a closed manifold M. Denote $\dim M$ by n. The set of critical points of f will be denoted by $S(f)$. A chart $\Phi_p\colon U_p \to B^n(0, r_p)$ (where $p \in S(f)$, U_p is a neighborhood of p, $r_p > 0$) is called the *standard chart for f around p of radius r_p* (or simply an *f-chart*) if there is an extension of Φ_p to a chart $\widetilde{\Phi}_p\colon V_p \to B^n(0, r'_p)$ (where $\overline{U_p} \subset V_p$ and $r'_p > r_p$) such that

$$(f \circ \widetilde{\Phi}_p^{-1})(x_1, \ldots, x_n) = f(p) + \sum_{i=1}^{n} \alpha_i x_i^2,$$

where $\alpha_i < 0$ for $i \leqslant \operatorname{ind}_f p$ and $\alpha_i > 0$ for $i > \operatorname{ind}_f p$. The domain U_p is called the *standard coordinate neighborhood.* Any such extension $\widetilde{\Phi}_p$ of Φ_p will be called the *standard extension of Φ_p.*

A family $\mathcal{U} = \{\Phi_p\colon U_p \to B^n(0, r_p)\}_{p \in S(f)}$ of f-charts is called the *f-chart-system* if the family $\{\overline{U_p}\}$ is disjoint.

Given an f-chart system $\mathcal{U} = \{\Phi_p\colon U_p \to B^n(0, r_p)\}_{p \in S(f)}$, we say that a vector field v on M is an *f-gradient with respect to $\mathcal{U}$* if

1) $\forall x \in M \setminus S(f)$ we have $df(v)(x) > 0$;

2) $\forall p \in S(f)$ we have $(\widetilde{\Phi}_p)_*(v) = (-x_1, \ldots, -x_k, x_{k+1}, \ldots, x_n)$, where $k = \operatorname{ind}_f p$, and $\widetilde{\Phi}_p$ is some standard extension of Φ_p.

We say that a vector field v is an *f-gradient* if there is an f-chart system $\mathcal{U}$ such that v is an f-gradient with respect to $\mathcal{U}$.

The notion of gradient-like vector field from [**16**] coincides essentially with that of [**26**], except that in [**26**] we considered the standard coordinate system around the critical point as a part of the data of a Morse function.

The following simple lemma shows that the notion of f-gradient in the sense of [**31**] and the notion of gradient-like vector field are the same.

LEMMA 2.7. *Let $f\colon W \to \mathbb{R}$ be a Morse function on a cobordism W and let v be a vector field on W. Then v is a gradient-like vector field if and only if v is an f-gradient in the sense of* [**31**].

PROOF. To prove that every f-gradient in the sense of [**31**] is a gradient-like vector field for f (in the other direction the implication is obvious), consider the standard coordinate system $\Phi\colon U \to V$ for f and v around a critical point $p \in S_k(f)$. We have

$$(f \circ \widetilde{\Phi}_p^{-1})(x_1, \dots, x_n) = f(p) + \sum_{i=1}^n \alpha_i x_i^2,$$

where $\alpha_i < 0$ for $i \leqslant k$ and $\alpha_i > 0$ for $i > k$, and

$$(\widetilde{\Phi}_p)_*(v) = (-x_1, \dots, -x_k, x_{k+1}, \dots, x_n).$$

Define a diffeomorphism

$$\Psi\colon \mathbb{R}^n \to \mathbb{R}^n, \qquad \Psi(x_1, \dots, x_n) = (\sqrt{|\alpha_1|}x_1, \dots, \sqrt{|\alpha_n|}x_n).$$

Then the chart $\Psi \circ \Phi$ is a standard chart for f around p, and it is obvious that $(\Psi \circ \Phi)_*(v)(x) = (-x_1, \dots, -x_k, x_{k+1}, \dots, x_n)$. □

Of course the notion of f-gradient suggested in the present paper is strictly wider than that of gradient-like vector field. In this paper we shall sometimes use gradient-like vector fields. For brevity we shall sometimes use the term "f-Gradient" instead of "gradient-like vector field". The reason for using, in this paper, the term "f-gradient" already used in the previous paper in a different sense (we apologize to the reader for any inconvenience caused by this) is that we try to find the widest possible class of "gradients" for which the usual Morse-theoretic constructions like "cellular-like decomposition, formed by the descending separatrix discs" are still valid. The extension which we propose in this paper (Definition 2.3) is motivated by the fact that the Riemannian gradient of a Morse function is now an f-gradient (this was not true in general with the definitions of [**26, 31**]).

2.1.2. *Local structure of a Morse function and its gradient in a neighborhood of a critical point.* In this subsection W is a cobordism of dimension n, $f\colon W \to [a, b]$ is a Morse function on W, p is a critical point of f of index k, and v is an f-gradient. Recall the famous Morse Lemma, asserting that f is equivalent to its Hessian $d^2f(p)$ locally in a neighborhood of p. The content of the next lemma is to show that the coordinate change providing the standard form for f can be chosen so as to satisfy some additional restrictions.

Denote by A the bilinear form on $\mathbb{R}^n$ with the following matrix:

$$A_{ij} = \begin{cases} 0 & \text{if } i \neq j, \\ -1 & \text{if } i = j \leqslant k, \\ 1 & \text{if } i = j > k. \end{cases} \tag{2.1}$$

We shall use the symbol $\mathbb{R}^k$ to denote also the subspace $\{x_j = 0 \mid k+1 \leqslant j \leqslant n\}$ of $\mathbb{R}^n$; similarly $\mathbb{R}^{n-k}$ will denote the subspace $\{x_j = 0 \mid 1 \leqslant j \leqslant k\}$.

LEMMA 2.8. *There is a chart $\varphi\colon U \to V$ at the point p such that $\varphi(p) = 0$, and*

$$f \circ \varphi^{-1}(x_1, \dots, x_n) = f(p) - \sum_{i=1}^k x_i^2 + \sum_{i=k+1}^n x_i^2. \tag{2.2}$$

A) *Assume that the Hessian $(\partial^2 f/\partial x_i \partial x_j)$ of f at 0 equals $2A$. Then the diffeomorphism φ can be chosen so that $\varphi'(0) = \mathrm{id}$.*

B) *Let L_-, L_+ be the subspaces of T_pW such that*

$$d^2 f(p)|L_- < 0, \qquad d^2 f(p)|L_+ > 0,$$

and L_- is orthogonal to L_+ with respect to the bilinear form $d^2 f(a)$. Then the diffeomorphism φ can be chosen so that

$$\varphi'(p)(L_-) = \mathbb{R}^k, \qquad \varphi'(p)(L_+) = \mathbb{R}^{n-k}.$$

PROOF. The very first assertion is the standard assertion of the Morse lemma. The proof can be found for example in [**15**, p. 6], or in [**12**, p. 59]. To obtain assertion A) of our lemma, check through either of those proofs and note that the diffeomorphism φ constructed there satisfies $\varphi'(0) = \mathrm{id}$ if the Hessian of f equals $2A$.

To prove B), let $K_+ = \Phi'(p)(L_+)$, $K_- = \Phi'(p)(L_-)$. It suffices to find a linear isomorphism $\xi\colon \mathbb{R}^n \to \mathbb{R}^n$ conserving A and satisfying $\xi(K_+) = \mathbb{R}^{n-k}$, $\xi(K_-) = \mathbb{R}^k$. Note that $A|K_+ > 0$, $A|K_- < 0$, $A(K_+, K_-) = 0$.

From these properties one can easily deduce the existence of such an isomorphism ξ. □

A chart $\varphi\colon U \to V$ satisfying the conclusions of 2.8 is called a *Morse chart for f at p*. If the Morse chart satisfies the additional restriction in B) of the lemma, it will be called a *Morse chart for f at p with respect to L_+, L_-*.

Morse charts provide a natural tool for the study of the local behavior of f and v near the critical point p.

Note that the linear subspaces $T_+(p,v), T_-(p,v) \subset T_pW$ satisfy

$$d^2 f(p)|T_+(p,v) > 0, \qquad d^2 f(p)|T_-(p,v) < 0,$$
$$d^2 f(p)(T_+(p,v), T_-(p,v)) = 0$$

(Indeed, the third property is already in the definition of f-gradient. Since the function $f|W^{\mathrm{st}}_{\mathrm{loc}}(p,v)$ has a local maximum in p, we have $d^2 f(p)|T_+(p,v) \geqslant 0$, and using the third property together with the nondegeneracy of $d^2 f$, it is not difficult to establish the strict positivity of $d^2 f|T_+$.)

Set $L_- = T_-(p;v)$, $L_+ = T_+(p;v)$, and let $\varphi\colon U \to V \subset \mathbb{R}^n$ be a Morse chart for f at p with respect to L_+, L_-. In the following definition we develop the notation for subsets, maps, etc., in $\mathbb{R}^n$ related to $\varphi_*(v)$ and to $f \circ \varphi^{-1}$ (see Figure 2).

DEFINITION 2.9. Denote $\varphi_* v$ by $\tilde{v}$, $f \circ \varphi^{-1}$ by $\tilde{f}$.

We write a point $x \in \mathbb{R}^n$ as (x_+, x_-), where $x_+ \in \mathbb{R}^{n-k}$ and $x_- \in \mathbb{R}^k$. The orthogonal projection of $\mathbb{R}^n$ onto $\mathbb{R}^k$ will be denoted by p_-. The orthogonal projection of $\mathbb{R}^n$ onto $\mathbb{R}^{n-k}$ will be denoted by p_+.

Set

$$Q_r = D^k(0,r) \times D^{n-k}(0,r), \qquad Q_r^\circ = B^k(0,r) \times B^{n-k}(0,r),$$
$$Q_r^- = Q_r \cap \{(x_+, x_-) \mid |x_-| \geqslant |x_+|\}, \qquad Q_r^+ = Q_r \cap \{(x_+, x_-) \mid |x_-| \leqslant |x_+|\},$$
$$C = \{(x_+, x_-) \mid |x_+| = |x_-|\}.$$

The boundary ∂Q_r is the union of two subsets:

$$\partial_- Q_r = \{x \in \partial Q_r \mid |x_-| = r\}, \qquad \partial_+ Q_r = \{x \in \partial Q_r \mid |x_+| = r\}.$$

Set

$$\partial_-^\circ Q_r = \partial_- Q_r \setminus \partial_+ Q_r, \qquad \partial_+^\circ Q_r = \partial_+ Q_r \setminus \partial_- Q_r.$$

Let W^- be the local stable manifold, and W^+ the local unstable manifold, with respect to a neighborhood $U(p)$ of p. We can assume $U(p) \subset U$. Denote

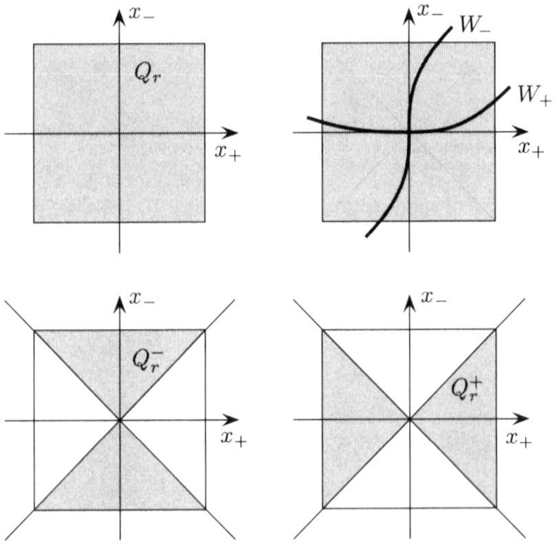

FIGURE 2

$\varphi(W^{\pm}(p,v))$ by $N_{\pm}$. Since $\tilde{v}$ is an $\tilde{f}$-gradient, we have $Q_r \cap N_{\pm} \subset Q_r^{\pm}$. Let $\tilde{v}'(0) = A$. By definition, A preserves $\mathbb{R}^{n-k}$ and $\mathbb{R}^k$, and there is $\nu > 0$ such that

$$\langle Ax_+, x_+\rangle \geqslant \nu\langle x_+, x_+\rangle, \qquad \langle Ax_-, x_-\rangle \leqslant -\nu\langle x_-, x_-\rangle.$$

We have $\tilde{v}(x) = Ax + \delta(x)$, where $\delta(x) = o(|x|)$.

In what follows we shall assume that the size of the set Q_r is sufficiently small. The following condition (R) is satisfied for each sufficiently small r.

(R)
- $\tilde{v}$ points into Q_r at the points of $\partial_-^{\circ} Q_r$ and out of Q_r in $\partial_+^{\circ} Q_r$;
- $p_-|(N_- \cap Q_r)$ is a diffeomorphism onto $D^k(0,r)$;
- $p_-|(N_+ \cap Q_r)$ is a diffeomorphism onto $D^{n-k}(0,r)$;
- for $x \in Q_r$ we have $|\delta(x)| \leqslant (\nu/10)x$;
- $Q_r \subset \varphi(U(p))$.

Up to the end of this subsection, we assume that r satisfies condition (R). Let $N_+(r)$, $N_-(r)$ be the local stable, resp. local unstable manifolds of $\tilde{v}|Q_r^{\circ}$. Obviously, $N_{\pm}(r) \subset N_{\pm} \cap Q_r^{\circ}$. Actually the reverse inclusion is also true. Indeed, let $x \in N_- \cap Q_r^{\circ}$. If the trajectory $\gamma(x, \cdot\,; \tilde{v})$ quits Q_r°, it must intersect $\partial_+^{\circ} Q_r^{\circ}$, and then at the moment of intersection the value $\tilde{f}(\gamma(x,t;\tilde{v}))$ will be strictly greater than $\tilde{f}(0)$, which contradicts $x \in N_-$. Thus $\gamma(x,t;\tilde{v}) \in Q_r^{\circ}$ for all $t \geqslant 0$, and $x \in N_{\pm}(r)$.

The next lemma will be useful in the sequel.

LEMMA 2.10. *Assume that a $\tilde{v}$-trajectory $\gamma(x,t;\tilde{v})$ stays in Q_r for all $t \geqslant 0$. Then $\gamma(x,t;\tilde{v}) \to p$ as $t \to \infty$.*

PROOF. Using the fact that $\tilde{f}(\gamma(x,t;\tilde{v}))$ is an increasing function of t, it is easy to prove that $\gamma(x,t;\tilde{v}) \in Q_r^-$ for every $t \geqslant 0$. Write

$$\gamma'(t) = (\gamma_+'(t), \gamma_-'(t)) = (A\gamma_+(t), A\gamma_-(t)) + \delta(\gamma(t)).$$

So $\langle \gamma_-', \gamma_- \rangle = \langle A\gamma_-, \gamma_- \rangle + \langle \delta(\gamma), \gamma_- \rangle$. Since $|\gamma| \leqslant 2\,|\gamma_-|$ in Q_r^-, we have

$$\langle \gamma_-', \gamma_- \rangle \leqslant -\frac{\nu}{2}\langle \gamma_-, \gamma_- \rangle.$$

Therefore $\gamma_-(t) \to 0$ as $t \to \infty$, and the same is true for $\gamma_+(t)$. □

2.1.3. *Global structure of v-trajectories.* In this subsection $f\colon W \to [a,b]$ is a Morse function on W, and v is an f-gradient.

Here we consider the behavior of v-trajectories on the whole of W, not just in a neighborhood of a critical point. Since $v|\partial W$ is transversal to ∂W and points into W on $\partial_0 W$ and out on $\partial_1 W$, there are 4 possible types of behavior for an integral curve of v (a, b in the formulas below are finite real numbers):

(1) γ is defined on $[a, \infty[$, $\gamma(a) \in \partial_0 W$, $\gamma(]a,\infty[) \subset W^\circ$;
(2) γ is defined on $]-\infty, a]$, $\gamma(a) \in \partial_1 W$, $\gamma(]-\infty, a[) \subset W^\circ$;
(3) γ is defined on $[a,b]$, $\gamma(a) \in \partial_0 W$, $\gamma(b) \in \partial_1 W$, $\gamma(]a,b[) \subset W^\circ$;
(4) γ is defined on $]-\infty,\infty[$, $\gamma(]-\infty,\infty[) \subset W^\circ$

(see, e.g., [**26**, Proposition 8.9]). Since v is a gradient of a Morse function, we can say more about the behavior of v-trajectories. First we study the integral curve in a neighborhood of a critical point. In the next lemma p is a critical point of f, and r is small enough for (R) to hold. Set $I_r = \{t \in \mathbb{R} \mid \gamma(t) \in \varphi^{-1}(Q_r)\}$.

LEMMA 2.11. *One of the following possibilities occurs:*

(1) $I_r = [a, \infty[$, $\gamma(t) \to p$ *as* $t \to \infty$;
(2) $I_r =]-\infty, a]$, $\gamma(t) \to p$ *as* $t \to -\infty$;
(3) $I_r = [a,b]$ (*where* $a = b$ *is not excluded*);
(4) $I_r =]-\infty,\infty[$, $\gamma(t) = p$ *for every* t;
(5) $I_r = \varnothing$.

PROOF. Assume that $I_r \neq \varnothing$, and let $\tau_0 \in I_r$. Consider the connected component J of τ_0 in I_r; then J is a closed interval, finite or infinite from one or both sides. The proof goes by considering these cases one after the other. We do three of them, and leave the rest to the reader.

(1) $J = \mathbb{R}$. Then $\operatorname{Im}\gamma \subset U(p) \cap (W^{\mathrm{st}}_{\mathrm{loc}} \cap W^{\mathrm{un}}_{\mathrm{loc}}) = \{p\}$ (by Lemma 2.10), and we obtain the case (4).

(2) $J =]-\infty, \alpha]$. Denote $\varphi(\gamma(t))$ by $\widetilde{\gamma}(t)$. Then $\widetilde{\gamma}(t) \in Q_r$ for every $t \in]-\infty,\alpha]$, and by Lemma 2.10 we have $\gamma(t) \in W^{\mathrm{un}}_{\mathrm{loc}}$. Note that $\widetilde{\gamma}(\alpha) \in \partial Q_r$, and since $\widetilde{v}$ points into Q_r on $\partial^\circ_- Q_r$, we have $\widetilde{\gamma}(\alpha) \in \partial_+ Q_r$. Therefore $f(\gamma(\alpha)) \geqslant f(p)$, and for every $t > \alpha$ we have $f(\gamma(t)) > f(p)$. Moreover, for $t > \alpha$ sufficiently close to α, we have $\widetilde{\gamma}(t) \notin Q_r$. Assume now that $I_r \neq J$, and let t_0 be the least number greater than α such that $\widetilde{\gamma}(t) \in Q_r$. Then $\widetilde{\gamma}(t_0) \notin \partial^\circ_+ Q_r$; therefore $\widetilde{\gamma}(t_0) \in \partial_- Q_r$, which is impossible since $f \mid \partial_- Q_r \leqslant f(p)$. Thus $J = I_r$.

(3) $J = [\alpha, \beta]$. The same reasoning as in the preceding item proves that $\widetilde{\gamma}(\beta) \in \partial^+ Q_r$, and that $\widetilde{\gamma}(\tau) \notin Q_r$ for $\tau > \beta$. By symmetry we have $\widetilde{\gamma}(\alpha) \in \partial_- Q_r$ and $\gamma(\tau) \notin Q_r$ for $\tau < \alpha$. □

From this lemma it follows easily that for a nonconstant v-trajectory γ the following four possibilities can occur:

(1) γ is defined on $[a, \infty[$, $\gamma(a) \in \partial_0 W$, $\gamma(t) \to p$ as $t \to \infty$, where $p \in S(f)$;
(2) γ is defined on $]-\infty, a[$, $\gamma(a) \in \partial_1 W$, $\gamma(t) \to q$ as $t \to -\infty$, where $q \in S(f)$;
(3) γ is defined on $[a,b]$, $\gamma(a) \in \partial_0 W$, $\gamma(b) \in \partial_1 W$, $\gamma(]a,b[) \subset W^\circ$;

(4) γ is defined on $]-\infty,\infty[$, $\gamma(t) \to p$ as $t \to \infty$, where $p \in S(f)$, and $\gamma(t) \to q$ as $t \to -\infty$, where $q \in S(f)$.

DEFINITION 2.12. Let p be a critical point of f. We denote by $D(p,v)$ the set of all $x \in W$ such that $\lim_{t\to\infty} \gamma(x,t;v) = p$, and we call it the *descending disc*.

We denote by $D(p,-v)$ the set of all $x \in W$ such that $\lim_{t\to\infty} \gamma(x,t;-v) = p$, and we call it the *ascending disc*.

REMARKS. 1. $D(p,v)$ is the analog of the global stable manifold in dynamical system theory. Note, however, that the flow generated by v is not defined on the whole of $\mathbb{R}$ in the case $\partial W \neq \varnothing$ (this is one reason to prefer this notation to the notation $W^{\mathrm{st}}(p)$ used in the literature on dynamical systems).

2. By the Grobman–Hartman theorem, the behavior of the v-trajectories belonging to $D(p,v)$ near p is *topologically* equivalent to that of the solutions of the standard system $x' = -x$ in $\mathbb{R}^n$. This is not the case in the differentiable category, and the behavior of trajectories near p can be complicated. The following lemmas show nevertheless that globally $D(p,v)$ is a nice geometrical object.

In the next three lemmas p is a critical point of f. In these lemmas we use the terminology from Definition 2.9, and we assume that r satisfies (R).

LEMMA 2.13. $D(p,v) \cap \varphi^{-1}(Q_r) = W^{\mathrm{st}}_{\mathrm{loc}}(p,v) \cap \varphi^{-1}(Q_r)$.

PROOF. Let $\tilde{x} = \varphi(x) \in Q_r$. If $\gamma(\tilde{x}, \cdot\,; \tilde{v})$ stays forever in Q_r, then $x \in W^{\mathrm{st}}_{\mathrm{loc}}(p,v)$ by Lemma 2.10. If $\gamma(\tilde{x}, \cdot\,; \tilde{v})$ leaves Q_r, then it must intersect $\partial_+ Q_r$ (since in $\partial_-^\circ Q_r \setminus C$ the vector field $\varphi_*(v)$ points into Q_r), say at the moment t_0. Then for $t > t_0$ we have $f(\gamma(x,t;v)) > f(p)$ and $x \notin D(p,v)$. □

Denote $f^{-1}(\lambda)$ by V_λ and $f^{-1}([\lambda,\mu])$ by $W_{[\lambda,\mu]}$.

LEMMA 2.14. *We have*:

(1) $D(p,v) \cap W^\circ$ *is a submanifold of* W° *of dimension* $\operatorname{ind} p$;
(2) $D(p,v) \cap V_\lambda$ *is a submanifold of* V_λ *of dimension* $\operatorname{ind} p - 1$.

PROOF. The second part follows from the first, since $D(p,v)$ is transversal to V_λ. To prove (1), note that it suffices to prove that $D(p,v) \cap W^\circ$ is locally a submanifold of W° in a neighborhood of a point $x \in \varphi^{-1}(Q_\varepsilon)$, where φ is a Morse chart. But this follows immediately from the preceding lemma. □

The next lemma justifies the using the term *descending disc* for $D(p,v)$.

LEMMA 2.15. *Let* $\lambda \in [a, f(p)]$ *be a regular value of* f. *Assume that for every* $x \in D(p,v) \cap W_{[\lambda,b]}$ *the* $(-v)$*-trajectory reaches* V_λ. *Then the pair* $(D(p,v), D(p,v) \cap V_\lambda)$ *is diffeomorphic to* $(D^k, \partial D^k)$, *where* $k = \operatorname{ind} p$.

PROOF. For $0 \leqslant \rho \leqslant r$ we set

$$\Delta_\rho = N_- \cap \{(x_+,x_-) \mid |x_-| \leqslant \rho\}, \qquad \Sigma_\rho = \partial\Delta_\rho = N_- \cap \{(x_+,x_-) \mid |x_-| = \rho\},$$
$$A(\rho,r) = N_- \cap \{(x_+,x_-) \mid \rho \leqslant |x_-| \leqslant r\}.$$

For $0 < \rho < r$ the set Δ_ρ is diffeomorphic to a k-disc and $A(\rho,r)$ is diffeomorphic to the annulus $S^{k-1} \times [0,1]$. Each $(-v)$-trajectory starting from a point $x \in A(\rho,r)$ reaches Σ_r at the moment $t(x)$ and reaches V_λ at the moment $T(x) > t(x)$, both $t(x)$ and $T(x)$ being C^∞ functions of x. It is now easy to construct a diffeomorphism $\Phi\colon \Delta_r \to D(p,v) \cap V_{[\lambda,f(p)]}$. Namely, set $\Phi|\Delta_\rho = \mathrm{id}$, and to define Φ outside Δ_ρ accelerate the motion along each $(-v)$-trajectory $\gamma(x,t;-v)$ for $x \in \Sigma_\rho$. We leave the details to the reader. □

2.1.4. *Thickenings of descending discs and compactness properties.* Let $f\colon W \to [a,b]$ be a Morse function on a Riemannian cobordism W of dimension n, and v an f-gradient.

DEFINITION 2.16. For $p \in S(f)$ set

$$B_\delta(p,v) = \{x \in M \mid \exists t \geqslant 0 : \gamma(x,t;v) \subset B_\delta(p)\},$$
$$D_\delta(p,v) = \{x \in M \mid \exists t \geqslant 0 : \gamma(x,t;v) \subset D_\delta(p)\}.$$

Recall from 1.3.1 that in the notation $B_\delta(p)$ and $D_\delta(p)$ it is implicitly understood that δ is sufficiently small.

We denote by $D_\delta(v)$, resp. by $B_\delta(v)$, the union of all the $D_\delta(p,v)$, resp. $B_\delta(p,v)$. We denote by $D(v)$ the union of all the descending discs.

For $s \in \mathbb{N}$, $0 \leqslant s \leqslant n$, we denote by $D(\text{ind}{\leqslant}s\,; v)$ the union of all the $D(p,v)$, where p ranges over the critical points of f of index $\leqslant s$. We denote by $B_\delta(\text{ind}{\leqslant}s\,; v)$, resp. by $D_\delta(\text{ind}{\leqslant}s\,; v)$, the union of all the $B_\delta(p,v)$, resp. $D_\delta(p,v)$, where p ranges over the critical points of f of index $\leqslant s$.

REMARK 2.17. We shall use similar notation like $D_\delta(\text{ind}{=}s\,; v)$ or $B_\delta(\text{ind}{\geqslant}s\,; v)$, which is now clear without special definition.

LEMMA 2.18. *$B_\delta(p,v)$ is open.*

PROOF. This follows immediately from the continuous dependence of the integral trajectories of vector fields on their initial values. □

Things are not that simple for $D_\delta(p,v)$ or for $D(p,v)$. These sets are not necessarily closed. The general reason for that is the existence of v-trajectories joining critical points: if there is a $(-v)$-trajectory joining two critical points p and q, then clearly $q \in \overline{D(p,v)} \setminus D(p,v)$. But for the union of all the $D(p,v)$ the compactness property is valid:

LEMMA 2.19. (1) *$D(v)$ is compact.*
(2) *$D_\delta(v)$ is compact.*
(3) $D_\delta(v) = \overline{B_\delta(v)}$.

PROOF. 1) The complement $W \setminus D(v)$ consists of all points $x \in W$ such that $\gamma(x, \cdot\,; v)$ reaches $\partial_1 W$. This set is open by Corollary 2.69.[4]
2) is similar to 1).
3) follows from the inclusions $B_\delta(v) \subset D_\delta(v) \subset \overline{B_\delta(v)}$. □

As δ goes to 0, these δ-thickenings, for example, $B_\delta(p,v)$, must collapse to their "skeleton" $D(p,v)$; that is, $B_\delta(p,v)$ must form a fundamental system of neighborhoods for $D(p,v)$.

Again in general this is not true, but the analog of this property is true for the union of all $D(p,v)$, which is the subject of the following proposition.

PROPOSITION 2.20. (1) *For every $\theta > 0$ we have $D_\theta(v) = \bigcap_{\delta>\theta} B_\delta(v)$.*
(2) $D(v) = \bigcap_{\delta>0} B_\delta(v)$.
(3) *The sets $\{B_\delta(v)\}_{\delta>\theta}$ form a fundamental system of neighborhoods for $D_\theta(v)$.*
(4) *The sets $\{B_\delta(v)\}_{\delta>0}$ form a fundamental system of neighborhoods for $D(v)$.*

[4]In this section we shall use the results of subsection 2.6. The results of 2.6 makes no use of results of the preceding sections; we have placed it at the end in order not to interrupt the exposition.

PROOF. 1) A point $x \in W$ *does not* belong to $D_\theta(v)$ if and only if the trajectory $\gamma(x, \cdot\,; v)$ reaches $\partial_1 W$ without intersecting the discs $D_\theta(p), p \in S(f)$. Then it does not intersect the discs $D_\delta(p)$ for some $\delta > \theta$. A similar argument proves 2).

3) Denote $B_\delta(v)$ by R_δ, and $D_\theta(v)$ by K (here δ is in some interval $I =]\theta, \delta_0[$). Then

(1) $\overline{R_\delta} \subset R_{\delta'}$, if $\delta < \delta'$;
(2) $\bigcap_{\delta > \theta} R_\delta = K$;
(3) K and W are compact.

These three properties imply in general that the sets $\{R_\delta\}_{\delta>\theta}$ form a fundamental system of neighborhoods of K. This is an exercise from elementary general topology, which we leave to the reader (see also Lemma 2.36, where a similar argument is presented). □

2.1.5. *Transversality conditions and the rearrangement lemma.* In this section $f\colon W \to [a, b]$ is a Morse function on a cobordism W and v is an f-gradient.

DEFINITION 2.21. We say that v satisfies the *transversality condition* if

$$(x, y \in S(f)) \implies (D(x, v) \cap W^\circ \pitchfork D(y, -v) \cap W^\circ).$$

We say that v satisfies the *almost transversality condition* if

$$(x, y \in S(f)\ \&\ \operatorname{ind} x \leqslant \operatorname{ind} y) \implies (D(x, v) \cap W^\circ \pitchfork D(y, -v) \cap W^\circ).$$

One of the basic instruments in Smale's theory of handlebodies is the construction, from any Morse function on a cobordism W, of a so-called self-indexing Morse function, i.e., a function such that $f(p) = \operatorname{ind} p$ for every $p \in S(f)$. The details of this construction can be found in [**16**, §4]. The construction actually gives a bit more: if we start with an arbitrary Morse function g and a gradient-like vector field v for g, satisfying the transversality condition, then it is possible to find a self-indexing Morse function f on the cobordism so that v is also a f-gradient. The same is true for f-gradients; the corresponding rearrangement lemma is the main subject of the present subsection.

DEFINITION 2.22. A Morse function $\phi\colon W \to [\alpha, \beta]$ is called *adjusted to* (f, v), if:

1) $S(\phi) = S(f)$;

2) the function $f - \phi$ is constant in a neighborhood of V_0, in a neighborhood of V_1, and in a neighborhood of each point of $S(f)$;

3) v is also a ϕ-gradient.

PROPOSITION 2.23. *Assume that $S(f) = \{p, q\}$ and $D(p, \pm v) \cap D(q, \mp v) = \varnothing$. Let $\alpha, \beta \in]a, b[$. Then there is a Morse function $g\colon W \to [a, b]$, adjusted to (f, v), such that $g(p) = \alpha$, $g(q) = \beta$.*

PROOF. Note first that we can assume the relation $f(p) \neq f(q)$ (this can always be achieved by a small perturbation of f with support in a neighborhood of $S(f)$). Assume, for example, that $f(p) < f(q)$. Denote $D_\delta(p, v) \cup D_\delta(p, -v)$ by $\mathcal{D}_\delta(p)$ and $B_\delta(p, v) \cup B_\delta(p, -v)$ by $\mathcal{B}_\delta(p)$. Define $\mathcal{D}_\delta(q)$ and $\mathcal{B}_\delta(q)$ similarly. Note that $\mathcal{B}_\delta(q)$ and $\mathcal{B}_\delta(p)$ are open.

LEMMA 2.24. *For $\delta > 0$ sufficiently small we have $\mathcal{B}_\delta(p) \cap \mathcal{B}_\delta(q) = \varnothing$.*

PROOF. Choose any λ such that $f(p) < \lambda < f(q)$. Denote the cobordism $f^{-1}([a,\lambda])$ by W_0, and the cobordism $f^{-1}([\lambda,b])$ by W_1. Applying Proposition 2.20 to W_0 and W_1, from $D(p,-v) \cap D(q,v) = \varnothing$ we deduce the following:

For $\delta > 0$ sufficiently small the sets $D_\delta(p,-v) \cap f^{-1}(\lambda)$ and $D_\delta(q,v) \cap f^{-1}(\lambda)$ are disjoint. This implies the lemma. □

It follows from the lemma that for $\delta > 0$ sufficiently small the sets $\mathcal{D}_\delta(p)$, $\mathcal{D}_\delta(q)$ are disjoint and compact. Denote their intersections with $\partial_0 W$ by K_p, resp. K_q. Choose a C^∞ function $\alpha\colon \partial_0 W \to [0,1]$ so that $\alpha | K_p = 1$, $\alpha | K_q = 0$. Note that $W' = W \setminus (D(v) \cup D(-v))$ is the union of all the v-trajectories starting at a point of $\partial_0 W \setminus D(v)$ and reaching $\partial_1 W$. Therefore there is a unique extension A of α to W' such that A is constant on each such trajectory. This function is smooth on W'. Note that $W = W' \cup \mathcal{B}_\delta(p) \cup \mathcal{B}_\delta(q)$, and it follows from the definition that $A | W' \cap \mathcal{B}_\delta(p) = 1$, $A | W' \cap \mathcal{B}_\delta(q) = 0$. Therefore A extends to C^∞ function on W which is constant on every v-trajectory. The rest of the proof is standard (see [**16**, §4]). We reproduce it briefly for the sake of completeness.

One first constructs an auxiliary real C^∞-function of two real arguments

$$G\colon [a,b] \times [0,1] \to [a,b],$$

which will be considered as a real function $[a,b] \to [a,b]$ depending on the real parameter $s \in [0,1]$. This function must satisfy the following conditions:

(1) $\partial G_s(x)/\partial x > 0$ for every s and x;
(2) $G_s(x) = x$ for x close to a or b and for every s;
(3) $G_0(f(q)) = \beta$ and $\partial G_0(x)/\partial x = 1$ in a neighborhood of $x = f(q)$;
(4) $G_1(f(p)) = \alpha$ and $\partial G_1(x)/\partial x = 1$ in a neighborhood of $x = f(p)$.

Then the function $x \mapsto G_{A(x)}(f(x))$ satisfies the conclusion of our proposition. □

COROLLARY 2.25. *Assume that v satisfies the almost transversality condition. Let $\prec$ be any partial order on the set $S(f)$ such that for every $p,q \in S(f)$ with* $\operatorname{ind} p < \operatorname{ind} q$ *we have $p \prec q$. Then there is a Morse function $q\colon W \to [a,b]$, adjusted to (f,v), such that $f(p) < f(q) \iff p \prec q$.*

COROLLARY 2.26. *Assume that v satisfies the almost transversality condition. Then for every s the set $D(\operatorname{ind}_{\leqslant s}, v)$ is compact.*

PROOF. By the preceding corollary, there exist a function $\phi\colon W \to [a,b]$ and a regular value λ of ϕ such that $\operatorname{ind} p \leqslant s \iff \phi(p) < \lambda$. Now apply Lemma 2.19 to the cobordism $\phi^{-1}([a,\lambda])$. □

The preceding corollary has an analog for δ-thickenings of descending discs. To formulate it we need two more definitions.

DEFINITION 2.27. We say that f is *ordered* with *ordering sequence* $a_0, \ldots, a_{n+1}$ if $a = a_0 < a_1 < \cdots < a_{n+1} = b$ are regular values of f such that $S_i(f) \subset f^{-1}(]a_i, a_{i+1}[)$.

DEFINITION 2.28. We say that v is δ*-ordered* (where $\delta > 0$) if there is an ordered Morse function $\phi\colon W \to [a,b]$, adjusted to (f,v), with ordering sequence $a_0, \ldots, a_{n+1}$, such that

$$\forall p \in S_k(f): \quad D_\delta(p) \subset \phi^{-1}(]a_k, a_{k+1}[). \tag{2.3}$$

REMARK 2.29. If the f-gradient v is δ-ordered, then it is δ'-ordered for some $\delta' > \delta$. Any f-gradient satisfying the almost transversality assumption is δ-ordered for some δ (this follows from Corollary 2.25).

Warning: Being δ-ordered does not imply the almost transversality condition.

PROPOSITION 2.30. *If v is ε-ordered, then for all δ, $0 < \delta \leqslant \varepsilon$, and for all s, $0 \leqslant s \leqslant n$, we have*

(1) *$D_\delta(\mathrm{ind}_{\leqslant s}\,; v)$ is compact;*
(2) *$D(\mathrm{ind}_{\leqslant s}\,; v)$ is compact;*
(3) $\bigcap_{\theta>0} B_\theta(\mathrm{ind}_{\leqslant s}\,; v) = D(\mathrm{ind}_{\leqslant s}\,; v)$;
(4) $\bigcap_{\theta>\delta} B_\theta(\mathrm{ind}_{\leqslant s}\,; v) = D_\delta(\mathrm{ind}_{\leqslant s}\,; v)$;
(5) $\overline{B_\delta(\mathrm{ind}_{\leqslant s}\,; v)} = D_\delta(\mathrm{ind}_{\leqslant s}\,; v)$.

PROOF. Let $\phi\colon W \to [a, b]$ be an ordered Morse function satisfying (2.3). Apply Lemmas 2.20 and 2.19 to the cobordism $\phi^{-1}([a_0, a_{s+1}])$. □

2.1.6. *Adding a horizontal component to a vector field near the boundary.* In this subsection we describe a (fairly standard) construction: modifying a vector field on a cobordism by adding a horizontal component near the boundary. See, for example, [**16**, p. 62]. The aim of this subsection is to give explicit estimates and to derive some properties for the resulting vector fields.

Let W be a cobordism, and v a vector field on W, transversal to $\partial_0 W$ (i.e., for $x \in \partial_0 W$ we have $v(x) \notin T_x(W)$). We shall describe here the procedure of changing v near $\partial_0 W$. (If v is transversal to $\partial_1 W$, a similar procedure can be performed near $\partial_1 W$; we leave the statement of the results to the reader.)

Let $a > 0$; assume that for every $x \in \partial_0 W$ the v-trajectory $\gamma(x, \cdot\,; v)$ is defined on $[0, a]$ and $\gamma(x, [0, a]; v) \subset W \setminus \partial_1 W$. The set $\gamma(\partial_0 W, [0, a]; v)$ will be called the *closed collar* of $\partial_0 W$ of height a with respect to v, and denoted by C_0. Consider a map $\Psi\colon \partial_0 W \times [0, a] \to W$ defined by $\Psi(x, t) = \gamma(x, t; v)$. Then Ψ is a diffeomorphism on its image C_0. The tangent space $T_{(x,t)}(\partial_0 W \times [0, a])$ being naturally isomorphic to $T_x(\partial_0 W) \oplus \mathbb{R}$, any vector field η on $\partial_0 W \times [0, a]$ can be written as $\eta = (\eta_1(x, t), \eta_2(x, t))$, where η_1 is a vector field on $\partial_0 W$ (it will be referred to as the *first component* of η) and η_2 is a function $\partial_0 W \to \mathbb{R}$ (it will be referred to as the *second component* of η). In this notation we have $(\Psi^{-1})_*(v) = (0, 1)$.

Let ξ be a vector field on $\partial_0 W$, and let $h\colon [0, a] \to \mathbb{R}$ be a positive C^∞-function such that $h(t) = 0$ for t in a neighborhood of 0 and in a neighborhood of a. Define a vector field v' on $\partial_0 W \times [0, a]$ by $v'(x, t) = (h(t)\xi(x), 1)$, and define a vector field u on W by the following formula:

$$u = v \quad \text{in } W \setminus C_0, \qquad u = \Psi_*(v') \quad \text{in } C_0.$$

We shall call this vector field u *the result of adding to v a horizontal component proportional to ξ near $\partial_0 W$* (or simply the result of the AHC-construction). The triple (a, ξ, h) will be called the *input data* for the AHC-construction. We shall also write $u = \mathrm{AHC}(a, \xi, h; v)$. We shall establish three basic properties of this construction. They are all easily checked, and we leave the proofs to the reader.

AHC1. *Estimate of $u - v$.*

Let E be the expansion constant of Ψ. We have

$$\|u - v\| \leqslant E\|\xi\|\|h\|$$

(where $\|\cdot\|$ is the C^0 norm). Moreover, the map $(\xi, h) \mapsto u$ is continuous with respect to C^∞ topology.

AHC2. *The behavior of $(-u)$-trajectories.*

Let γ be a $(-v)$-trajectory starting at $x_0 \in W \setminus C_0$ and such that we have $y = \gamma(x, t_0; -u) \in \partial_0 W$. Then the $(-u)$-trajectory starting at x_0 reaches $\partial_0 W$ at the moment t_0, and

$$\gamma(x_0, t_0; -u) = \Phi(-\xi, T_0)(y), \quad \text{where } T_0 = \int_0^a h(\tau)\, d\tau.$$

Thus

$$(-u)^{\leadsto}_{[b,a]} = \Phi(-\xi, T_0) \circ (-v)^{\leadsto}_{[b,a]}$$

AHC3. *Gradient property.*

Assume that v is an f-gradient, where $f\colon W \to [a, b]$ is a Morse function on W. Set

$$m = \min_{x \in C_0} df(v)(x), \qquad M = \max_{x \in C_0} \|df(x)\|.$$

If $\|\xi\| \|h\| < m/(EM)$, then u is also an f-gradient.

2.2. s-submanifolds and gfn-systems. This section is an extended version of §§2.1 and 2.2 of [**35**].

2.2.1. *s-submanifolds.* Here we define and study s-submanifolds. The basic example of an s-submanifold is the set of all the descending discs of an f-gradient v satisfying the almost transversality condition (where $f\colon M \to \mathbb{R}$ is a Morse function on a closed manifold.) The notion of an s-submanifold of a manifold M is close to the notion of a stratified manifold in the sense of Whitney and Thom. We have chosen to introduce the notion of s-submanifold and to deal with these objects, rather than to stick to the classical definition, since s-submanifolds carry less structure, and are more flexible than the classical stratified manifolds. It seems that if we impose additional restriction on v, these s-manifolds are actually Whitney-stratified manifolds. We hope to return to these questions in future publications.

Proceeding to the precise definitions, let $\mathbb{A} = \{A_0, \ldots, A_k\}$ be a finite sequence of subsets of a topological space X. We denote A_s also by $\mathbb{A}_{(s)}$, and we denote $A_0 \cup \cdots \cup A_s$ by $A_{\leqslant s}$ and also by $\mathbb{A}_{(\leqslant s)}$. We say that $\mathbb{A}$ is a *compact family* if $\mathbb{A}_{(\leqslant s)}$ is compact for every s.

DEFINITION 2.31. Let M be a manifold without boundary. A finite sequence $\mathbb{X} = \{X_0, \ldots, X_k\}$ of subsets of M is called an *s-submanifold of M* (s for stratified) if

(1) $\mathbb{X}$ is disjoint and each X_i is a submanifold of M of dimension i with trivial normal bundle;[5]
(2) $\mathbb{X}$ is a compact family.

For an s-submanifold $\mathbb{X} = \{X_1, \ldots, X_s\}$ the largest k such that $X_k \neq \varnothing$ is called the *dimension* of $\mathbb{X}$, and denoted by $\dim \mathbb{X}$. For a diffeomorphism $\Phi\colon M \to N$ and an s-submanifold $\mathbb{X}$ of M we denote by $\Phi(\mathbb{X})$ the s-submanifold of N defined by $\Phi(\mathbb{X})_{(i)} = \Phi(\mathbb{X}_{(i)})$.

If V is a submanifold of M and $\mathbb{X}$ is an s-submanifold of M, then we say that V is *transversal to* $\mathbb{X}$ (notation: $V \pitchfork \mathbb{X}$) if $V \pitchfork \mathbb{X}_{(i)}$ for each i. If V is a compact

[5]We impose the triviality of the normal bundle condition, since it makes some of the proofs easier. See for example the proof of Proposition 2.34.

submanifold of M, transversal to an s-submanifold $\mathbb{X}$, then the family $\{\mathbb{X}_{(i)} \cap V\}$ is an s-submanifold of V that will be denoted by $\mathbb{X} \cap V$.

If $\mathbb{X}, \mathbb{Y}$ are two s-submanifolds of M, we say, that $\mathbb{X}$ is *transversal to* $\mathbb{Y}$ (notation: $\mathbb{X} \pitchfork \mathbb{Y}$) if $\mathbb{X}_{(i)} \pitchfork \mathbb{Y}_{(j)}$ for every i, j; we say that $\mathbb{X}$ is *almost transversal to* $\mathbb{Y}$ (notation: $\mathbb{X} \nmid \mathbb{Y}$) if $\mathbb{X}_{(i)} \pitchfork \mathbb{Y}_{(j)}$ for $i + j < \dim M$.

REMARK 2.32. Note that $\mathbb{X} \nmid \mathbb{Y}$ if and only if $\mathbb{X}_{(\leqslant i)} \cap \mathbb{Y}_{(\leqslant j)} = \varnothing$ whenever $i + j < \dim M$.

Let $f\colon W \to [a, b]$ be a Morse function on a cobordism W, and v an f-gradient. Denote by $\mathbb{D}(v)$ the family $\{D(\text{ind}{=}i\,; v)\}_{0 \leqslant i \leqslant \dim W}$ of subsets of W. For $\lambda \in \mathbb{R}$ denote by $\mathbb{D}_\lambda(v)$ the family $\{D(\text{ind}{=}i{+}1\,; v) \cap f^{-1}(\lambda)\}_{0 \leqslant i \leqslant \dim W - 1}$ of subsets of $V_\lambda = f^{-1}(\lambda)$.

LEMMA 2.33. *Assume that v satisfies the almost transversality assumption. Then:*

(1) *$\mathbb{D}(v)$ is a compact family.*
(2) *If M is a closed manifold, then $\mathbb{D}(v)$ is an s-submanifold, transversal to $f^{-1}(\lambda)$ for every regular value λ of f.*
(3) *If M is any cobordism, then for every regular value λ of f the family $\mathbb{D}_\lambda(v)$ is an s-submanifold of $f^{-1}(\lambda)$.*

PROOF. (1) follows immediately from Proposition 2.19. Note that the normal bundle to $D(p, v)$ is always trivial, which implies (2) and (3). □

We proceed to an analog for s-submanifolds of the Thom theorem on the density of the transversality property. In Proposition 2.34 we shall prove that if $\mathbb{X}$, $\mathbb{Y}$ are two s-submanifolds, it is always possible to perturb one of them, say $\mathbb{X}$, a little, so that the result is almost transversal to $\mathbb{Y}$. First we need a definition.

Let M be a manifold without boundary. Denote by $\text{Vectt}(M)$ the subset of $\text{Vect}(M \times [0, 1])$, formed by the vector fields $v \in \text{Vect}(M \times [0, 1])$, satisfying

(1) v vanishes together with all partial derivatives in $M \times \{0\} \cup M \times \{1\}$;
(2) the coordinate of v corresponding to the factor $[0, 1]$ is equal to zero.

One can consider the elements of $\text{Vectt}(M)$ as C^∞ vector fields on M depending smoothly on the parameter $t \in [0, 1]$. There is a natural topology on $\text{Vect}(M \times [0, 1])$, and $\text{Vectt}(M)$ is a closed subspace of $\text{Vect}(M \times [0, 1])$ (see, e.g., [**26**, §8]). From now on, until the end of the subsection, we assume that M is compact. In this case the space $\text{Vect}(M \times [0, 1])$, as well as the space $\text{Vectt}(M)$, has a sup-norm (with respect to a Riemannian metric), which induces the C^0 *topology* on this space.

For every $v \in \text{Vectt}(M)$ the integral curve $\gamma(x, t; v)$ is defined for all $t \in \mathbb{R}$. Recall that $\gamma(x, t; v)$ is also denoted by $\Phi(v, t)(x)$, and that for a fixed t the map $x \mapsto \Phi(v, t)(x)$ is a diffeomorphism of M.

PROPOSITION 2.34. *Let $\mathbb{X}$, $\mathbb{Y}$ be s-submanifolds of a closed manifold M. Then the set of $v \in$* Vectt(M) *with $\Phi(v, 1)(\mathbb{X}) \nmid \mathbb{Y}$ is open and dense in* Vectt(M).

PROOF. We shall give a sketch of the proof. A reader familiar with differential topology will easily reconstruct the details. Denote by $\mathcal{T}(\mathbb{X}, \mathbb{Y})$ the subset of all $v \in \text{Vectt}(M)$ such that $\Phi(v, 1)(\mathbb{X}) \nmid \mathbb{Y}$. Note first that $\mathcal{T}(\mathbb{X}, \mathbb{Y})$ is open in the C^0-topology, which follows easily from the results of §5 and Remark 2.32. To prove its C^∞-density, we proceed by induction. Consider two auxiliary assertions:

$A(n)$: $\mathcal{T}(\mathbb{X}, \mathbb{Y})$ is dense for every $\mathbb{X}$ and $\mathbb{Y}$ with $\dim \mathbb{X} + \dim \mathbb{Y} \leqslant n$.

$A_0(n)$: For every $\mathbb{X}$, $\mathbb{Y}$ with $\dim \mathbb{X} + \dim \mathbb{Y} \leqslant n$, the closure of $\mathcal{T}(\mathbb{X}, \mathbb{Y})$ contains 0.

It is easy to show that $A_0(n) \implies A(n)$. Therefore it is sufficient to show that $A(n-1) \implies A_0(n)$. For this, let $\mathbb{X} = \{X_0, \dots, X_k\}$, $\mathbb{Y} = \{Y_0, \dots, Y_l\}$ be s-submanifolds of M, $\dim \mathbb{X} = k$, $\dim \mathbb{Y} = l$, $k + l = n$. Denote by $\mathbb{X}'$, $\mathbb{Y}'$ the s-submanifolds $\{X_0, \dots, X_{k-1}\}$, resp. $\{Y_0, \dots, Y_{l-1}\}$. By the induction assumption, there is a small vector field $w \in \text{Vectt}(M)$, such that $\Phi(w,1)(\mathbb{X}') \pitchfork \mathbb{Y}$ and $\Phi(w,1)(\mathbb{X}) \pitchfork \mathbb{Y}'$. The manifolds which are not yet in general position are $X'_k = \Phi(w,1)(X_k)$ and Y_l. It suffices to consider the case $k + l < \dim M$. Consider three disjoint compact sets: $\mathbb{X}'_{(\leqslant k-1)}$, $X'_k \cap Y_l$, $\mathbb{Y}_{(\leqslant l-1)}$. The set $X'_k \cap Y_l$ is in X_k. Recall that the normal bundle to X'_k is trivial. Let U_0 be a neighborhood of $X'_k \cap Y_l$ in X'_k. Choose a tubular neighborhood U of U_0, fiberwise diffeomorphic to the product $U_0 \times \mathbb{R}^{n-k}$, such that $\overline{U} \cap (\mathbb{X}'_{(k-1)} \cup \mathbb{Y}_{(l-1)}) = \varnothing$. A standard argument using the Sard lemma (see, e.g., [**16**, p. 46]) implies that there is a small isotopy Ψ_t with support in U, such that $\Psi_1(X'_k) \pitchfork Y_l$. □

2.2.2. *Good fundamental systems of neighborhoods.* The aim of this subsection is to introduce the notions of gfn-systems and ts-submanifolds. The ts-submanifolds are "thickenings of s-submanifolds" in the same sense as $B_\delta(v)$ is a thickening of $D(v)$. The notion of gfn-system is a generalization of the notion of ts-submanifold to the situation in which we replace s-submanifolds by compact families of subsets of topological spaces. They appear in our setting when we deal with Morse functions on cobordisms with nonempty boundary, so that the descending discs are manifolds with boundary, noncompact in general.

Proceeding to the definitions, we start with good fundamental systems. Recall that a *fundamental system of neighborhoods* of a subset of A in a topological space X is a family $\{U_i\}_{i \in I}$ of open subsets of X such that for every open set $U \supset A$ there is an index $i \in I$ with $U_i \subset U$.

DEFINITION 2.35. Let X be a topological space, $A \subset X$ a closed subset, I an open interval $]\delta_0, \delta_1[$. A *good fundamental system of neighborhoods of* A (abbreviation: gfn-system for A) is a family $\{A(\delta)\}_{\delta \in I}$ of open subsets of X, satisfying the following conditions:

(1) For each $\delta \in I$ we have $A \subset A(\delta)$; moreover, $\delta_1 < \delta_2 \Rightarrow A(\delta_1) \subset A(\delta_2)$.
(2) For each $\delta \in I$ we have $\overline{A(\delta)} = \bigcap_{\theta > \delta} A(\theta)$.
(3) $A = \bigcap_{\theta > \delta_0} A(\theta)$.

LEMMA 2.36. *Assume that X is compact. Let $I =]\delta_0, \delta_1[$. Let $\{A(\delta)\}_{\delta \in I}$ be a gfn-system for A. Then:*

(1) *the family $\{A(\theta)\}_{\theta > \delta_0}$ is a fundamental system of neighborhoods of A;*
(2) *for every $\delta \in I$ the family $\{A(\theta)\}_{\theta > \delta}$ is a fundamental system of neighborhoods of $\overline{A(\delta)}$.*

PROOF. (1) Let U be an open neighborhood of A. The sets $\{X \setminus \overline{A(\theta)}\}_{\theta > \delta_0}$ form an open covering of the compact set $X \setminus U$. There is a finite subcovering, and so there is a $\theta > \delta_0$ such that $\overline{A(\theta)} \subset U$. Condition (2) is proved similarly. □

Next we define good fundamental systems of families of subsets.

DEFINITION 2.37. Let X be a topological space, $\mathbb{A} = \{A_0, \dots, A_k\}$ a compact family of subsets of X, I an open interval $]\varepsilon_0, \varepsilon_1[$. A *good fundamental system of neighborhoods of* $\mathbb{A}$ (abbreviation: *gfn*-system for $\mathbb{A}$) is a family $\mathbf{A} = \{A_s(\delta)\}_{\delta\in I, 0\leqslant s\leqslant k}$ of open subsets of X, satisfying the following conditions:

(FS1) For every s, $0 \leqslant s \leqslant k$, and every $\delta \in I$ we have $A_s \subset A_s(\delta)$.

(FS2) For every s, $0 \leqslant s \leqslant k$, we have $\delta_1 < \delta_2 \implies A_s(\delta_1) \subset A_s(\delta_2)$.

(FS3) For every $\delta \in I$ and every j with $0 \leqslant j \leqslant k$ we have

$$\overline{A_{\leqslant j}(\delta)} = \bigcap_{\theta>\delta} (A_{\leqslant j}(\theta)).$$

(FS4) For every j with $0 \leqslant j \leqslant k$ we have

$$A_{\leqslant j} = \bigcap_{\theta>0} (A_{\leqslant j}(\theta)).$$

I is called the *interval of definition* of the system. The family $\mathbb{A}$ is called the *core* of $\mathbf{A}$.

We shall denote $A_s(\delta)$ also by $\mathbf{A}_{(s)}(\delta)$ and $A_{\leqslant i}(\delta)$ also by $\mathbf{A}_{(\leqslant i)}(\delta)$. We shall denote A_s also by $A_s(\varepsilon_0)$.

REMARK 2.38. Assume that X is compact. Let $I =]\varepsilon_0, \varepsilon_1[$. Let

$$\mathbf{A} = \{A_s(\delta)\}_{\delta\in I, 0\leqslant s\leqslant k}$$

be a *gfn*-system for a compact family $\mathbb{A} = \{A_0, \dots, A_k\}$ of subsets of X. Then for every $\delta \in [\varepsilon_0, \varepsilon_1[$ and every $0 \leqslant s \leqslant k$ the family $\{A_{\leqslant s}(\theta)\}_{\theta\in]\delta,\varepsilon_1[}$ is a *gfn*-system for $\overline{A_{\leqslant s}(\delta)}$.

If M is a manifold without boundary, $\mathbb{X}$ is an s-submanifold of M, and $\mathbf{X}$ is a *gfn*-system for $\mathbb{X}$, then we say that $\mathbf{X}$ is a *ts-submanifold with the core* $\mathbb{X}$. For a *ts*-submanifold $\mathbf{X} = \{X_s(\delta)\}_{\delta\in I, 0\leqslant s\leqslant k}$ we shall denote $X_i(\delta)$ by $\mathbf{X}_{(i)}(\delta)$, and $X_{\leqslant i}(\delta)$ by $\mathbf{X}_{(\leqslant k)}(\delta)$.

The next lemma follows immediately from Proposition 2.30. It provides the basic example of a *gfn*-system and of a *ts*-manifold.

LEMMA 2.39. *Let W be a Riemannian cobordism, and let $n = \dim M$. Let $f\colon W \to [a,b]$ be a Morse function and v an ε-ordered f-gradient. Then the family $\{B_\delta(\mathrm{ind}{=}s\,; v)\}_{\delta\in]0,\varepsilon[, 0\leqslant s\leqslant n}$ is a gfn-system for $\mathbb{D}(v)$. If W is a closed manifold, this family is a ts-submanifold with the core $\mathbb{D}(v)$.*

The *gfn*-system introduced in the lemma will be denoted by $\mathbf{D}(v)$. Note that there is no canonical choice for the interval of definition of this system.

2.3. Handle-like filtrations. It is well known since Smale (see e.g. [**43, 42**]) that to any Morse function on a closed manifold M one can associate the so-called *handle decomposition* of M. This section contains a version of this construction, which suits our purposes better than the original one. The object in our setting which replaces the classical "handle corresponding to a critical point p" is the set $D_\delta(p, v)$ (see Definition 2.16). With our definition, a systematic treatment of "δ-thin handles with $\delta \to 0$" is possible. On the other hand, these objects are not very well suited for studying the diffeomorphism type of manifolds (which was the original aim of the seminal papers of Smale, cited above). Indeed, we do not even introduce a smooth structure on $D_\delta(\mathrm{ind}{\leqslant}s\,; v)$. We only identify the homotopy type of $D_\delta(\mathrm{ind}{\leqslant}s\,; v)$ (under some mild restrictions on v).

2.3.1. *δ-separated gradients: definition.* Recall from [**8**, Chapter 5], that a *filtration* of a topological space X is a family $\{X_i\}_{i\in\mathbb{Z}}$ of subsets of X such that $i \leqslant j \implies X_i \subset X_j$. We say that a filtration $\{X_i\}$ is *complete* if $\bigcup_{i\in\mathbb{Z}} X_i = X$.

Let $f: W \to [a,b]$ be a Morse function on a Riemannian cobordism W of dimension n, and let v be an f-gradient. The very first filtration of W to introduce is of course the following: $W_s = D_\delta(\text{ind}{\leqslant}s\,; v)$ (where $\delta > 0$ is some fixed number). In general these W_s do not form a good filtration; W_s can well be noncompact. To obtain a filtration with "handle-like" properties, we shall impose more restrictions on f and on v.

Let $\phi: W \to [a,b]$ be an ordered Morse function with an ordering sequence $(a = a_0 < a_1 < \cdots < a_{n+1} = b)$. Let $\delta > 0$ and let v be a ϕ-gradient. Denote $\phi^{-1}([a_i, a_{i+1}])$ by W_i.

DEFINITION 2.40. We say that v is *δ-separated with respect to the ordering sequence* $(a_0, \ldots, a_{n+1})$ if

i) for every i and every $p \in S_i(f)$ we have $D_\delta(p) \subset W_i^\circ$;

ii) for every i and every $p \in S_i(f)$ there is a Morse function $\psi: W_i \to [a_i, a_{i+1}]$, adjusted to $(\phi|W_i, v)$, and a regular value λ of ψ such that
$$D_\delta(p) \subset \psi^{-1}(]a_i, \lambda[)$$
and for every $q \neq p$, $q \in S_i(f)$, we have
$$D_\delta(q) \subset \psi^{-1}(]\lambda, a_{i+1}[).$$

If there is no possibility of confusion, we shall say that v is a δ-separated ϕ-gradient.

DEFINITION 2.41. Let $f: W \to [a,b]$ be an arbitrary Morse function, and v an f-gradient. We say that v is *δ-separated* if it is δ-separated with respect to some ordered Morse function $\phi: W \to [a,b]$, adjusted to (f, v).

It is obvious that each f-gradient satisfying the almost transversality condition is δ-separated for some $\delta > 0$.

2.3.2. *Definition of handle-like filtrations.* Now let $\phi: W \to [a,b]$ be an ordered Morse function with an ordering sequence $(a_0, \ldots, a_{n+1})$, where $n = \dim W$, let v be a δ-separated ϕ-gradient, and let $\nu \in\,]0, \delta]$. Let s be an integer between 0 and $n+1$. Set
$$W^{\{\leqslant s\}} = \phi^{-1}([a_0, a_{s+1}]), \qquad W^{\{\geqslant s\}} = \phi^{-1}([a_s, a_{n+1}]), \tag{2.4}$$
$$W^{[\leqslant s]}(\nu) = D_\nu(\text{ind}{\leqslant}s\,; v) \cup \partial_0 W, \tag{2.5}$$
$$W^{\lceil s \rceil}(\nu) = \phi^{-1}([a_0, a_s]) \cup D_\nu(\text{ind}{=}s\,; v). \tag{2.6}$$

Note that
$$W^{\{\leqslant 0\}} = \phi^{-1}([a_0, a_1]), \qquad W^{\{\geqslant 0\}} = W,$$
$$W^{[\leqslant 0]}(\nu) = \Big(\bigcup_{p \in S_0(f)} D_\delta(p) \Big) \cup \partial_0 W = W^{\lceil 0 \rceil}.$$

Extend the definition to negative values of s as follows. For $s \leqslant -2$ define all these sets (except $W^{\{\geqslant -s\}}$) to be empty. For $s = -1$ the definitions of $W^{\{\leqslant s\}}$ and $W^{[\leqslant s]}(\nu)$ are good as they are; we have $W^{\{\leqslant -1\}} = \partial_0 W = W^{[\leqslant -1]}(\nu)$. For $s \leqslant 0$ set $W^{\{\geqslant 0\}} = W$. Finally, set $W^{\lceil -1 \rceil}(\nu) = W^{[\leqslant -1]}(\nu) = \partial_0 W$.

There is an ambiguity in this notation, since the $W^{\{\leqslant s\}}$ depend on ϕ, and the $W^{[\leqslant s]}(\nu)$ and $W^{\lceil s\rceil}(\nu)$ depend on v. Whenever we use this notation, it will be clear from the context what ϕ and v are. For the rest of subsection 2.3 we fix both ϕ and v, so there is no possibility of confusion. We obviously have

$$W^{\{\leqslant s\}} \supset W^{\lceil s\rceil}(\nu) \supset W^{[\leqslant s]}(\nu).$$

All three filtrations will be referred to as *handle-like filtrations*.

We introduce one more family of subsets, setting

$$U_s = \phi^{-1}([a_0, a_s]) \cup \Big(\bigcup_{\operatorname{ind} p = s} D(p, v) \Big)$$

for $s \geqslant 0$ and setting $U_{-1} = \partial_0 W$, $U_s = \varnothing$ for $s \leqslant -2$. The space U_s is obtained from $\phi^{-1}([a_0, a_s])$ by attaching the following family of s-dimensional cells:

$$\{D(p, v) \cap \phi^{-1}([a_s, a_{s+1}])\}_{p \in S_s(\phi)}.$$

2.3.3. *Homotopy properties of the filtrations.*

PROPOSITION 2.42. *Let $s \in \mathbb{Z}$ and $\nu \in]0, \delta]$.*

(1) *The inclusions $W^{[\leqslant s]}(\nu) \subset W^{\lceil s\rceil}(\nu) \subset W^{\{\leqslant s\}}$ are homotopy equivalences.*
(2) *The inclusion $U_s \subset W^{\lceil s\rceil}(\nu)$ is a homotopy equivalence.*
(3) *The inclusions of pairs*

$$(W^{[\leqslant s]}(\nu), W^{[\leqslant s-1]}(\nu)) \subset (W^{\lceil s\rceil}, W^{\{\leqslant s-1\}}) \subset (W^{\{\leqslant s\}}, W^{\{\leqslant s-1\}})$$

are homotopy equivalences.
(4) *The inclusion*

$$(U_s, W^{\{\leqslant s-1\}}) \subset (W^{\lceil s\rceil}(\nu), W^{\{\leqslant s-1\}})$$

is a homotopy equivalence.

PROOF. We consider only the case $s \geqslant 0$; the rest is trivial. Assertions (1) and (2) follow immediately from (3) and (4). The proofs of (3) and (4) are based on the following lemma.

LEMMA 2.43. *Let $(Y, B) \subset (X, A)$ be pairs of topological spaces. Assume that there is a homotopy $\Phi_t \colon X \to X$, $t \in [0, a]$, with the following three properties:*

(i) $\Phi_0 = \mathrm{id}$;
(ii) *the subsets Y, A, and B are invariant under Φ_t for all t;*
(iii) $\Phi_a(X) \subset Y$, $\Phi_a(A) \subset B$.

Then the inclusion $i \colon (Y, B) \hookrightarrow (X, A)$ is a homotopy equivalence.

PROOF. The map $\Phi_a \colon (X, A) \to (Y, B)$ is the homotopy inverse for i. Indeed, $\Phi_a \circ i \colon (Y, B) \to (Y, B)$ equals $\Phi_a | (Y, B)$, which is homotopic to the identity map $(Y, B) \to (Y, B)$ via the homotopy $\Phi_t | (Y, B)$. Further, the map $i \circ \Phi_a \colon (X, A) \to (X, A)$ is homotopic to the identity via the homotopy Φ_t. □

To deduce assertion (3) from Lemma 2.43, define a homotopy $\Phi_t \colon W \to W$ by $\Phi_t(x) = \Gamma(x, t; -v)$ (see Definition 2.70 and Corollary 2.71 for the definition and properties of $\Gamma(\cdot, \cdot; \cdot)$). Here $t \geqslant 0$; note that for $T > 0$ sufficiently large and every s we have $\Phi_T(W^{\{\leqslant s\}}) \subset W^{[\leqslant s]}(\nu)$. (Indeed, there is a $\lambda > 0$ such that $df(v)(x) \geqslant \lambda$ for

$$x \in W^{\{\leqslant s\}} \setminus \bigcup_{p \in S(f)} B_\nu(p).$$

Therefore if T is sufficiently large, there exists a $t_0 \leqslant T$ such that $\Gamma(x, t_0; -v) \in \partial_0 W \cup B_\nu(\text{ind}\leqslant s\,; v)$.)

Assertion (4) of Proposition 2.42 is a bit more delicate: we cannot manage using only deformations of the type "shift along the $(-v)$-trajectories". Note that the compact sets $D_\delta(p, v) \cap W_{[a_s, a_{s+1}]}$, where $\operatorname{ind} p = s$, are disjoint; therefore it suffices to prove that for each $p \in S_s(f)$ the space $D(p, v) \cup W_{[a, a_s]}$ is a strong deformation retract of $D_\delta(p, v) \cup W_{[a, a_s]}$.[6] We shall proceed as if there were only one critical point p of index s; the general case differs only by complications in the notation.

Let φ be a Morse chart for f at p. We shall use the terminology from Definition 2.9. We assume that condition (R) from Definition 2.9 is true, and also that $\varphi^{-1}(Q_r) \subset D_\nu(p)$. Recall that the vector field $(-\tilde{v})$ points out from Q_r at the points of $\partial_-^\circ Q_r$ and into Q_r at the points of $\partial_+^\circ Q_r$. Set

$$R = \{z \in W_{[a_s, a_{s+1}]} \mid z = \gamma(x, t; -v), \text{ where } x \in \varphi^{-1}(Q_r),\ t \geqslant 0\},$$
$$R_- = \{z \in W_{[a_s, a_{s+1}]} \mid z = \gamma(x, t; -v), \text{ where } x \in \varphi^{-1}(\partial_- Q_r),\ t \geqslant 0\}.$$

The sets R and R_- are weakly $(-v)$-invariant. Set $W_0 = W_{[a, a_s]}$. We have the inclusions

$$(U_s, W_0) \subset (R_- \cup U_s, W_0) \subset (R \cup W_0, W_0) \subset (W^{\lceil s \rceil}(\nu) \cup W_0, W_0).$$

An argument similar to the one used above shows that the first and the third inclusions are homotopy equivalences. To prove that the second inclusion is a homotopy equivalence, it suffices to prove that $R_- \cup U_s$ is a strong deformation retract of $R \cup W_0$. For this it suffices in turn to prove that $\partial_- Q_r \cup (D(p, v) \cap Q_r)$ is a strong deformation retract of Q_r. The last assertion is obvious since the pair $(Q_r, \partial_- Q_r \cup (D(p, v) \cap Q_r))$ is homeomorphic to $(Q_r, \partial_- Q_r \cap \mathbb{R}^k)$. □

2.4. Morse complex.

2.4.1. *Definition of the Morse complex.* In this section W is a Riemannian cobordism of dimension n, $f\colon W \to [a, b]$ is a Morse function on W, and v is an f-gradient. Let s be an integer between 0 and n. Consider the free abelian group C_s generated by all critical points of f of index s. Assuming that v satisfies the transversality condition, we shall introduce for every s a homomorphism $\partial_s\colon C_s \to C_{s-1}$ satisfying $\partial_{s-1} \circ \partial_s = 0$, obtaining a chain complex over $\mathbb{Z}$, called the *Morse complex.* The construction of ∂_s mainly makes use of the orbits of v joining critical points of f, but it depends also on an additional arbitrary choice (namely on the choice of orientations of the descending discs). The homology of the complex is isomorphic to $H_*(W, \partial_0 W)$.

Proceeding to precise definitions, choose an ordered Morse function $\phi\colon W \to [a, b]$ with an ordering sequence $(a_0 = a, a_1, \dots, a_{n+1} = b)$, adjusted to (f, v). In the preceding subsection we constructed a filtration $W^{\{\leqslant s\}}$. The filtration $W^{\{\leqslant s\}}$ is complete; we thus obtain a complete filtration in the singular chain complex of $(W, \partial_0 W)$. From Proposition 2.42 we know that the pair $(W^{\{\leqslant s\}}, W^{\{\leqslant s-1\}})$ is homotopy equivalent to $(U_s, W^{\{\leqslant s-1\}})$, and U_s is obtained from $W^{\{\leqslant s-1\}}$ by attaching the following family of s-dimensional cells: $\{D(p, v) \cap \phi^{-1}([a_s, a_{s+1}])\}_{p \in S_s(\phi)}$.

[6]Recall that $A \subset X$ is a *strong deformation retract of* X if there is a homotopy $f_t\colon X \to X$ such that $f_0 = \mathrm{id}$, $f_t | A = \mathrm{id}$, $f_1(X) \subset A$.

Therefore we have

$$H_*(W^{\{\leqslant s\}}, W^{\{\leqslant s-1\}}) = \begin{cases} 0 & \text{if } * \neq s, \\ (\mathbb{Z})^{\# S_s(\phi)} & \text{if } * = s. \end{cases} \tag{2.7}$$

We shall denote $H_s(W^{\{\leqslant s\}}, W^{\{\leqslant s-1\}})$ by C_s. The property (2.7) implies that the filtration $\{W^{\{\leqslant s\}}\}$ is a *cellular filtration* in the sense of [**8**, Chapter 5, §1]. (Note that in 2.3.2 we defined $W^{\{\leqslant s\}}$ for all $s \in \mathbb{Z}$; in particular, $W^{\{-1\}} = \partial_0 W$). Therefore, by [**8**, Chapter 5, Proposition 1.3], there is an isomorphism

$$H_*(W, \partial_0 W) \approx H_*(C_*, \partial_*).$$

Here

$$\partial_s \colon C_s = H_s(W^{\{\leqslant s\}}, W^{\{\leqslant s-1\}}) \to C_{s-1} = H_{s-1}(W^{\{\leqslant s-1\}}, W^{\{\leqslant s-2\}})$$

is the composition of the boundary homomorphism of the exact sequence of the pair $(W^{\{\leqslant s-1\}}, W^{\{\leqslant s-2\}})$ and the projection

$$H_{s-1}(W^{\{\leqslant s-1\}}) \to H_{s-1}(W^{\{\leqslant s-1\}}, W^{\{\leqslant s-2\}}).$$

The complex (C_*, ∂_*) is called a *Morse complex*, and denoted by $C_*(\phi, v)$.

By Theorem 2.45 and Remark 2.44 below, this complex depends only on v, but not on the choice of ordered Morse function ϕ, of which v is the gradient. So we shall often denote it by $C_*(v)$.

It turns out that the boundary operator ∂_* is computable from the geometric data (see Theorem 2.45 below).

HISTORICAL REMARK. The essentials of this construction were discovered by S. Smale and used in his proof of the Poincaré conjecture in dimensions $\geqslant 5$ (see [**43**]). In the book [**16**, §7] one finds a statement almost equivalent to Theorem 2.45. Much later, interest in the subject was stimulated by the paper [**45**], where one finds Theorem 2.45 in its present form (for the case of Riemannian gradients). The proof suggested in [**45**] is, however, based on different ideas than those of Smale. A proof of Theorem 2.45 is contained in the Appendix to my paper [**25**] (only the case of pseudo-gradients is treated there, but we shall see a bit later that this does not diminish the generality).

For each $p \in S_s(f)$, choose an orientation of the tangent space $T_-(p, v)$ to the local stable manifold. Then the manifold $D(p, v) \cap W^\circ$ acquires an orientation. Further, the manifold $D(p, -v)$ gets a coorientation (recall that a *coorientation* of a submanifold N of a manifold M is an orientation of the normal bundle to N in M). The fundamental class of $D(p, v)$ modulo its boundary defines an element $d(p) \in C_s$. These elements form a free base in C_s. Write

$$\partial_s d(p) = \sum_{q \in S_{s-1}(f)} n(p, q) \cdot q,$$

where $n(p, q) \in \mathbb{Z}$. We call $n(p, q)$ the *homological incidence coefficient* corresponding to p and q. We shall also write $n(p, q; v)$ to stress that a priori this index depends on v (we shall see later that often it does not).

Assume now that v satisfies the transversality condition. Denote $D(p, v) \cap \phi^{-1}(a_s)$ by $S(p, v)$, and $D(q, -v) \cap \phi^{-1}(a_s)$ by $S(q, -v)$. For each $p \in S_s(f)$ and $q \in S_{s-1}(f)$ the sets $S(p, v)$ and $S(q, -v)$ are compact submanifolds of complementary dimension in V_{a_s}, and they are transversal to each other. Therefore their intersection $N(p, q; v)$ is a finite set. Its points are in one-to-one correspondence with the orbits of v joining q to p. Due to our choice of orientations, each point $x \in N(p, q; v)$ acquires a sign $+$ or $-$, denoted by $\varepsilon(x)$. (Namely, the manifold $S(p, v)$ is oriented as the boundary of $D(p, v)$. Further, $S(q, -v)$ is a cooriented submanifold of V_{a_s}, since $D(q, -v)$ is cooriented. Then at each point of $x \in S(p, v) \cap S(q, -v)$ we obtain an orientation of $T_x(S(p, v))$ and a coorientation of $T_x(S(q, -v))$. Since these two subspaces of $T_x V_{a_s}$ are complementary, we obtain a sign.)

Set

$$\nu(p, q; v) = \sum_{x \in N(p,q;v)} \varepsilon(x).$$

REMARK 2.44. The number $\nu(p, q; v)$ depends only on v, not on the choice of the ordered Morse function ϕ adjusted to (f, v). Indeed, the above definition is easily reformulated in terms of orbits of v: the number $\nu(p, q; v)$ is the algebraic number of v-orbits joining p with q.

2.4.2. *Properties of incidence coefficients.*

THEOREM 2.45. *For all p, q with $\operatorname{ind} p = \operatorname{ind} q + 1$ we have $n(p, q; v) = \nu(p, q; v)$.*

PROOF. By Remark 2.44, we can assume that all critical points of ϕ of index s lie on the same critical level of ϕ, say $c_s \in \,]a_s, a_{s+1}[$.

We shall say that two ϕ-gradients v and w are *tangent* if for every critical point $p \in S(\phi)$ we have $T_-(p, v) = T_-(p, w)$, $T_+(p, v) = T_+(p, v)$. Using partitions of unity, it is not difficult to show that every f-gradient is tangent to an f-Gradient.

PROPOSITION 2.46. *Let w be another ϕ-gradient, satisfying the transversality condition. Assume that v and w are tangent. Then for every $p, q \in S(f)$ with $\operatorname{ind} p = \operatorname{ind} q + 1$ we have*

1) $\nu(p, q; v) = \nu(p, q; w)$;
2) $n(p, q; v) = n(p, q; w)$.

PROOF. 1) The number $\nu(p, q; v)$ is the intersection index of the submanifolds $S(p, v)$ and $S(q, -v)$ in V_{a_s} (the first being oriented, the second cooriented). To prove that $\nu(p, q; v) = \nu(p, q; w)$ it suffices to show that $S(p, v)$ is homotopic to $S(p, w)$ and $S(q, -v)$ is homotopic to $S(q, -w)$.

The proof of $S(p, v) \sim S(p, w)$ makes use of the Morse chart φ for ϕ around p. Since v and w are tangent, we can use the same Morse chart φ for both v and w. Therefore here we use the terminology of Definition 2.9. Let r be so small that the condition (R) from Definition 2.9 is valid for v and for w. Choose $t \in [0, 1]$ and consider the vector field $v_t = tw + (1 - t)v$. In general it is not a ϕ-gradient, but it satisfies $d\phi(v_t)(x) > 0$ for every $x \notin S(\phi)$. This implies that every $(-v_t)$-trajectory starting in $\partial_- Q_r$ reaches V_{a_s}.

Consider the continuous map $\psi_t \colon \partial_- Q_r \to \partial_0 W$ that associates to each point y of $\partial_- Q_r$ the intersection point of $\gamma(y, \cdot\,; -v_t)$ with V_{a_s}. Set

$$\sigma(p, v) = \partial_- Q_r \cap D(p, v), \qquad \sigma(p, w) = \partial_- Q_r \cap D(p, w).$$

These are two oriented embedded spheres in $\partial_- Q_r$, which are homotopic to each other; denote the corresponding homotopy by H_t. Then $\psi_t \circ H_t$ is the homotopy that we were seeking. The proof of $S(q, -v) \sim S(q, -w)$ is similar.

2) The proof is similar to 1): deforming w to v one shows that the discs $D(p, v)$ and $D(p, w)$ define the same homology classes in $H_*(W^{\{\leqslant s\}}, W^{\{\leqslant s-1\}})$. We leave the details to the reader. □

In view of this proposition, we can assume that the image of v in the Morse chart corresponding to p is the standard vector field $\boldsymbol{v}_s$ (where $s = \operatorname{ind} p$). Denote $\phi^{-1}([a, a_{s-1}])$ by W_0, $\phi^{-1}([a, a_s])$ by W_1, $\phi^{-1}([a, c_{s-1}[)$ by W'.

Fix some $p \in S_s(f)$ and denote $S(p, v)$ by M. Consider its fundamental class $[M]$ as an element of $H_{s-1}(W_1, W_0)$. The statement of our proposition describes the decomposition of $[M]$ with respect to the base $\{d(q)\}_{q \in S_{s-1}(f)}$ of $H_{s-1}(W_1, W_0)$ in terms of the algebraic intersection index of M with the spheres $S(q, -v)$. In order to find this decomposition, we shall perform a homotopy of M in W_1. The inclusion $(W_1, W_0) \hookrightarrow (W_1, W')$ is a homotopy equivalence; therefore it suffices to calculate the image of $[M]$ in the group $H_{s-1}(W_1, W')$. To simplify the notation, we shall assume that there is only one critical point of f of index $s-1$, say, q (in the general case the proof is similar).

Let $x_1, \dots, x_l \in M$ be the points of intersection of M with $S(q, -v)$ and let ε_i be the intersection sign of $S(p, v)$ with $S(q, -v)$ at x_i. Let Δ denote the closed $(s-1)$-disc $D^{s-1}(0, 1)$. For each i, $1 \leqslant i \leqslant l$, choose an embedding $E_i \colon \Delta \to M$ such that $E_i(0) = x_i$ and E_i preserves the orientation. We assume also that the images of the E_i are disjoint. The set $M' = M \setminus \bigcup_i \operatorname{Int}(E_i(\Delta))$ is a compact subset of the domain of definition of $v^{\rightsquigarrow}_{[a_s, a_{s-1}]}$; therefore the gradient descent will push M' below c_{s-1}. That is, denote $\Gamma(x, t; -v)$ by $H_t(x)$. Then for $t > 0$ sufficiently large we shall have $H_t(M') \subset \phi^{-1}([a, c_{s-1}[)$. Thus the image of $[M]$ in $H_*(W_1, W')$ equals the sum (over all i) of the fundamental classes modulo boundary of singular discs $H_t(E_i(\Delta))$.

Therefore the proof will be completed if we show that for every $1 \leqslant i \leqslant l$ and for t sufficiently large the homology class of $H_t(E_i(\Delta))$ modulo the boundary is equal to $\varepsilon_i \cdot d(q)$. For example let $i = 1$. Let $\varphi \colon U \to V$ be a Morse chart for f around q. We refer again to Definition 2.9 for the notation and definitions. Let T be so large that $\gamma(x_1, T; -v)$ belongs to $\operatorname{Int} \varphi^{-1}(Q_r)$. The map $\rho \colon y \mapsto \varphi(\gamma(E_1(y), T; -v))$ restricted to a sufficiently small neighborhood of 0 is a diffeomorphism onto its image, which is in $\operatorname{Int} Q_r$. Let $a = \varphi(\gamma(x_1, T; -v))$. The manifold $\varphi(H_t(M))$ is transversal to the positive subspace $\mathbb{R}^{n-s+1}$ at a. The gradient descent $\Phi(t, -\boldsymbol{v}_{s-1})$ is given inside Q_r by the formula $x = (x_-, x_+) \mapsto (e^t x_-, e^{-t} x_+)$; it shrinks the second coordinate and expands the first one, so that the germ of the submanifold $(\varphi \circ H_t \circ E_1)(\Delta)$ at the point a becomes very close to the germ of the subspace $\mathbb{R}^s$ for $t > 0$. In precise terms, for every sufficiently large $\tau \geqslant 0$ one can construct an embedding $\widetilde{E}_i \colon \Delta \to \operatorname{Int} Q_r \cap H_t(M)$ with $\widetilde{E}_i(0) = a$, sending Δ to a small neighborhood of a, and such that the disc $\widetilde{\Delta} = \Phi(\tau; -\boldsymbol{v}_{s-1}) \circ (\widetilde{E}_i(\Delta))$ has the following properties. First,

$$\partial \widetilde{\Delta} \subset \operatorname{Int} Q_r^- = \{(x_-, x_+) \in Q_r \mid |x_+| < |x_-|\}.$$

Second, the fundamental class of the pair $(\widetilde{\Delta}, \partial \widetilde{\Delta})$ in the homology of the pair $(Q_r, \operatorname{Int} Q_r^-)$ is equal to the fundamental class of $(B^s(0, r) \times \{0\}, \partial B^s(0, r) \times \{0\})$ multiplied by ε_i.

Therefore, for every $T \geqslant \tau$ the homology class of $H_t \circ \widetilde{E}_i(\Delta)$ modulo its boundary in $H_*(W_1, W')$ is equal to $\varepsilon_i \cdot d(q)$. □

2.5. Gradient descent map. In this subsection we deal with the following situation. Let W be a cobordism, $f \colon W \to [a, b]$ a Morse function on W, and v an f-gradient. Let $U_1 \subset \partial_1 W$ be the set of all x such that the $(-v)$-trajectory $\gamma = \gamma(x, t; -v)$ starting at x reaches $\partial_0 W$. For $x \in U_1$ the point of intersection of γ with $\partial_0 W$ will be denoted by $(-v)^{\rightsquigarrow}(x)$. We obtain a C^∞ diffeomorphism $(-v)^{\rightsquigarrow} : U_1 \to U_0$, where U_0 is an open subset of $\partial_0 W$. In general, of course, $U_i \neq \partial_i W$. For $A \subset \partial_1 W$ denote $(-v)^{\rightsquigarrow}(A \cap U_1)$ by $(-v)^{\rightsquigarrow}(A)$. Note that if $L \subset \partial_1 W$ is a submanifold of $\partial_1 W$, then $(-v)^{\rightsquigarrow}(L)$ is a submanifold of $\partial_0 W$, but the compactness of L does not imply the compactness of $(-v)^{\rightsquigarrow}(L)$. To make $(-v)^{\rightsquigarrow}(L)$ compact, we must add the soles of the descending discs, and the resulting object will not be a manifold. So one can think that the gradient descent map is well defined on the category of stratified manifolds. This is indeed so, and the corresponding construction is given in 2.5.2 (2.5.1 contains some preliminaries). To explain it briefly, let $\mathbb{A} = \{A_0, \dots, A_k\}$ be an s-submanifold of V_1 (see the definition in 2.2.1). Let A_i' denote the union of all $(-v)$-trajectories starting in A_i. Set $B_i = A_i' \cup D(\text{ind}{=}i\,, v)$. The main observation is that $\{B_i\}_{0 \leqslant i \leqslant k}$ is a compact family, provided that $\mathbb{A} \pitchfork \mathbb{D}(-v)$ (see Lemma 2.51). It is therefore natural to define an s-submanifold $(-v)^{\rightsquigarrow}(\mathbb{A})$ of $\partial_0 W$ by setting $(-v)^{\rightsquigarrow}(\mathbb{A})_{(i)} = B_{i+1} \cap \partial_0 W$. The most technical part of the section is 2.5.3. There we deal with thickenings. We construct a gfn-system for $\{B_i\}$ starting with a gfn-system for $\mathbb{A}$ and the gfn-system $\mathbf{D}(v)$.

2.5.1. *General properties of tracks.* In this subsection W is a cobordism, f is a Morse function on W, and v is an f-gradient.

DEFINITION 2.47. Let $X \subset W$. The set $\{\gamma(x, t; -v) \mid x \in X, t \geqslant 0\}$ is denoted by $T(X, -v)$ and called *the track of X with respect to $(-v)$*.

The aim of this subsection is to establish the basic properties of $T(X, -v)$. The next lemma is obvious.

LEMMA 2.48. (1) *For every $A \subset W$ we have $T(\overline{A}, -v) \subset \overline{T(A, -v)}$.*

(2) *If $\{X_i\}_{i \in I}$ is a family of subsets of $\partial_1 W$, then*

$$\bigcap_i T(X_i, -v) = T\Big(\bigcap_i X_i, -v\Big).$$

In the following lemma we list the compactness properties of the sets $T(X, -v)$. It can be regarded as an analog of 2.19.

LEMMA 2.49. (1) *If X is compact, then $T(X, -v) \cup D(v)$ is compact.*

(2) *If X is compact, and every $(-v)$-trajectory starting from a point of X reaches $\partial_0 W$, then $T(X, -v)$ is compact.*

(3) *For any X we have*

$$\overline{T(X, -v) \cup D(v)} = T(\overline{X}, -v) \cup D(v).$$

(4) *For any X and $\delta > 0$ we have*

$$\overline{T(X, -v) \cup B_\delta(v)} = T(\overline{X}, -v) \cup D_\delta(v).$$

PROOF. (1) A point $y \in W$ *does not* belong to $T(X, -v) \cup D(v)$ if and only if the trajectory $\gamma(x, \cdot\,; -v)$ reaches $\partial_1 W$ without intersecting X. Therefore the complement of $T(X, -v) \cup D(v)$ in W is open.

(2) For $x \in X$ let $\tau(x) \geqslant 0$ be the moment when $\gamma(x, t; -v)$ intersects $\partial_0 W$. Then $\tau(x)$ is continuous in x, so it is bounded by a constant, say, $C > 0$. Now let $y_i = \gamma(x_i, t_i; -v) \in T(X, -v)$, where $x_i \in X, 0 \leqslant t_i \leqslant C$. Choose a subsequence i_k such that $\{x_{i_k}\}$ and $\{t_{i_k}\}$ converge, and we are done.

(3) Since $\overline{X}$ is compact, $T(\overline{X}, -v) \cup D(v)$ is also compact, and the inclusion

$$\overline{T(X, -v) \cup D(v)} \subset T(\overline{X}, -v) \cup D(v)$$

holds. The reverse inclusion follows from the preceding lemma.

(4) Similar to (3). □

2.5.2. *Tracks of s-submanifolds.* In this and in the following subsection $f\colon W \to [a,b]$ is a Morse function on a cobordism W, v is an f-gradient satisfying the almost transversality condition, and $\mathbb{A} = \{A_0, \dots, A_k\}$ is an s-submanifold of $\partial_1 W$ satisfying $\mathbb{A} \nmid \mathbb{D}_b(-v)$.

DEFINITION 2.50. For $0 \leqslant i \leqslant k+1$ set

$$TA_i(-v) = T(A_{i-1}, -v) \cup D(\text{ind}{=}i\,; v)$$

(we set $A_{-1} = \varnothing$ by definition). Introduce the family $\mathbb{T}(\mathbb{A}, -v)$ of subsets of W by means of the formula

$$\mathbb{T}(\mathbb{A}, -v) = \{TA_i(-v)\}_{0 \leqslant i \leqslant k+1}.$$

For $\lambda \in [a,b]$ introduce the family $(-v)^{\rightsquigarrow}_{[b,\lambda]}(\mathbb{A})$ of subsets of V_λ as follows:

$$(-v)^{\rightsquigarrow}_{[b,\lambda]}(\mathbb{A}) = \{TA_{i+1}(-v) \cap f^{-1}(\lambda)\}_{0 \leqslant i \leqslant k}.$$

If the values b, λ are clear from the context, we shall abbreviate $(-v)^{\rightsquigarrow}_{[b,\lambda]}(\mathbb{A})$ to $\overset{\rightsquigarrow}{(-v)}(\mathbb{A})$. The family $\mathbb{T}(\mathbb{A}, -v)$ will be called the *track* of $\mathbb{A}$, and the family $v^{\rightsquigarrow}_{[b,\lambda]}(\mathbb{A})$ will be called the $\overset{\rightsquigarrow}{v}$*-image* of $\mathbb{A}$.

LEMMA 2.51. (1) *$\mathbb{T}(\mathbb{A}, -v)$ and $(-v)^{\rightsquigarrow}_{[b,\lambda]}(\mathbb{A})$ are compact families.*

(2) *If λ is a regular value of f, then $(-v)^{\rightsquigarrow}_{[b,\lambda]}(\mathbb{A})$ is an s-submanifold of $f^{-1}(\lambda)$.*

PROOF. (1) Let $0 \leqslant s \leqslant k+1$. Denote $T(A_{\leqslant s-1}, -v) \cup D(\text{ind}{\leqslant}s\,; v)$ by $Q(s)$. Let $\phi\colon W \to [a,b]$ be an ordered Morse function, adjusted to (f, v), and $(a_0, \dots, a_{n+1})$ the ordering sequence for ϕ. The all the critical points of ϕ of indices $> s$ are above a_s and those of indices $\leqslant s$ are below a_s. Since $\mathbb{A} \nmid (\mathbb{D}(-v) \cap V_1)$, every $(-v)$-trajectory starting at a point of $A_{\leqslant s-1}$ reaches $\phi^{-1}(a_{s+1})$. Then $Q(s) \cap \phi^{-1}([a_{s+1}, a_{n+1}])$ is compact by 2.49(2), and $Q(s) \cap \phi^{-1}([a_0, a_{s+1}])$ is compact by 2.49(1); therefore $Q(s)$ is compact. Part (2) is obvious. □

The assumption $\mathbb{A} \nmid \mathbb{D}_b(-v)$ adopted at the beginning of this subsection is not very restrictive: given $\mathbb{A}$, we can always perturb v so that this assumption will hold. A more general statement is the subject of the following proposition.

PROPOSITION 2.52. *Let $\mathbb{B}$ be an s-submanifold of $\partial_1 W$ and $\mathbb{C}$ an s-submanifold of $\partial_0 W$. Let U be a neighborhood of ∂W. Then there is an f-gradient w satisfying the almost transversality condition and such that*

(1) $\mathbb{D}_b(-w) \nmid \mathbb{B}$;

(2) $(-w)^{\rightsquigarrow}(\mathbb{B}) \nmid \mathbb{C}$, $\overset{\rightsquigarrow}{w}(\mathbb{C}) \nmid \mathbb{B}$.

Moreover, w can be chosen arbitrarily C^∞ close to v, and so that

$$\operatorname{supp}(u - w) \subset U, \qquad \operatorname{supp}(u - v) \cap \partial W = \varnothing.$$

PROOF. Choose $a > 0$ so that the collars

$$C_0 = \gamma(\partial_0 W, [0,a]; v), \qquad C_1 = \gamma(\partial_1 W, [0,a]; -v)$$

are disjoint. Our new f-gradient w will satisfy $\operatorname{supp}(w - v) \subset \operatorname{Int}(C_0 \cup C_1)$, which immediately implies the almost transversality condition. Apply to v the AHC-construction near the upper boundary $\partial_1 W$ of W. We choose the input data (a, ξ, h) as follows. First fix a function h so as to satisfy

$$\int_0^a h(\tau)\, d\tau = 1.$$

Then choose a vector field ξ on $\partial_1 W$ so that $\Phi(\xi, 1)(\mathbb{D}_b(-v)) \pitchfork \mathbb{B}$ (this is possible by 2.34), and, moreover, ξ can be chosen so C^∞ small that $v_0 = \mathrm{AHC}(a, \xi, h; v)$ is C^∞ close to v and v_0 is still an f-gradient. Now AHC2 from subsection 2.1.6 implies that $\Phi(\xi, 1)(\mathbb{D}_b(-v)) = \mathbb{D}_b(-v_0)$, and point (1) of our conclusions is true with v_0 instead of w. Therefore $\mathbb{T}(\mathbb{B}, -v_0)$ is defined and the s-submanifold $\mathbb{L} = (-v_0)^{\leadsto}(\mathbb{B})$ of $\partial_0 W$ is defined. Applying a similar construction to $\mathbb{L}$ and to $\mathbb{C}$, we obtain an f-gradient w satisfying item (1) and the first half of (2) of our conclusion. Note that the first half of (2) implies the second, and the proof is finished. $\square$

2.5.3. *Tracks of ts-submanifolds.* In this subsection we assume that our f-gradient v is δ_1-ordered.

DEFINITION 2.53. Let $\mathbf{A} = \{A_s(\delta)\}_{\delta \in]0,\delta_0[, 0 \leqslant s \leqslant k}$ be a ts-submanifold of V_1 with core $\mathbb{A}$. For $0 < \delta < \min(\delta_0, \delta_1)$ and $0 \leqslant s \leqslant k+1$ set

$$TA_s(\delta, -v) = T(A_{s-1}(\delta), -v) \cup B_\delta(\mathrm{ind}{=}s\,; v)$$

(where we set by definition $A_{-1}(\delta) = \varnothing$).

We shall prove that for sufficiently small $\varepsilon > 0$ the system

$$\{TA_s(\delta, -v)\}_{\delta \in]0,\varepsilon[, 0 \leqslant s \leqslant k}$$

is a gfn-system for $\mathbb{T}(\mathbb{A}, -v)$. Up to the end of this section, $Q(s, \delta)$ stands for

$$T(\overline{A_{\leqslant s-1}(\delta)}, -v) \cup D_\delta(\mathrm{ind}{\leqslant}s\,; v).$$

PROPOSITION 2.54. *There is an $\varepsilon \in]0, \min(\delta_0, \delta_1)[$ such that*

$$\{TA_s(\delta, -v)\}_{\delta \in]0,\varepsilon[, 0 \leqslant s \leqslant k}$$

is a gfn-system for $\mathbb{T}(\mathbb{A}, -v)$.

PROOF. First we shall prove two lemmas that establish certain properties of the family $\{TA_s\}$.

LEMMA 2.55. *For every δ with $0 < \delta \leqslant \min(\delta_0, \delta_1)$ and for every s we have*
(1) $TA_{\leqslant s}(-v) = \bigcap_{\theta > 0} TA_{\leqslant s}(\theta, -v)$,
(2) $Q(s, \delta) = \bigcap_{\theta > \delta} TA_{\leqslant s}(\theta, -v)$,
(3) $TA_{\leqslant s}(\delta, -v) \subset Q(s, \delta) \subset \overline{TA_{\leqslant s}(\delta, -v)}$.

PROOF. To prove (1), note first that

$$TA_{\leqslant s}(-v) = \bigcap_{\theta > 0} TA_{\leqslant s}(\theta, -v)$$

(by 2.48) and

$$\bigcap_{\theta > 0} B_\theta(\mathrm{ind}{\leqslant}s\,; v) = D(\mathrm{ind}{\leqslant}s\,; v),$$

(by 2.30). Since both $TA_{\leqslant s}(\theta, -v)$ and $B_\theta(\text{ind}_{\leqslant s}\,; v)$ are decreasing (in θ) families of subsets, item (1) holds. Item (2) is proved similarly. To prove (3), note that the first inclusion is straightforward, while the second follows from 2.48. □

LEMMA 2.56. *Let* $0 < \varepsilon \leqslant \min(\delta_0, \delta_1)$. *Then the following two conditions are equivalent*:

(i) *For every* $0 < \delta < \varepsilon$ *and every* s, $0 \leqslant s \leqslant k+1$, *the set* $Q(s, \delta)$ *is compact.*
(ii) $TA_s(\delta, -v)_{0<\delta<\varepsilon, 0\leqslant s \leqslant k+1}$ *is a* gfn*-system for* $\mathbb{T}(\mathbb{A}, -v)$.

PROOF. Properties (FS1), (FS2), (FS4) hold for the system

$$TA_s(\delta, -v)_{0<\delta<\varepsilon, 0\leqslant s\leqslant k+1}$$

for every ε. So we must check that (i) is equivalent to (FS3). But this follows immediately from the preceding lemma. □

Now the question is reduced to the compactness of the subsets $Q(s, \delta)$. Since v is δ_1-ordered, there is a Morse function $\phi\colon W \to [a, b]$, adjusted to (f, v) with an ordering sequence $(a_0, a_1, \dots, a_{n+1})$, such that for every $p \in S_s(f)$ we have $D_{\delta_1}(p) \subset \phi^{-1}(]a_s, a_{s+1}[)$.

LEMMA 2.57. *Let* $0 < \delta \leqslant \min(\delta_0, \delta_1)$. *Then the following two conditions are equivalent*:

(i) $Q(s, \delta)$ *is compact.*
(ii) *Every* $(-v)$*-trajectory starting at a point of* $\overline{A_{\leqslant s-1}(\delta)}$ *reaches* $\phi^{-1}(a_{s+1})$.

PROOF. Assume that (ii) holds. Then $Q(s,\delta) \cap \phi^{-1}([a_{s+1}, a_{n+1}])$ is compact by 2.49 (2) and $Q(s, \delta) \cap \phi^{-1}([a_0, a_{s+1}])$ is compact by 2.49 (1).

Assume now that (ii) does not hold. Then there is a point $x \in \overline{A_{\leqslant s-1}(\delta)}$ such that $\gamma(x, t; -v)$ converges to $p \in S(f)$ as $t \to \infty$, and $p \in \phi^{-1}(]a_{s+1}, a_{n+1}[)$. But in this case $p \in \overline{Q(s, \delta)} \setminus Q(s, \delta)$. □

This lemma implies in particular that if $0 < \delta' \leqslant \delta \leqslant \min(\delta_0, \delta_1)$ and $Q(s, \delta)$ is compact, then $Q(s, \delta')$ is also compact.

DEFINITION 2.58. Let $\lambda > 0$. Let $\mathbf{A}$ be a ts-submanifold of $\partial_1 W$ with interval of definition $I =]0, \delta_0[$. We say that the pair $(\mathbf{A}, v)$ is λ*-ordered* if

(1) v is λ-ordered;
(2) λ is in the interval of definition of $\mathbf{A}$;
(3) $Q(s, \lambda)$ is compact for every s.

The three preceding lemmas imply that if $(\mathbf{A}, v)$ is λ-ordered, then

$$\{TA_s(\delta, -v)\}_{\delta \in]0,\lambda[, 0\leqslant s \leqslant k}$$

is a gfn-system. Therefore to prove our proposition it suffices to show that there is a λ such that $(\mathbf{A}, v)$ is λ-ordered. But this is obvious. Indeed, every $(-v)$-trajectory starting in $A_{\leqslant s-1}$ reaches $\phi^{-1}(a_{s+1})$. Recall that $\{A_{\leqslant s-1}(\theta)\}_{\theta > 0}$ is a fundamental system of neighborhoods of the compact set $A_{\leqslant s-1}$. An easy compactness argument shows that for some $\theta > 0$ every $(-v)$-trajectory starting in $A_{\leqslant s-1}(\theta)$ reaches $\phi^{-1}(a_{s+1})$ as well. □

We shall denote the gfn-system introduced in 2.54 by $\mathbf{T}(\mathbf{A}, -v)$, and call it the *track* of $\mathbf{A}$ (with respect to $-v$). Note that there is no canonical choice of the interval of definition for this system.

2.6. Properties of continuity of integral curves of vector fields with respect to C^0 small perturbations of vector fields. Let v be a C^1 vector field on a manifold M, and let γ be an integral curve of v. It is well known that, given $t \geqslant 0$, the value $\gamma(t)$ depends continuously on the initial value $\gamma(0)$ and on the vector field itself (where the topology on the space of vector fields is the *C^1-topology*). Actually this continuity property is still true if we replace the C^1 topology by the *C^0-topology.*[7] It is due to this fact that the C^0 topology plays an important role in our work. The present section contains the statements of the continuity theorems we shall need in the sequel. We start by citing two propositions from [**35**]. Afterwards we deduce from them some properties (see Corollary 2.64, Proposition 2.65, Proposition 2.62) in order to obtain the results in a form suitable for use in later sections. The results which are used in the sequel are gathered in 2.6.2. Essentially all the content of the present section was already present in [**35**]; we have only rearranged the material.

2.6.1. *Manifolds without boundary.* For a manifold M (without boundary) we denote by $\mathrm{Vect}^1(M)$ (resp. $\mathrm{Vect}^1_0(M)$) the vector space of C^1 vector fields on M (resp. the vector space of C^1 vector fields on M with compact support). In this subsection M is a Riemannian manifold without boundary, $v \in \mathrm{Vect}^1_0(M)$, and $n = \dim M$.

PROPOSITION 2.59 ([**35**, Proposition 5.1]). *Suppose $a, b \in M$, $t_0 \geqslant 0$, and $\gamma(a, t_0; v) = b$. Then for every open neighborhood U of $\gamma(a, [0, t_0]; v)$ and every open neighborhood R of b there exist a $\delta > 0$ and an open neighborhood $S \subset U$ of a such that for all $x \in S$ and all $w \in \mathrm{Vect}^1_0(M)$ with $\|w - v\| < \delta$ we have $\gamma(x, t_0; w) \in R$ and $\gamma(x, [0, t_0]; w) \subset U$.*

PROPOSITION 2.60 ([**35**, Proposition 5.1]). *Suppose $a, b \in M$, $t_0 \geqslant 0$, and $\gamma(a, t_0; v) = b$. Let E be a submanifold without boundary of M of codimension 1 such that $b \in E$ and $v(b) \notin T_bE$.*

Let U be an open neighborhood of $\gamma(a, [0, t_0]; v)$ and R an open neighborhood of b. Then there is a $\theta_0 > 0$ such that for every $\theta \in]0, \theta_0[$ there exist $\delta > 0$ and an open neighborhood $S \subset U$ of a such that for all $x \in S$ and all $w \in \mathrm{Vect}^1_0(M)$ with $\|w - v\| < \delta$ we have

(1) *$\gamma(x, [-\theta, t_0 + \theta]; w) \subset U$ and $\gamma(x, [t_0 - \theta, t_0 + \theta]; w) \subset R$;*
(2) *there is a unique $\tau_0 = \tau_0(w, x) \in [t_0 - \theta, t_0 + \theta]$ such that $\gamma(x, \tau_0; w) \in E$;*
(3) *if E is compact and t_0 is the unique t from $[0, t_0]$ such that $\gamma(x, t_0; v) \in E$, then for all $y \in S$ the number $\tau_0(w, y)$ is the unique τ from $[-\theta, t_0 + \theta]$ such that $\gamma(y, \tau; w) \in E$.*

For further use we shall need a version of Proposition 2.59; this version (Proposition 2.62 below) is proved by the same method, so we shall describe the proof only briefly.

DEFINITION 2.61. We denote by $\|\cdot\|$ the sup-norm on $\mathrm{Vect}^1_0(M)$. Let $\delta > 0$ and $v \in \mathrm{Vect}^1_0(M)$. We denote by $B(v, \delta)$ the set of all $w \in \mathrm{Vect}^1_0(M)$ such that $\|w - v\| < \delta$. The topology induced in $\mathrm{Vect}^1_0(M)$ by the sup-norm will be called the *C^0-topology.*

[7] At the time the preprint [**31**] was written, I had no references for this property at hand. I am grateful to O. Viro for the indication that this "C^0 continuity property" is known and must appear somewhere in the literature.

PROPOSITION 2.62. *Denote by* $\mathcal{E}\colon M \times \mathbb{R} \times \mathrm{Vect}_0^1(M) \to M$ *the map defined by*

$$(x, t, v) \mapsto \gamma(x, t; v).$$

Then $\mathcal{E}$ *is continuous with respect to the* C^0 *topology in* $\mathrm{Vect}_0^1(M)$.

PROOF. Let

$$(a, t_0, v_0) \in M \times \mathbb{R} \times \mathrm{Vect}_0^1(M), \qquad b = \gamma(a, t_0; v_0).$$

The continuity of $\mathcal{E}$ at the point (a, t_0, v_0) is equivalent to the following assertion: for every neighborhood U of b there are a neighborhood S of a, a number $\delta > 0$, and a number $\theta > 0$ such that

$$(x \in S,\ w \in B(v, \delta),\ |t - t_0| < \theta) \implies (\gamma(x, t; w) \in U).$$

Subdividing the interval $[0, t_0]$ into small intervals, it is easy to see that we can restrict our assertion to the case in which the curve $\gamma(a, [0, t_0]; v_0)$ belongs to the domain of a single chart $\Phi\colon V \to V' \subset \mathbb{R}^n$. This case is reduced to the case of vector fields with compact support in $\mathbb{R}^n$. This last case is settled via the next lemma, which in turn is a direct consequence of [**4**, §4, Theorem 3].

For $v \in \mathrm{Vect}_0^1(\mathbb{R}^n)$ we shall denote by $\|v\|_1$ the sup-norm of the derivation map $dv\colon \mathbb{R}^n \to L(\mathbb{R}^n, \mathbb{R}^n)$.

LEMMA 2.63. *Let* $u, w \in \mathrm{Vect}_0^1(\mathbb{R}^n)$, $\|u - w\| < \alpha$, $\|u\|_1 \leqslant D$, *where* $D > 0$. *Let* γ, η *be trajectories of, respectively,* u, w, *and assume that* $|\gamma(0) - \eta(0)| \leqslant \varepsilon$. *Then for every* $t \geqslant 0$ *we have*

$$|\gamma(t) - \eta(t)| \leqslant \varepsilon e^{Dt} + \frac{\alpha}{D}(e^{Dt} - 1).$$

□

COROLLARY 2.64. *Let* $a, b \in M$, $t_0 \geqslant 0$, $\gamma(a, t_0; v) = b$. *Let* E *be a submanifold without boundary of* M *of codimension* 1 *such that* $b \in E$ *and* $v(b) \notin T_bE$. *Let* U *be an open neighborhood of* b. *There is a* $\theta_0 > 0$ *such that for every* $\theta \in \,]0, \theta_0[$ *we can find a* $\delta > 0$, *a neighborhood* S *of* a, *and a function* $\tau_0\colon B(v, \delta) \times S \to \mathbb{R}$ *so that*

(1) *for all* $(w, x) \in B(v, \delta) \times S$ *we have*

$$y \in \gamma(x, \tau_0(w, x); w) \in E \cap U, \qquad w(y) \notin T_yE;$$

(2) $\tau_0(w, x) \in \,]t_0 - \theta, t_0 + \theta[$, *and* $\tau_0(w, x)$ *is the unique number* τ *in this interval such that* $\gamma(x, \tau; w) \in E$.

PROOF. Choose a neighborhood $R \subset U$ of b and a $\delta_0 > 0$ so that for every $w \in B(v, \delta_0)$ and every $y \in R \cap E$ we have $w(y) \notin T_yE$. Then apply Proposition 2.60 with this R. There is a $\theta_0 > 0$ such that for every $\theta \in \,]0, \theta_0[$ there exist a $\delta > 0$, a neighborhood S of a, and a function $\tau_0\colon B(v, \delta) \times S \to \mathbb{R}$, satisfying the conclusion of 2.60. For $(w, x) \in B(v, \delta) \times S$, set $y(w, x) = \gamma(x, \tau_0(w, x); w)$. It follows from our choice of R that $w(y(w, x)) \notin T_{y(w,x)}E$. It remains only to diminish δ and S in order to force $\tau_0(w, x)$ to belong to $]t_0 - \theta, t_0 + \theta[$ instead of the interval $[t_0 - \theta, t_0 + \theta]$ from (2). For this just take δ' and S' corresponding to $\theta/2$. □

PROPOSITION 2.65. *Let* $a, b \in M$, $t_0 \geqslant 0$, $\gamma(a, t_0; v) = b$. *Let* E *be a submanifold without boundary of* M *of codimension* 1 *such that* $b \in E$ *and* $v(b) \notin T_bE$. *Let* $\delta > 0$ *and* $\theta > 0$, *and let* S *be a neighborhood of* a *and assume that* $\tau_0\colon B(v, \delta) \times S \to$

$\mathbb{R}$ *is a function satisfying* (1) *and* (2) *in Corollary* 2.64. *Then* τ_0 *is continuous on its domain.*

PROOF. Let $x_0 \in S$, $w_0 \in B(v,\delta)$, and consider the w_0-trajectory $\gamma(x_0,\,\boldsymbol{\cdot}\,;w_0)$. Set $t_0' = \tau_0(w_0,x_0)$ and $y_0 = \gamma(x_0,t_0';w_0)$. Then $w(y_0) \notin T_{y_0}E$. Thus we can apply the preceding corollary to the trajectory $\gamma(x_0,\,\boldsymbol{\cdot}\,;w_0)$. Let $\varepsilon > 0$. We apply Corollary 2.64 with θ_0' satisfying

$$\theta_0' < \varepsilon, \qquad [t_0' - \theta', t_0' + \theta'] \subset]t_0 - \theta, t_0 + \theta[.$$

We obtain a number $\delta' > 0$, a neighborhood S' of x_0, and a function $\tau'\colon B(w,\delta') \times S' \to \mathbb{R}$ such that

(1) for every $(w,x) \in B(w_0,\delta') \times S'$ we have $y = \gamma(x,\tau'(w,x);w) \in E$ and $w(y) \notin T_yE$;
(2) $\tau'(w,x) \in]t_0' - \theta', t_0' + \theta'[$, and $\tau'(w,x)$ is the unique number τ of this interval such that $\gamma(x,t;w) \in E$.

We can make δ' and S' so small that $B(w_0,\delta') \subset B(v,\delta)$ and $S' \subset S$. Then the domain of definition of τ' is in the domain of definition of τ_0, and our choice of θ' implies that $\tau'(w,x) = \tau_0(w,x)$. Then condition (2) implies that $\tau_0(w,x) \in]t_0 - \varepsilon, t_0 + \varepsilon[$ for $(w,x) \in B(w_0,\delta') \times S'$. Since $\varepsilon > 0$ was arbitrary, this proves continuity. □

In Proposition 2.66 we assume that the submanifold E and the vector field v satisfy the following assumptions:

($\mathcal{E}1$) there is an $\alpha > 0$ such that for every $w \in B(v,\alpha)$ every w-trajectory intersects E at most once;
($\mathcal{E}2$) for every $a \in E$ we have $v(a) \notin T_aE$.

PROPOSITION 2.66. *Let $K \subset M$ be a compact set. Let U be an open subset of E. Assume that for every $x \in K$ the trajectory $\gamma(x,\,\boldsymbol{\cdot}\,;v)$ reaches E at the moment $\tau(x)$ and that $\gamma(x,\tau(x);v) \in U$. Then there exist a $\delta > 0$ and a continuous function $\tau\colon B(v,\delta) \times K \to \mathbb{R}_+$ such that the following assertions are true:*

(1) *For every $(w,x) \in B(v,\delta) \times K$ we have $\gamma(x,\tau(w,x);w) \in U$.*
(2) *Denote by $A\colon B(v,\delta) \times K \to U$ the map defined by*

$$A(w,x) = \gamma(w,\tau(x,w);w).$$

For $w \in B(v,\delta)$ denote by A_w the map $K \to U$ defined by $A_w(x) = A(w,x)$. Then A is continuous, A_w is continuous for every w, and A_w is homotopic to A_v for every W.

PROOF. Apply Corollary 2.64 and for every $a \in K$ choose a neighborhood $S(a)$ of a, a number $\delta(a)$, and a function $\tau_{(a)}\colon B(v,\delta(a)) \times S(a) \to \mathbb{R}_+$ so that for every $x \in S(a)$ and $w \in B(v,\delta(a))$ we have $\gamma(w,\tau_a(x,w);w) \in U$. Obviously, the function $\tau_{(a)}$ is continuous at (v,a). We assume that $\delta(a) < \alpha$ for every a. Note that if a pair (w,x) is in the intersection of $B(v,\delta(a)) \times S(a)$ and $B(v,\delta(b)) \times S(b)$ (where $a,b \in K$), then $\tau_a(w,x) = \tau_b(w,x)$ (this follows from ($\mathcal{E}1$)). Choose a finite covering of K by sets $S(a_i)$, $1 \leqslant i \leqslant N$, and put $\delta = \min_i \delta(a_i)$, $S = \bigcup_i S(a_i)$. We obtain a function $\tau\colon B(v,\delta) \times S \to \mathbb{R}_+$. This function is continuous. Now the continuity of the map A defined in part (2) of our proposition is obvious: just apply Proposition 2.62, obtaining that A is a composition of continuous maps. To obtain a homotopy between A_w and A_v, set $H_t = A_{v+t(w-v)}$. □

2.6.2. *Manifolds with boundary.* Let W be a compact manifold with boundary ∂W. Let $\phi\colon \partial W \times [0,\infty[\to W$ be a collar of ∂W (see [**11**, p. 113] for the definition of collar and the existence theorem). Consider the smooth manifold W' without boundary obtained by gluing W to $\partial W \times]-1,1[$ via the diffeomorphism $\phi|\partial W \times [0,1[$. The manifolds W and $\partial W \times]-1,1[$ are embedded in the obvious way in W', so we shall consider W and $\partial W \times]-1,1[$ as submanifolds of W'. Let

$$\xi\colon \partial W \times]-1,1[\to]-1,1[$$

be the projection onto the second factor. Denote by $\mathrm{Vect}^1(W,\perp)$ the set of all C^1 vector fields v on W such that for every $x \in \partial W$ we have $v(x) \notin T_x(\partial W)$. Let $v \in \mathrm{Vect}^1(W,\perp)$. Denote by $\partial_v^+ W$, resp. by $\partial_v^- W$, the set of $x \in \partial W$ such that $v(x)$ points into W, resp. out. (If the value of v is clear from the context, we shall write simply $\partial^- W$ and $\partial^+ W$.) Since W is compact, the set $\mathrm{Vect}^1(W,\perp)$ is open in $\mathrm{Vect}^1(W)$ with respect to the C^0 topology. Moreover, for a given $v \in \mathrm{Vect}^1(W,\perp)$ there is a $\delta > 0$ such that for $\|v-w\| < \delta$ we have $\partial_v^+ W = \partial_w^+ W$, $\partial_v^- W = \partial_w^- W$.

A vector field u on W' will be called a *good extension* of v if

(E1) $u|W = v$;
(E2) $\operatorname{supp} u$ is compact;
(E3) the restriction of u to $\partial_0 W \times [0,1[$ satisfies

$$\xi'(x,t)(u(x,t)) \geqslant 0, \qquad \xi'(x,t)(u(x,t)) > 0$$

for every x and t in some interval $[0,\beta]$;
(E4) the restriction of u to $\partial_1 W \times [0,1[$ satisfies

$$\xi'(x,t)(u(x,t)) \leqslant 0, \qquad \xi'(x,t)(u(x,t)) < 0$$

for every x and t in some interval $[0,\beta]$.

REMARK 2.67. It is not difficult to show, using partitions of unity, that any $v \in \mathrm{Vect}^1(W,\perp)$ has good extensions. Moreover, one can show that if $v, w \in \mathrm{Vect}^1(W,\perp)$ and $\partial_v^\pm W = \partial_w^\pm W$, then good extensions $\tilde v$, $\tilde w$ of v, w, can be chosen so that $\|\tilde v - \tilde w\|_{W'} \leqslant 2\|v-w\|_W$.

LEMMA 2.68. *Let $v \in \mathrm{Vect}^1(W,\perp)$ and let $\tilde v$ be a good extension of v. Then the vector field $\tilde v$ and the submanifold $\partial^+ W \subset W'$ satisfy $(\mathcal{E}1)$ and $(\mathcal{E}2)$ from 2.6.1.*

PROOF. The condition $(\mathcal{E}2)$ holds by definition. To prove $(\mathcal{E}1)$, note first that every $\tilde v$-trajectory γ that intersects $\partial^+ W$ at some moment, say t_0, necessarily leaves the domain $\partial^+ W \times [0,\beta]$ passing through a point of $\partial^+ W \times \{\beta\}$. Afterwards it stays forever in $\partial^+ W \times [\beta, 1[$. So $(\mathcal{E}1)$ holds for v. If w is sufficiently C^0 close to $\tilde v$, the property $(E3)$ above is still true for w, and thus $(\mathcal{E}1)$ holds for w. □

COROLLARY 2.69. *Let $v \in \mathrm{Vect}^1(W,\perp)$. Let $x \in W$, and assume that the trajectory $\gamma(x,\cdot\,;v)$ reaches $\partial^+ W$. Then there is an open neighborhood S of x such that for any $y \in S$ the trajectory $\gamma(y,\cdot\,;v)$ reaches $\partial^+ W$.*

DEFINITION 2.70. For $x \in W$ and $t \geqslant 0$, define the element $\Gamma(x,t;v) \in W$ as follows:

(1) if $\gamma(x,\cdot\,;v)$ is defined on $[0,t]$, then $\Gamma(x,t;v) = \gamma(x,t;v)$;
(2) if $\gamma(x,t_0;v) \in \partial^+ W$ for some $t_0 \leqslant t$, then $\Gamma(x,t;v) = \gamma(x,t_0;v)$.

COROLLARY 2.71. *The map $(x,t) \mapsto \Gamma(x,t;v)$ is continuous.*

PROOF. Let $x \in W$. If $\gamma(x, \cdot\,; v)$ does not reach $\partial^+ W$, then the continuity of Γ in (x,t) for every t is obvious. If $\gamma(x, \cdot\,; v)$ reaches $\partial^+ W$, then the same is true for any y in a neighborhood of x; denote the moment of intersection of $\gamma(x, \cdot\,; v)$ with $\partial^+ W$ by $\tau_0(x)$. Then $\Gamma(x, \cdot\,; v) = \gamma(x, \min(t, \tau_0(x)); \tilde{v})$, and the continuity is proved. □

COROLLARY 2.72. *Let $v \in \mathrm{Vect}^1(W, \perp)$. Let U be an open subset of $\partial^+ W$, and let $K \subset W$ be compact. Assume that for every $x \in K$ the v-trajectory $\gamma(x, \cdot\,; v)$ starting at K reaches $\partial^+ W$ and intersects it at $\alpha_v(x) \in U$.*

Then the map $x \mapsto \alpha_v(x)$ is a continuous map $K \to U$. There is a $\lambda > 0$ such that for every w with $\|w - v\| < \lambda$ every w-trajectory starting at K reaches $\partial^+ W$, intersects it at a point $\alpha_w(x) \in U$, and the map $x \mapsto \alpha_w(x)$ is a continuous map $K \to U$ homotopic to α_v.

PROOF. Let $\tilde{v}$ be a good extension of v. It follows from Lemma 2.68 that the vector fields $\tilde{v}$ and the submanifold $\partial^+ W$ of W' satisfy $(\mathcal{E}1)$ and $(\mathcal{E}2)$. Therefore Proposition 2.66 applies, and we obtain a number $\delta > 0$ and a continuous function $A\colon B(\tilde{v}, \delta) \times K \to U$ such that $A(u, x) = \gamma(x, \tau(x, u); u)$. Set $u = \tilde{v}$; we obtain a continuous function $A_{\tilde{v}}\colon K \to U$ and a continuous function $x \mapsto \tau(x, \tilde{v})$ that we shall denote by $\tau_{\tilde{v}}$. Since $K \subset W$ for every $x \in K$, we have $\tau_{\tilde{v}}(x) \geqslant 0$ and $\gamma(x, [0, \tau_{\tilde{v}}(x)]; \tilde{v}) \subset W$. Therefore $\alpha_v(x) = A_{\tilde{v}}(x)$, and the map α_v is continuous.

Assume now that δ is so small that for every $\xi \in \mathrm{Vect}^1(W)$ with $\|\xi - v\|_W < \delta$ we have $\xi \in \mathrm{Vect}^1(W, \perp)$ and $\partial_\xi^\pm W = \partial_v^\pm W$. Let $\lambda < \delta/2$; let $w \in \mathrm{Vect}^1(W)$ and $\|v - w\|_W < \lambda$. Let $\widetilde{w}$ be a good extension of v satisfying $\|\tilde{v} - \widetilde{w}\|_{W'} < \delta$. We have $\gamma(x, \tau(\widetilde{w}, x); w) \in U \subset E$ for $x \in K$. Note that $\tau(x, \widetilde{w}) \geqslant 0$ (indeed, if $\tau(x, \widetilde{w}) < 0$, then for some $\theta \in]\tau(x, \widetilde{w}), 0[$ we have $\gamma(x, \theta; \widetilde{w}) \in \partial^+ W \times]0,1[$, and the relation $\gamma(x, 0; \widetilde{w}) \in W$ is no longer possible). Further, $\gamma(x, [0, \tau(x, \widetilde{w})]; \widetilde{w}) \subset W$; therefore $\gamma(x, \tau(\widetilde{w}, x); w) = A(\widetilde{w}, x) = \alpha_w(x) \in U$, and α_w is continuous in x.

To prove that α_w is homotopic to α_v for $\|v - w\|_W < \lambda$, consider the vector field $\xi(t) = \tilde{v} + t(\widetilde{w} - \tilde{v})$, where $t \in [0,1]$. Note that $\|\xi(t) - \tilde{v}\|_{W'} < \delta$, and so for every $t \in [0,1]$ the vector field $\xi(t)$ is in $B(\tilde{v}, \delta)$. The map $[0,1] \times K \to U$ given by the formula $(t, x) \mapsto A(\xi(t), x)$ is then a homotopy between α_w and α_v. □

For further applications, we shall need a generalization of Corollary 2.72 to the case of pairs of compact sets (K_1, K_2). For future reference we shall formulate it as a separate statement. The proof is an immediate generalization of the proof of 2.72.

COROLLARY 2.73. *Assume the hypotheses and terminology of Corollary 2.72. Let K' be a compact subset in K, and U' an open subset of U. Assume that $\alpha_v(K') \subset U'$.*

There is a $\lambda > 0$ such that for every w with $\|w - v\| < \lambda$ every w-trajectory starting at K', resp. at K, reaches $\partial^+ W$, intersects it at a point $\alpha_w(x) \in U'$, resp. $\alpha_w(x) \in U$, and the map $x \mapsto \alpha_w(x)$ defines a continuous map of pairs $(K, K') \to (U, U')$ which is homotopic to α_v regaded as a map of pairs $(K, K') \to (U, U')$.

2.6.3. *Gradients of Morse functions.* In this subsection $f\colon W \to [a, b]$ is a Morse function on a Riemannian cobordism W, and v is an f-gradient. We start with an elementary lemma.

LEMMA 2.74. *Let $\phi\colon W \to [\alpha, \beta]$ be a Morse function adjusted to (f, v). Then there is a $\delta > 0$ such that every f-gradient w satisfying $\|w - v\| < \delta$ is also a ϕ-gradient.*

PROOF. To check condition (1) in Definition 2.3 for the pair (ϕ, w) (the other two conditions obviously hold), for each $x \in S(f)$ choose a neighborhood $U(x)$ such that $(f - \phi)|U = \text{const}$. Set

$$U = \bigcup_x U(x), \qquad K = W \setminus U.$$

We have $d\phi(x)(v(x)) > 0$ for every $x \in K$. Since K is compact, this implies that for w sufficiently close to v in the C^0-topology we have $d\phi(x)(w(x)) > 0$ for every $x \in K$. For every $x \in U \setminus S(f)$ we have $d\phi(x)(w(x)) = df(x)(w(x)) > 0$, and condition (1) is proved. □

COROLLARY 2.75. (1) *Assume that v is a δ-ordered f-gradient. Then every f-gradient w sufficiently close to v in the C^0-topology is δ-ordered.*

(2) *Assume that v is a δ-separated f-gradient. Then every f-gradient w sufficiently close to v in the C^0-topology is δ-separated.*

In the following two propositions $\lambda \in [a, b]$ is a regular value of f, p is a critical point of f satisfying $f(p) \geqslant \lambda$, and U is an open subset of $f^{-1}(\lambda)$. We denote $f(p)$ by c, and $f^{-1}(\lambda)$ by V'.

PROPOSITION 2.76. *Assume that for every $x \in D(p, v) \cap f^{-1}([\lambda, c])$ the $(-v)$-trajectory $\gamma(x, \cdot\,; -v)$ reaches V'. Assume that $D(p, v) \cap V' \subset U$. Then there is a $\delta > 0$ such that for every f-gradient w satisfying $\|v - w\| < \delta$ the following statement holds*:

For every $x \in D(p, v) \cap f^{-1}([\lambda, c])$ the $(-w)$-trajectory $\gamma(x, \cdot\,; -w)$ reaches V', and $D(p, v) \cap V' \subset U$.

PROOF. Let $W' = f^{-1}([\lambda, b])$. For every $q \in S(f) \cap W'$ we have

$$D(q, -v) \cap D(p, v) = \varnothing.$$

Therefore, applying Rearrangement Lemma 2.23 several times, we can find a Morse function $\phi\colon W' \to [\lambda, b]$ adjusted to (f, v) and a regular value μ of ϕ such that p is the unique critical point of ϕ in the domain $\phi^{-1}([\lambda, \mu])$. Consider the cobordism $W'' = \phi^{-1}([\lambda, \mu])$. Recall from Lemma 2.74 that any f-gradient w sufficiently close to v in the C^0-topology is also a ϕ-gradient. Therefore for w sufficiently close to v in the C^0-topology a $(-w)$-trajectory starting at p necessarily reaches V', since there is only one critical point of ϕ in W''. Further, let $K = V' - U$. Every v-trajectory starting at K reaches $\phi^{-1}(\mu) = \partial_1 W''$. Therefore the same is true for every vector field w sufficiently close to v in the C^0-topology (by Corollary 2.72). Thus $D(p, w) \cap V' \subset V' \setminus K = U$. □

Assume the hypotheses of the preceding proposition. Choose an orientation of $D(p, v)$. Set $\Sigma(v) = D(p, v) \cap V'$. Then $\Sigma(v)$ is an embedded s-dimensional oriented sphere in V', where $s + 1 = \operatorname{ind} p$ (see Lemma 2.15). Therefore its fundamental class defines an element $\sigma(v) \in H_s(U)$. By Proposition 2.76, the same is true for any f-gradient w, provided w is sufficiently close to v in C^0-topology.

PROPOSITION 2.77. *Assume the hypotheses of Proposition 2.76. Let U be a neighborhood of p. Then there is a $\delta > 0$ such that for every f-gradient w satisfying $v|U = w|U$ and $\|v - w\| < \delta$ we have $\sigma(v) = \sigma(w)$.*

PROOF. We assume the terminology from the proof of Proposition 2.76. Let $\nu = \phi(p)$, and let $\varepsilon > 0$ be so small that $D(p,v) \cap \phi^{-1}([\nu-\varepsilon, \nu]) \subset U$. Set

$$\Sigma_0(v) = D(p,v) \cap \phi^{-1}(\nu - \varepsilon), \qquad \Sigma_0(v) = D(p,v) \cap \phi^{-1}(\nu - \varepsilon).$$

Then

$$\Sigma(v) = (-v)^{\leadsto}_{[\nu-\varepsilon,\lambda]}(\Sigma_0(v)), \qquad \Sigma(w) = (-w)^{\leadsto}_{[\nu-\varepsilon,\lambda]}(\Sigma_0(w)).$$

Note that the condition $v|U = w|U$ implies $\Sigma_0(v) = \Sigma_0(w)$. Therefore, in order to obtain our proposition, we can simply apply Corollary 2.72 to the cobordism $\phi^{-1}([\lambda, \nu - \varepsilon]) = W''$ and the vector field $(-v)|W''$. □

3. Quick Flows

Let M be a closed manifold of dimension n, f a Morse function on M, and v an f-gradient satisfying the transversality condition. Consider the union of all the descending discs of indices $\leqslant k$, and denote it by M_k. These descending discs form a filtration of M, which is a close analog of a cellular decomposition. (In many cases the descending discs *do* give a real cell decomposition, but we neither study this subject in our paper, nor use the corresponding results.) The ascending discs form a "dual decomposition", denoted by $\{\widehat{M}_k\}_{0\leqslant k\leqslant n}$. The transversality condition implies that $M_k \cap \widehat{M}_{n-k-1} = \varnothing$. Let U be an open neighborhood of M_k, and V an open neighborhood of $\widehat{M}_{n-k-1}$. It is clear that the gradient descent map $\Phi(T, -v)$ pushes $M \setminus V$ inside U, provided T is large enough. The main aim of this section is to find, for any M, a function f and its gradient v, and the neighborhoods U, V, so that the above number T and the C^0 norm of v are small. We require also that V and U be sufficiently small neighborhoods of M_k, resp. of $\widehat{M}_{n-k-1}$. The flows generated by such gradients are called *rapid flows*, since they push $M \setminus V$ to U rapidly.

The precise statement of the result is Theorem 3.4, which is the main result of this section. This theorem will be used in the proof of the C^0-density of condition ($\mathfrak{C}$) (Theorem 4.3), which in turn is essential for the proof of Theorem 5.5. Later we shall use only the statement of Theorem 3.4, but not the proof, so the reader interested only in the applications of the theorem can read the present introduction and 3.1.1, skipping the rest of the section.

One may try to give a construction of rapid flows using triangulations of M. Indeed, choose a smooth triangulation of M. Choose a Morse function f on M such that the critical points of f of index p are the barycenters of the simplices of the triangulation. (The existence of such a function seems to be known from folklore since the sixties; there is a recent paper [**37**] of Pozniak on this subject.) It seems that if the triangulation of M is sufficiently fine (that is, the simplices are of small diameter), the function f and some f-gradient would give a solution to our problem.

I have chosen another approach, in which it is easier to obtain explicit estimates for times and norms. I do not make any use of triangulations of the manifold. Instead, my approach suggests an alternative construction of a "small handle decomposition" of a manifold, which may prove useful in other domains of topology.

Now I shall give a brief informal description of the construction. Start with an arbitrary Morse function $f \colon M \to \mathbb{R}$ such that the values of f at all the critical points are different. Then choose a finite sequence of regular values $\alpha_1, \ldots, \alpha_N$ of f such that between α_i and α_{i+1} there is at most one critical value of f. We then think of M as the union of slices $W_i = f^{-1}([\alpha_i, \alpha_{i+1}])$, which we choose to be very thin. Consider a slice W_i that has no critical points inside it. It looks like a cylinder $V_i \times [0,1]$, where $V_i = f^{-1}(\alpha_i)$ (see Figure 3A). Now modify f inside W_i so as to make a fold (see Figure 3B).[8]

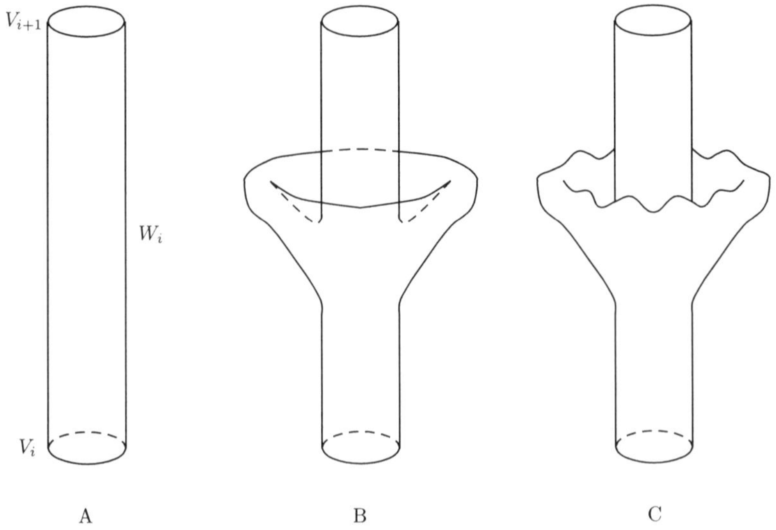

FIGURE 3

Glue together the functions $\tilde{f}_i$ and obtain a smooth function on W. Note that the trajectories of its gradient have become much shorter: they can intersect no more than three successive slices W_i. However, each function $\tilde{f}_i$ is not a Morse function: there are two level surfaces, consisting only of critical points. This must be repaired by perturbing $\tilde{f}_i$ near these level surfaces with the help of an auxiliary Morse function on V_i (see Figure 3C). Gluing together the perturbed functions, we obtain the required Morse function.

What is going on in one slice is the subject of subsection 4.1; "gluing back" is done in subsection 3.3, which contains also the end of the proof of Theorems 3.12 and 3.13. Subsection 3.1 contains all the new terminology we need. Unfortunately we do need a lot, since the precise estimates of times that a trajectory of a gradient of a Morse function can spend in different domains is not a standard item. In particular, we shall distinguish between *rapid flows* (Definition 3.1) and *quick flows* (Definition 3.8). In both definitions the main item is the time that a trajectory of a flow can spend outside a neighborhood U of $S(f)$; but in the definition of rapid flows, U is the union of some Riemannian balls around the critical points of f, and in the definition of quick flows, U is the union of coordinate neighborhoods corresponding to Morse charts. In subsection 3.1 we also state the main result of this section (Theorem 3.4) and perform the first steps in its proof.

[8]"fold" is "skladka" in Russian, which explains the term "S-construction".

3.1. Introduction and statement of results.

3.1.1. *Theorem on the existence of rapid gradient flows.* In this subsection $f\colon M \to \mathbb{R}$ is a Morse function on a closed manifold of dimension m.

DEFINITION 3.1. Let v be an f-gradient satisfying the almost transversality condition. Let $\varepsilon > 0$ and $t \geqslant 0$. We say that the flow generated by v is *(t, ε)-rapid* if for every s, $0 \leqslant s \leqslant m$, we have

$$\Phi(v,t)(M \setminus B_\varepsilon(\mathrm{ind}_{\leqslant s}\,; v)) \subset B_\varepsilon(\mathrm{ind}_{\leqslant m-1-s}\,; -v,), \tag{3.1}$$

$$\Phi(-v,t)(M \setminus B_\varepsilon(\mathrm{ind}_{\leqslant s}\,; -v)) \subset B_\varepsilon(\mathrm{ind}_{\leqslant m-1-s}\,; v). \tag{3.2}$$

Of course the definition would make sense for an arbitrary f-gradient, not necessarily satisfying the almost transversality condition. The reason to restrict ourselves to gradients satisfying this condition is the following lemma.

LEMMA 3.2. *For every $\varepsilon > 0$ there is a $t \geqslant 0$ such that the flow generated by v is (t, ε)-rapid.*

PROOF. If the lemma is true for some ν, then it is true for every $\nu' \geqslant \nu$, so it is sufficient to prove it for ν such that v is ν-ordered. For such ν consider the compact subset

$$K = M \setminus \Big(B_\varepsilon(\mathrm{ind}_{\leqslant m-s-1}\,, -v) \cup \bigcup_{\mathrm{ind}\, p \leqslant s} B_\varepsilon(p) \Big).$$

There are no critical points in K; therefore $df(v)(x) > C > 0$ for $x \in K$. This implies the existence of a $D > 0$ such that for every $x \in K$ there is $t_0 \leqslant D$ such that $\gamma(x, t_0; -v) \in M \setminus K$. Since $M \setminus B_\varepsilon(\mathrm{ind}_{\leqslant m-s-1}\,, -v)$ is $(-v)$-invariant, we have

$$\gamma(x, t_0, -v) \in \bigcup_{\mathrm{ind}\, p \leqslant s} B_\varepsilon(p),$$

and since $B_\varepsilon(\mathrm{ind}_{\leqslant s}\,, v)$ is also $(-v)$-invariant, we have $\gamma(x, D; -v) \subset B_\varepsilon(\mathrm{ind}_{\leqslant s}\,, v)$. $\square$

Similarly, given t and ε, one easily constructs a (t, ε)-rapid f-gradient which may have a very large C^0 norm. (Indeed, multiply a given f-gradient satisfying the almost transversality condition by a large constant outside $B = \bigcup_{\mathrm{ind}\, p \leqslant s} B_\varepsilon(p)$ and smooth the result near the boundary ∂B.) The aim of the construction of rapid flows is to produce flows v that have small C^0 norm and are (t, ε)-rapid for a small t and ε. This is achieved by Theorem 3.4, which is the main result of this section. We need a definition.

DEFINITION 3.3. Let $f\colon W \to [a, b]$ be a Morse function on a compact cobordism and let u, v be f-gradients. We say that u and v are *equivalent* if $u(x) = \phi(x)v(x)$, where $\phi\colon W \to \mathbb{R}$ is a C^∞ function such that $\phi(x) > 0$ for every x and $\phi(x) = 1$ for x in a neighborhood of $S(f)$.

THEOREM 3.4. *Let M be a closed Riemannian manifold. Let $C > 0$, $t > 0$. There exist a Morse function $f\colon M \to \mathbb{R}$ and an f-gradient v satisfying the almost transversality condition, and for every $\varepsilon > 0$ there is a (t, ε)-rapid f-gradient u with $\|u\| \leqslant C$ equivalent to v.*

We shall deduce Theorem 3.4 from Theorem 3.12. See page 78.

3.1.2. *Terminology.* The rapid f-gradients that we shall construct will be gradient-like vector fields. (The time that a v-trajectory spends in some domain near a critical point of f is much easier to compute or estimate for a gradient-like vector field than for a general f-gradient.)

Let n, s be positive integers and $0 \leqslant s \leqslant n$; set

$$\mathcal{Q}_s(x) = -\sum_{i=1}^{s} x_i^2 - \sum_{i=s+1}^{n} x_i^2, \tag{3.3}$$

$$\mathfrak{n}_s(x) = (-x_1, \dots, -x_s, x_{s+1}, \dots, x_n). \tag{3.4}$$

DEFINITION 3.5. **1. Quasi-Morse charts (qM-charts) for Morse functions.**

Let p be a critical point of f of index s.

A chart $\Phi\colon U \to B^n(0,r)$, where U is a neighborhood of p and $r > 0$, is called a *quasi-Morse chart for f around p of radius r* (or briefly a *qM-chart for f*) if $\Phi(p) = 0$ and there is an extension of Φ to a chart $\widetilde{\Phi}\colon V \to B^n(0,r')$, where $\overline{U} \subset V$ and $r' > r$, such that

$$(f \circ \widetilde{\Phi}^{-1})(x_1, \dots, x_n) = f(p) + \sum_{i=1}^{n} \alpha_i x_i^2$$

with $\alpha_i < 0$ for $i \leqslant s$ and $\alpha_i > 0$ for $i > s$. The domain U is called a *quasi-Morse coordinate neighborhood.* Any such extension $\widetilde{\Phi}$ of Φ will be called a *quasi-Morse extension of* Φ. We shall often abbreviate "quasi-Morse" to qM. The set $\Phi^{-1}(\mathbb{R}^s \times \{0\})$, resp. $\Phi^{-1}(\{0\} \times \mathbb{R}^{n-s})$, is called the *negative disc*, resp. the *positive disc*.

2. Quasi-Morse chart systems. Assume that for every critical point $p \in S(f)$ we are given a quasi-Morse chart $\{\Phi_p\colon U_p \to B^n(0,r_p)\}_{p \in S(f)}$. The family

$$\mathcal{U} = \{\Phi_p\colon U_p \to B^n(0,r_p)\}_{p \in S(f)}$$

is called a *quasi-Morse chart system* if the family $\{\overline{U_p}\}$ is disjoint and none of the sets $\overline{U_p}$ intersect ∂W. We denote $\min_p r_p$ by $d(\mathcal{U})$ and $\max_p r_p$ by $D(\mathcal{U})$. If all the r_p are equal to r, we shall say that *$\mathcal{U}$ is of radius r*. The set $\Phi_p^{-1}(B^n(0,\lambda))$, where $\lambda \leqslant r_p$, will be denoted by $U_p(\lambda)$. For $\lambda \leqslant d(\mathcal{U})$ we denote $\bigcup_{p \in S(f)} U_p(\lambda)$ by $\mathcal{U}(\lambda)$. In the case when W is Riemannian, we denote $\max_p \mathcal{G}(U_p, \Phi_p)$ by $\mathcal{G}(\mathcal{U})$. Let

$$\mathcal{U} = \{\Phi_p\colon U_p \to B^n(0,r_p)\}_{p \in S(f)}, \qquad \mathcal{U}' = \{\Phi'_p\colon U'_p \to B^n(0,r'_p)\}_{p \in S(f)}$$

be two qM-chart systems for f. We say that $\mathcal{U}'$ is *a restriction of $\mathcal{U}$* if for every $p \in S(f)$ we have $r'_p \leqslant r_p$, $U'_p \subset U_p$, $\Phi'_p = \Phi_p | U'_p$.

3. f-Gradients. Given a qM-chart system $\mathcal{U} = \{\Phi_p\}$, we say that a vector field v on W is an *f-Gradient with respect to $\mathcal{U}$* if

1) for all $x \in W \setminus S(f)$ we have $df(v)(x) > 0$;

2) for all $p \in S(f)$ we have $(\widetilde{\Phi}_p)_*(v)(x) = (-x_1, \dots, -x_k, x_{k+1}, \dots, x_n) = \mathfrak{n}_k(x)$, where k is the index of p and $\widetilde{\Phi}_p$ is some qM extension of Φ_p.

Note that a vector field v is a gradient-like vector field for f if and only if it is an f-Gradient with respect to some qM-chart system.

4. $\mathcal{M}_0$-flows, $\mathcal{M}$-flows, $\mathcal{AM}_0$-flows, $\mathcal{AM}$-flows. If

$$\mathcal{U} = \{\Phi_p\colon U_p \to B^n(0,r_p)\}_{p \in S(f)}$$

is a qM-chart system for f and v is an f-Gradient of f with respect to $\mathcal{U}$, we shall say that the triple $\mathcal{V} = (f, v, \mathcal{U})$ is an *$\mathcal{M}_0$-flow*. Further we say that $\mathcal{V}$ *is of radius* r if $\mathcal{U}$ is of radius r. We set $d(\mathcal{V}) = d(\mathcal{U})$, $D(\mathcal{V}) = D(\mathcal{U})$.

If W is Riemannian, we shall say that the triple $\mathcal{V} = (f, v, \mathcal{U})$ is an *$\mathcal{M}$-flow* if for each chart $\Phi_p : U_p \to B^n(0, r_p)$ from $\mathcal{U}$ the coordinate frame in p is orthonormal with respect to the Riemannian metric.

An $\mathcal{M}_0$-flow $(f, v, \mathcal{U})$ (resp. an $\mathcal{M}$-flow) will be called an *$\mathcal{AM}_0$-flow* (respectively, an $\mathcal{AM}$-flow) if v satisfies the almost transversality condition.

Let $\mathcal{V} = (f, v, \mathcal{U})$ be an $\mathcal{M}$-flow on W, where $\mathcal{U} = \{\Phi_p : U_p \to B^n(0, r_p)\}_{p \in S(f)}$. We denote $\max_p \mathcal{G}(U_p, \Phi_p)$ by $\mathcal{G}(\mathcal{V})$.

5. Subordinate flows. Let $\mathcal{V}_1 = (f, v_1, \mathcal{U}_1)$ and $\mathcal{V}_2 = (f, v_2, \mathcal{U}_2)$ be two $\mathcal{M}_0$-flows on W (the Morse function is the same for both), where

$$\mathcal{U}_1 = \{\Phi_p : U_p \to B^n(0, r_p)\}_{p \in S(f)}, \qquad \mathcal{U}_2 = \{\Psi_p : U_p \to B^n(0, r'_p)\}_{p \in S(f)}$$

are the corresponding qM-chart systems. We say that $\mathcal{V}_2$ is *subordinate to* $\mathcal{V}_1$ if

(1) $\mathcal{U}_2$ is a restriction of $\mathcal{U}_1$;
(2) $v_2 = \phi \cdot v_1$, where $\phi : W \to \mathbb{R}$ is a strictly positive C^∞ function equal to 1 in a neighborhood of the closure of $\bigcup_p U_q$.

(Note that if $\mathcal{V}_2$ is subordinate to $\mathcal{V}_1$, then v_1 and v_2 are equivalent in the sense of Definition 3.3.)

REMARK 3.6. Here are two observations that may serve as a justification for having introduced so many definitions.

1. As we have already noted (Remark 2.6), each f-gradient is actually a gradient-like vector field. Still we keep the qM-charts in our terminology, since we shall often consider different Morse functions with the same f-gradient (for example, we shall multiply Morse functions by a constant).

2. The condition of orthonormality in the definition of $\mathcal{M}$-flow seems artificial, but it makes some constructions easier. Namely, it implies that the Riemannian δ-balls around critical points and the δ-balls with respect to standard charts are very close to each other. The basic invariants of such flows, like $N(\mathcal{V})$ and $T(\mathcal{V})$ (see Definition 3.8), can of course be defined for any $\mathcal{M}_0$-flow.

LEMMA 3.7. *Let $f : \mathbb{R}^n \to \mathbb{R}$ be a quadratic form $f(x) = \sum_{i=1}^n \alpha_i x_i^2$, where $\alpha_i < 0$ for $i \leqslant s$ and $\alpha_i > 0$ for $i > s$. (Note that $\mathfrak{n}_s$ is an f-gradient.) Let $R > R' > r' > r > 0$. Then there is a Morse function $g : \mathbb{R}^n \to \mathbb{R}$ such that*

(1) *$g(x) = f(x)$ outside $B^n(0, R')$;*
(2) *$g(x) = C\mathcal{Q}_s(x)$ in $B^n(0, r)$, where $C > 0$;*
(3) *$S(g) = \{0\}$, and $g'(x)(v(x)) > 0$ for every $x \neq 0$.*

PROOF. Choose a C^∞ function $h : \mathbb{R}^n \to [0, 1]$ such that $\operatorname{supp} h \subset B^n(0, R')$ and $h(x) = 1$ for $x \in B^n(0, r')$. It is obvious that the function

$$x \mapsto Ch(x)\mathcal{Q}_s(x) + (1 - h(x))f(x)$$

satisfies the conclusion of the lemma provided C is sufficiently small. $\square$

It follows from this lemma that if v is an f-Gradient with respect to a qM-chart system $\mathcal{U} = \{\Phi_p : U_p \to B^n(0, r_p)\}_{p \in S(f)}$ and γ is a v-trajectory, then for every $p \in S(f)$ and every $\lambda \leqslant r_p$ the set $I = \{t \in \mathbb{R} \mid \gamma(t) \in U_p(\lambda)\}$ is an open interval, empty, or finite, or infinite from one side.

DEFINITION 3.8 (Quick flows). Let $\mathcal{V} = (f, v, \mathcal{U})$ be an $\mathcal{M}_0$-flow on W. Let γ be a v-trajectory.

A) The number of sets $U_p = U_p(r_p)$ intersected by γ will be denoted by $N(\gamma)$. The number $\max_\gamma N(\gamma)$ will be denoted by $N(\mathcal{V})$. The set

$$\Big\{t \in \mathbb{R} \mid \gamma(t) \notin \bigcup_{p \in S(f)} U_p\Big\}$$

is the finite union of disjoint closed intervals; the sum of their lengths will be denoted by $T(\gamma)$. The number $\max_\gamma T(\gamma)$ is denoted by $T(\mathcal{V})$.

B) Let $\beta > 0$, $C > 0$. We say that $\mathcal{V}$ is (C, β)-*quick* if $\|v\| \leqslant C$ and $T(\mathcal{V}) \leqslant \beta$.

C) Let $\delta \leqslant d(\mathcal{U})$. The number of sets $U_p(\delta)$ intersected by γ will be denoted by $N(\gamma, \delta)$. The number $\max_\gamma N(\gamma, \delta)$ will be denoted by $N(\mathcal{V}, \delta)$. The set

$$\{t \in \mathbb{R} \mid \gamma(t) \notin \mathcal{U}(\delta)\}$$

is the finite union of disjoint closed intervals; the sum of their lengths will be denoted by $T(\gamma, \delta)$. The number $\max_\gamma T(\gamma, \delta)$ will be denoted by $T(\mathcal{V}, \delta)$.

D) Let $\beta > 0$, $C > 0$, $\delta \leqslant d(\mathcal{U})$. We say that $\mathcal{V}$ is (C, β, δ)-*quick* if $\|v\| \leqslant C$ and $T(\mathcal{V}, \delta) \leqslant \beta$.

3.1.3. *Two lemmas on standard gradients in* $\mathbb{R}^n$. In this subsection n, k are positive integers, and $0 \leqslant k \leqslant n$. We regard $\mathbb{R}^n$ as the product $\mathbb{R}^k \times \mathbb{R}^{n-k}$; a point $z \in \mathbb{R}^k$ is therefore written as (x, y), where $x \in \mathbb{R}^k$, $y \in \mathbb{R}^{n-k}$. The Euclidean norm in $\mathbb{R}^n$ is denoted by $|\cdot|$.

We have $\mathfrak{n}_k(x, y) = (-x, y)$, $\mathcal{Q}_k(x, y) = -|x|^2 + |y|^2$. The $\mathfrak{n}_s$-trajectories are of the form $(x_0 e^{-t}, y_0 e^t)$. Using this fact, it is not difficult to prove the following two lemmas.

LEMMA 3.9. *Let $R > r > 0$, and let γ be an $\mathfrak{n}_s$-trajectory. Then the time that γ spends in $B^n(0, R) \setminus B^n(0, r)$ is no greater than* $\ln((R/r)^2 + \sqrt{(R/r)^4 - 1})$, *and the length of the corresponding part of γ is not more than $2R$.*

PROOF. Let $\gamma(t) = (x(t), y(t))$ be an $\mathfrak{n}_s$-trajectory. We can assume that both x and y are never equal to 0. (Indeed, if $x(t_0) = 0$, then $x(t) = 0$ for all t and the time, resp. length, cited in the statement of the lemma is equal to $\ln(R/r)$, resp., to $R - r$.) Set $\mathcal{L}(x) = \ln(x + \sqrt{x^2 - 1})$. There is a unique t_0 such that $|x(t_0)| = |y(t_0)|$. This point is the minimum of $|\gamma(t)|$. Reparametrize γ so as to have $t_0 = 0$, and denote $|x(0)| = |y(0)|$ by α. Assuming $\sqrt{2}\alpha \leqslant R$, we easily calculate that γ quits $B(0, R)$ at the moment $\mathcal{L}(R^2/2\alpha^2)/2$. If $r < \sqrt{2}\alpha$, we are done. If not, the time spent between two spheres is equal to

$$\mathcal{L}\left(\frac{R^2}{2\alpha^2}\right) - \mathcal{L}\left(\frac{r^2}{2\alpha^2}\right),$$

and it suffices to prove that $\mathcal{L}(x) - \mathcal{L}(y) \leqslant \mathcal{L}(x/y)$ for $x > y > 1$; but this is obvious (take the derivative in x). To obtain the estimate of the length, just note that the length cannot be more than the length of the intersection of the one-dimensional hyperbola $H = \{(x, y) \in \mathbb{R}^2 \mid xy = \alpha\}$ with $B^2(0, R)$. □

LEMMA 3.10. *Let $r > 0$, and let γ be an $\mathfrak{n}_k$-trajectory. Then the time that γ spends in the set $\mathcal{Q}_k^{-1}([-r^2, r^2]) \setminus B^n(0, r)$ is not more than 2.*

PROOF. The function $\mathcal{Q}_k'(x)(v(x))$ is bounded below in the domain by r^2, and the variation of $\mathcal{Q}_k$ in this domain is not more than 2. □

COROLLARY 3.11. *Let $\mathcal{V} = (f, v, \mathcal{U})$ be an $\mathcal{M}$-flow on a cobordism. Assume that $\mathcal{V}$ is (C, β, α)-quick. Then $\mathcal{V}$ is $(C, \beta + 8N(\mathcal{V}), \alpha/2)$-quick.*

PROOF. By Lemma 3.9 the time that a v-trajectory can spend in the set $\mathcal{U}(\alpha) \setminus \mathcal{U}(\alpha/2)$ is not more than $8N(\mathcal{V})$. □

3.1.4. *Theorems on the existence of quick flows.* In this subsection W is a Riemannian cobordism of dimension n ($\partial_0 W$ and $\partial_1 W$ can both be empty).

THEOREM 3.12. *Let $C > 0$, $\beta > 0$, $A > 1$. There is an $\mathcal{AM}$-flow $\mathcal{V} = (f, v, \mathcal{U})$ on W of radius r, and for every $\mu \in]0, r[$ there is an $\mathcal{AM}$-flow $\mathcal{W} = (f, w, \mathcal{U}')$ subordinate to $\mathcal{V}$, and such that*
(1) $\mathcal{W}$ is of radius μ and is (C, β)-quick;
(2) $N(\mathcal{W}) \leqslant n2^n$;
(3) $\mathcal{G}(\mathcal{W}) \leqslant A$.

THEOREM 3.13. *Let $B > 0$, $\mu_0 > 0$. Then there is an $\mathcal{AM}$-flow $\mathcal{V} = (f, v, \mathcal{U})$ on W with the following properties:*
(1) $\mathcal{G}(\mathcal{V}) \leqslant 2$ and $D(\mathcal{V}) < \mu_0$;
(2) $N(\mathcal{V}) \leqslant n2^n$;
(3) $\mathcal{V}$ is $(B, S(n))$-quick, where $S(n) = 30 + n2^{n+5}$.

The outline of the proof of 3.12 and 3.13 is as follows. We shall use the notation 3.12(n), which means "Theorem 3.12 for manifolds of dimension n". The proof of the implication 3.13 $\Longrightarrow$ 3.12 is given below. The proof of 3.12(n) $\Longrightarrow$ 3.13($n+1$) is done in §2 and §3 with the help of the S-construction. The same argument also proves 3.13 for 1-dimensional manifolds.

PROOF OF THEOREM 3.4. Let $(f, v, \mathcal{U})$ be the $\mathcal{AM}$-flow of radius r satisfying the conclusion of 3.12 with respect to the manifold M, the numbers C, t, and $A = 2$. Let $\varkappa > 0$ be any number so small that
(1) $\varkappa < r$ and $\varkappa < \varepsilon$;
(2) v is $\varkappa$-ordered;
(3) for every $p \in S(f)$ the disc $D_\varepsilon(p)$ is in U_p.
Let $\mathcal{W} = (f, w, \mathcal{U}')$ be the $\mathcal{M}$-flow subordinate to $\mathcal{V}$ and satisfying 3.12 with respect to the number $\varkappa/2$. I claim that it satisfies our conclusions. Indeed, $\|w\| \leqslant C$ by definition. To show that the flow generated by u is (t, ε)-rapid, it suffices to check (3.1), (3.2) with $\varkappa$ instead of ε. The set $M \setminus B_\varkappa(\mathrm{ind}{\leqslant}n-s-1\,; -v)$ is a $(-v)$-invariant and a $(-w)$-invariant, so we need only check that for every $M \setminus B_\varkappa(\mathrm{ind}{\leqslant}n-s-1\,; -v)$ there are $p \in S(f)$ and $t_0 \leqslant t$ such that $\gamma(x, t_0; -w) \in B_\varkappa(p)$. This is possible, since $U_p(\varkappa/2) \subset B_\varkappa(p)$ and $\mathcal{W}$ is $(C, t, \varkappa/2)$-quick. The proof of (3.2) is similar. □

Before proceeding to the proof of the implication 3.13 $\Longrightarrow$ 3.12, let us informally explain the main idea of the proof. The principal difference between the statements of these two theorems is that 3.13 asserts the existence of *one* flow with prescribed quickness properties, and 3.12 asserts the existence of a whole family of such flows, subordinate to the given one. All these flows must have the prescribed quickness properties, and in this family there are flows of arbitrary small radius. Assume for example that we have a flow $\mathcal{V} = (f, v, \mathcal{U})$ of radius r satisfying the conclusions of 3.13. The problem is of course that any v-trajectory which passes very close to a critical point p spends a lot of time near p, in particular in the spherical slice $U_p(r) \setminus U_p(\mu)$, if μ is small. But if we seek a flow of radius μ with good quickness properties, we can multiply v in $U_p(r) \setminus U_p(\mu)$ by a large positive real function, so that the (Euclidean) norm of the resulting field becomes constant in this domain, which will dramatically reduce the time spent in $U_p(r) \setminus U_p(\mu)$.

These considerations are formalized in the following proposition. In this proposition we present a construction, producing, from a given $\mathcal{M}$-flow $\mathcal{V}$, another flow, denoted $\mathcal{V}'$, for which $D(\mathcal{V}')$ is as small as we like, but $T(\mathcal{V}')$ is still under control.

PROPOSITION 3.14. *Let* $\mathcal{V} = (f, v, \mathcal{U})$ *be an* $\mathcal{M}$*-flow on* W *with* $\mathcal{G}(\mathcal{V}) \leqslant 2$, $\|v\| \leqslant B$. *Write its atlas as* $\mathcal{U} = \{\Phi_p : U_p \to B^n(0, r_p)\}_{p \in S(f)}$. *Let* $C \geqslant B$. *Assume that to each* $p \in S(f)$ *a number* $\mu_p \in]0, r_p[$ *is assigned. Then there is an* $\mathcal{M}$*-flow* $\mathcal{V}' = (f, w, \mathcal{U}')$, *subordinate to* $\mathcal{U}$, *with* $\mathcal{U}' = \{\Phi'_p : U'_p \to B^n(0, \mu_p)\}_{p \in S(f)}$ *and* $\Phi'_p = \Phi_p | U'_p$, *which satisfies* $\|w\| \leqslant C$ *and*

$$C \cdot T(\mathcal{V}') \leqslant B \cdot T(\mathcal{V}) + 5N(\mathcal{V}) \cdot D(\mathcal{V}). \tag{3.5}$$

PROOF. Outside $\bigcup_p U_p$ we set $w = (C/B)v$. We shall construct w in each U_p separately with the help of the standard chart. Namely, $(\Phi_p)_*(v)$ is the standard vector field $\mathfrak{n}_k$ on $\mathbb{R}^k \times \mathbb{R}^{n-k}$, where $k = \operatorname{ind} p$. Let $\lambda : \mathbb{R}_+ \to \mathbb{R}$ be a C^∞ function such that

$$\begin{cases} \lambda(t) = 1 & \text{for } 0 \leqslant t \leqslant \mu_p, \\ \min(1, C/(2t)) \leqslant \lambda(t) \leqslant \max(1, C/(2t)) & \text{for } t \in [\mu_p, \mu_p + \delta], \\ \lambda(t) = C/(2t) & \text{for } \mu_p + \delta \leqslant t \leqslant r_p - \delta, \\ \min(C/B, C/(2t)) \leqslant \lambda(t) \leqslant \max(C/B, C/(2t)) & \text{for } t \in [r_p - \delta, r_p], \\ \lambda(t) = C/B & \text{for } r_p \leqslant t \end{cases}$$

(see Figure 4). The number $\delta \in]0, (r_p - \mu_p)/2[$ will be chosen later.

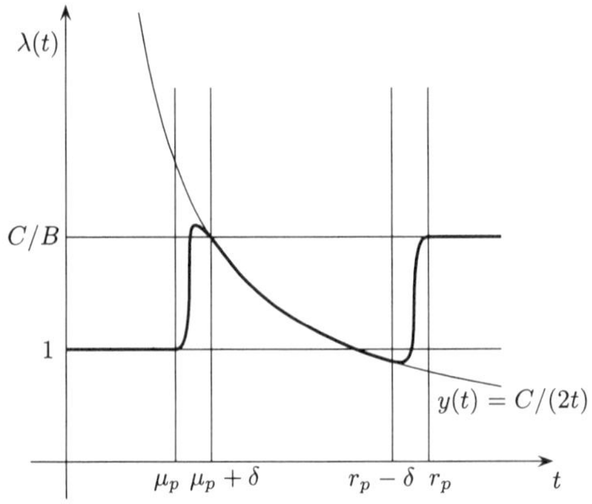

FIGURE 4

Define a vector field w_0 in $\mathbb{R}^n$ by $w_0(x) = \lambda(|x|)\mathfrak{n}_k(x)$. Note that $(\Phi_p^{-1})_*(w_0)$ can be glued to w, previously defined on $M \setminus \bigcup_p U_p$ (with the help of a standard extension); performing this construction for each p, we obtain an $\mathcal{M}$-flow $\mathcal{V}' = (f, w, \mathcal{U}')$, subordinate to $\mathcal{V}$.

The Euclidean norm of $w_0(x)$ equals $\lambda(|x|)|x|$, so that, using the inequality $\mathcal{G}(\mathcal{V}) \leqslant 2$, it is easy to check that $\|w\| \leqslant C$. To obtain an estimate for $T(\mathcal{V})$, note that the time that a w-trajectory spends outside $\bigcup_p U_p$ is not more than

$T(\mathcal{V})(B/C)$. Therefore to obtain (3.5) it suffices to prove that for each $p \in S(f)$ the time that a w-trajectory can spend in $U_p \setminus U_p(\mu_p)$ is not more than $(5/C)r_p$. This is a consequence of the two following observations (in which γ stands for a w_0-trajectory).

(1) The time spent by γ in each of the annuli

$$\mathcal{A}_1 = \{\mu_p \leqslant |x| \leqslant \mu_p + \delta\}, \qquad \mathcal{A}_2 = \{r_p - \delta \leqslant |x| \leqslant r_p\}$$

tends to zero as $\delta \to 0$ (this follows from 3.9).

(2) The Euclidean length of the part of γ in the annulus

$$\mathcal{A}_3 = \{\mu_p + \delta \leqslant |x| \leqslant r_p - \delta\}$$

does not exceed $2r_p$. The Euclidean norm of γ' in $\mathcal{A}_3$ equals $C/2$. □

PROOF OF THEOREM 3.12 FROM THEOREM 3.13. We are given C, β, and A. Choose $B > 0$ and $\mu_0 > 0$ so that

$$B \leqslant C, \quad B \leqslant C\beta \frac{1}{3S(n)}, \quad \mu_0 \leqslant C\beta \frac{1}{15n2^n},$$

and let $\mathcal{V}_0 = (f, v, \mathcal{U}_0)$ be an $\mathcal{M}$-flow satisfying the conclusions of 3.13 with respect to B and μ_0. Write $\mathcal{U}_0 = \{\Psi_p \colon U_p \to B^n(0, r_p)\}_{p \in S(f)}$.

Choose $r < d(\mathcal{V}_0)$ so small that $\mathcal{G}(U_p(r), \Psi_p) < A$ for every $p \in S(f)$. (This is possible since the coordinate frame of Ψ_p at p is orthonormal with respect to the Riemannian metric of W.) For each chart $\{\Psi_p \colon U_p \to B^n(0, r_p)\}_{p \in S(f)}$ of $\mathcal{U}_0$ consider its restriction $\Phi_p = \Psi_p | U_p(r)$, form the corresponding qM-chart system $\mathcal{U} = \{\Phi_p\}$, and denote by $\mathcal{V}$ the $\mathcal{M}$-flow $\mathcal{V} = (f, v, \mathcal{U})$.

I claim that $\mathcal{V}$ satisfies the conclusions of 3.12. Indeed, let $0 < \mu < r$. Apply Proposition 3.5 to the flow $\mathcal{V}_0$ and to the set of numbers $\mu_p = \mu$ for each $p \in S(f)$. We obtain an $\mathcal{M}$-flow $\mathcal{W} = (f, w, \mathcal{U}')$, subordinate to $\mathcal{V}_0$, where $\mathcal{U} = \{\Phi'_p\}_{p \in S(f)}$ and $\Phi'_p = \Phi_p | U_p(\mu)$. This $\mathcal{M}$-flow is (C, β)-quick, since $\|w\| \leqslant C$ and (by 3.14)

$$T(\mathcal{V}') \leqslant \left(\frac{\beta T(\mathcal{V}_0)}{3S(n)} + \frac{5}{C} n 2^n \mu_0 \right) \leqslant \frac{2}{3}\beta.$$

Properties (2) and (3) hold by construction. □

3.2. The S-construction. The aim of this section is to present the S-construction. Given a Morse function without critical points on a cobordism W, the S-construction produces another Morse function on W. This new function behaves like a "fold" (see Figure 3 and the informal introduction at the beginning of the present section).

The corresponding theorem is stated below (Theorem 3.15), and its proof occupies the rest of the subsection. In 3.2.1 we describe some auxiliary constructions and give some definitions. The output of the construction, namely, the Morse function F and its gradient u, are introduced in 3.2.2. The proof of the properties of F and u, listed in Theorem 3.15, occupies the end of 3.2.2 and 3.2.3.

Proceeding to the formulation of Theorem 3.15, let W be a Riemannian cobordism of dimension $n + 1$, where $n \geqslant 0$. In this section we denote by $|\cdot|$ the norm induced by the metric on the tangent spaces, and by $\|\cdot\|$ the corresponding C^0 norm in the space of vector fields.

THEOREM 3.15. *Let $g: W \to [a,b]$ be a Morse function without critical points, $g^{-1}(a) = \partial_0 W$, $g^{-1}(b) = \partial_1 W$. Let $C > 0$ and let w be a g-gradient such that $\|w\| \leqslant C$. Denote*

$$V_{1/3} = g^{-1}\Big(\frac{2a+b}{3}\Big), \qquad V_{2/3} = g^{-1}\Big(\frac{2b+a}{3}\Big), \qquad V_{1/2} = g^{-1}\Big(\frac{a+b}{2}\Big),$$

$$W_0 = g^{-1}\Big(\Big[a, \frac{2a+b}{3}\Big]\Big), \qquad W_1 = g^{-1}\Big(\Big[\frac{2b+a}{3}, b\Big]\Big),$$

$$W_{1/2} = g^{-1}\Big(\Big[\frac{2a+b}{3}, \frac{2b+a}{3}\Big]\Big).$$

Denote by $\mathrm{grad}(g)$ *the Riemannian gradient of g and by* $\mathrm{grd}(g)$ *the vector field* $\mathrm{grad}(g)/|\mathrm{grad}(g)|$. *Denote by T the maximal length of the interval of definition of a* $\mathrm{grd}(g)$*-trajectory. Then there is a $\nu_0 > 0$ such that:*

For every s-submanifold $\mathbb{X}$ of $\partial_0 W$, s-submanifold $\mathbb{Y}$ of $\partial_1 W$, every $\mathcal{AM}$-flow $\mathcal{V}_1 = (F_1, u_1, \mathcal{U}_1)$ on $V_{1/3}$ and $\mathcal{AM}$-flow $\mathcal{V}_2 = (F_2, u_2, \mathcal{U}_2)$ on $V_{2/3}$, and every $\mu \leqslant \min(\nu_0, d(\mathcal{V}_1), d(\mathcal{V}_2))$, there exists an $\mathcal{AM}$-flow $\mathcal{V} = (F, u, \mathcal{U})$ on W with the following properties.

1. *$F: W \to [a,b], S(F) = S(F_1) \cup S(F_2)$, and in a neighborhood of ∂W we have $F = g$, $u = w$.*
2. *$\mathcal{V}$ is of radius μ, and $\mathcal{G}(\mathcal{V}) \leqslant \frac{3}{2}\max(\mathcal{G}(\mathcal{V}_1), \mathcal{G}(\mathcal{V}_2))$.*
3. *$V_{1/3}$, $V_{2/3}$, $W_{1/2}$ are $(\pm u)$-invariant, W_0 is u-invariant and weakly $(-u)$-invariant, and W_1 is $(-u)$-invariant and weakly u-invariant.*
4. *Assume that $\mathcal{V}_1$ is $(C_1, \beta_1, \mu/2)$-quick and $\mathcal{V}_2$ is $(C_2, \beta_2, \mu/2)$-quick. Then $\mathcal{V}$ is $(\frac{3}{2}(C + C_1 + C_2), \beta_1 + \beta_2 + 5 + 4T/C, \mu)$-quick.*
5. *Let γ be a u-trajectory. If $\mathrm{Im}\,\gamma \subset W_{1/2}$, then $N(\gamma, \mu) \leqslant N(\mathcal{V}_1, \mu) + N(\mathcal{V}_2, \mu)$. If $\mathrm{Im}\,\gamma \subset W_0$, then $N(\gamma, \mu) \leqslant N(\mathcal{V}_1, \mu)$. If $\mathrm{Im}\,\gamma \subset W_1$, then $N(\gamma, \mu) \leqslant N(\mathcal{V}_2, \mu)$.*
6. *$\mathbb{D}_a(u) \pitchfork \mathbb{X}$, and $\mathbb{D}_b(-u) \pitchfork \mathbb{Y}$.*

COMMENTS. To visualize the constructions we suggest looking at three figures (see Figure 3) that we have already mentioned. What is drawn exactly is the case $\partial_0 W = \partial_1 W = S^1$. In part A we see the cobordism $W \approx S^1 \times [0,1]$, embedded in $\mathbb{R}^3$, the function g being the vertical projection onto $[0,1]$. Part B presents the manifold W' diffeomorphic to W but embedded differently into $\mathbb{R}^3$. The projection π onto the vertical, restricted to W', has the same behavior as the functions $\chi \circ g$ and $\chi \circ f$ (see 3.2.1). The function $\pi|W'$ has two critical levels, $2/3$ and $1/3$, and the level surfaces correspond to $V_{1/3}$ and $V_{2/3}$. Finally, in part C we find a further perturbed embedding W'' of $S^1 \times [0,1]$ into $\mathbb{R}^3$, and our resulting function F is the vertical projection restricted to W''.

The submanifolds $V_{1/3}$ and $V_{2/3}$ correspond to the critical level surfaces of $\tilde{f}_i$ on part B. The flows $\mathcal{V}_1$ and $\mathcal{V}_2$ correspond to the perturbations of $\tilde{f}_i$ necessary to turn $\tilde{f}_i$ into a Morse function.

In the proof we shall assume that $a = 0$ and $b = 1$, since the general case is easily reduced to this one by an affine transformation of $\mathbb{R}$.

3.2.1. *Auxiliary constructions and the choice of ν_0.*

1. *The function f and its gradient v.*

LEMMA 3.16. *There exist a Morse function $f: W \to [0,1]$ without critical points and an f-gradient v such that:*

(1) $\|v\| \leqslant C$;
(2) *in a neighborhood of* ∂W *we have* $f = g$ *and* $v = w$;
(3) $f^{-1}(1/3) = V_{1/3}$, $f^{-1}(2/3) = V_{2/3}$, $f^{-1}(1/2) = V_{1/2}$;
(4) *for* x *in a neighborhood of* $V_{1/3} \cup V_{1/2} \cup V_{2/3}$ *we have* $|v(x)| = C$ *and* $df(v)(x) = C$;
(5) *for* $\lambda = 1/3, 1/2, 2/3$ *and* $x \in g^{-1}(\lambda)$ *we have* $v(x) \bot g^{-1}(\lambda)$;
(6) *the maximal length of the domain of a* v*-trajectory is not more than* $2T/C$.

PROOF. Let U be an open neighborhood of ∂W such that

$$U \cap (V_{1/3} \cup V_{2/3} \cup V_{1/2}) = \varnothing$$

and let $h\colon W \to [0,1]$ be a C^∞ function such that $\operatorname{supp} h \subset U$ and for x in a neighborhood of ∂W we have $h(x) = 1$. Set

$$v(x) = h(x)w(x) + (1 - h(x))C \operatorname{grd}(g).$$

It is obvious that v is a g-gradient satisfying (1) and (5). We also have $v(x) = w(x)$ near ∂W as well as $|v(x)| = C$ near $V_{1/3} \cup V_{2/3} \cup V_{1/2}$. Part (6) holds also, provided U was chosen sufficiently small. To construct f, define it first of all in a neighborhood of $V_{1/3} \cup V_{2/3} \cup V_{1/2}$ directly by (4), and in a neighborhood of ∂W as equal to g. Now extend f to each of the cobordisms W_0, W_1, $g^{-1}(1/3, 1/2])$, $g^{-1}([1/2, 2/3])$ using the usual procedure of gluing two functions having the same gradient (this procedure is described in detail in [**26**, Corollary 8.14]). □

For $\lambda \in [0,1]$ we denote $f^{-1}(\lambda)$ by V_λ. Fix some $\varepsilon \in]0, 1/12[$. For $\nu > 0$ sufficiently small the map $(x, \tau) \mapsto \gamma(x, \tau; v/C)$ is defined on $V_i \times [-\nu, \nu]$, where $i = 1/3, 1/2, 2/3$, on $V_0 \times [0, \nu]$ and on $V_1 \times [-\nu, 0]$. The corresponding embeddings will be denoted by

$$\Psi_0(\nu)\colon V_0 \times [0, \nu] \to W, \tag{3.6}$$

$$\Psi_1(\nu)\colon V_1 \times [-\nu, 0] \to W, \tag{3.7}$$

$$\Psi_i(\nu)\colon V_i \times [-\nu, \nu] \to W, \qquad \text{for } i = 1/3, 1/2, 2/3. \tag{3.8}$$

The image of $\Psi_i(\nu)$ will be denoted by $\mathrm{Tb}_i(\nu)$ (for $i = 0, 1/3, 1/2, 2/3, 1$).

2. *The choice of* ν_0.

Let ν_0 satisfy the following condition:

$(\mathfrak{R})$
- (1) $2\nu_0 < C$;
- (2) for $i = 0, 1/3, 1/2, 2/3, 1$ we have
  - ($2A$) $\mathrm{Tb}_i(2\nu_0) \subset]i - \varepsilon, i + \varepsilon[$,
  - ($2B$) the Riemannian metric induced by $\Psi_i(2\nu_0)$ from W and the product metric on the domain of $\Psi_i(2\nu_0)$ are $(3/2)$-equivalent;
- (3) for $i = 1/3, 1/2, 2/3$ we have $|v(x)| = df(v(x)) = C$ for x in a neighborhood of $\mathrm{Tb}_i(\nu_0)$.

We shall prove that ν_0 satisfies the conclusions of our theorem. So let

$$\mathbb{X}, \quad \mathbb{Y}, \quad \mathcal{V}_1 = (F_1, u_1, \mathcal{U}_1), \quad \mathcal{V}_2 = (F_2, u_2, \mathcal{U}_2)$$

be as in the statement of the theorem, where

$$\mathcal{U}_1 = \{\Phi_{1p}\colon U_{1p} \to B^n(0, r_{1p})\}_{p \in S(F_1)}, \qquad \mathcal{U}_2 = \{\Phi_{2q}\colon U_{2q} \to B^n(0, r_{2q})\}_{q \in S(F_2)}.$$

Let $\mu \leqslant \min(\nu_0, d(\mathcal{V}_1), d(\mathcal{V}_2))$. We shall denote $\Psi_i(\mu)$ by Ψ_i. Using $(\mathfrak{R})$, it is not difficult to check the following property:

$$\text{for } |\lambda| \leqslant 2\mu \text{ and } i = 1/3, 1/2, 2/3 \text{ we have } f^{-1}(i+\lambda) = \Psi_i(V_i \times \{\lambda\}). \tag{3.9}$$

3. *Auxiliary functions, vector fields and s-submanifolds.*

For $\lambda \in [0, 1/3]$ we denote by $\mathbb{L}_\lambda(u_1)$ the s-submanifold $v^{\rightsquigarrow}_{[1/3,\lambda]}(\mathbb{D}(u_1))$ of V_λ. For $\lambda \in [1/3, 1]$ we denote by $\mathbb{L}_\lambda(u_1)$ the s-submanifold $(-v)^{\rightsquigarrow}_{[1/3,\lambda]}\mathbb{D}(u_1)$ of V_λ. For $\lambda \in [2/3, 1]$ we denote by $\mathbb{L}_\lambda(-u_2)$ the s-submanifold $(-v)^{\rightsquigarrow}_{[2/3,\lambda]}(\mathbb{D}(-u_2))$ and for $\lambda \in [0, 2/3]$ we denote by $\mathbb{L}_\lambda(-u_2)$ the s-submanifold $v^{\rightsquigarrow}_{[2/3,\lambda]}(\mathbb{D}(-u_2))$. Note that since v has no zeros, $v^{\rightsquigarrow}_{[\alpha,\beta]}$ is a diffeomorphism of V_α onto V_β for any $\beta < \alpha$. Note also that $\mathbb{L}_{2/3}(-u_2)$ equals $\mathbb{D}(-u_2)$ and $\mathbb{L}_{1/3}(u_1)$ equals $\mathbb{D}(u_1)$.

Let $\chi\colon [0,1] \to [0,1]$ be a C^∞ function, with the following properties:

$$\begin{aligned}
&\chi(x) = x && \text{for } x \in [0,\varepsilon] \cup [1-\varepsilon, 1],\\
&\chi(x) = 1 - x && \text{for } x \in [1/2-\varepsilon, 1/2+\varepsilon],\\
&\chi(x) = 2/3 - (x-1/3)^2 && \text{for } x \in [1/3-\varepsilon, 1/3+\varepsilon],\\
&\chi'(x) > 0 && \text{for } x \in [0, 1/3[\cup]2/3, 1],\\
&\chi(x) = 1/3 + (x-2/3)^2 && \text{for } x \in [2/3-\varepsilon, 2/3+\varepsilon],\\
&\chi'(x) < 0 && \text{for } x \in]1/3, 2/3[
\end{aligned}$$

(see the graph of χ in Figure 5).

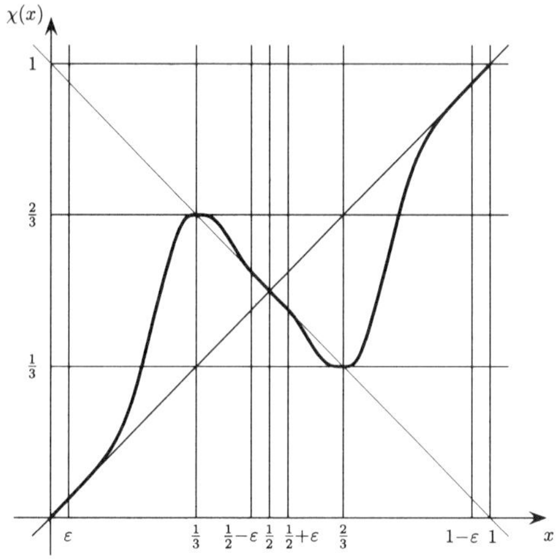

FIGURE 5

Denote $\chi \circ f$ by ψ. The following property follows immediately from $(\mathfrak{R})$:

$$\begin{aligned}
(\psi \circ \Psi_{1/3})(x,\tau) &= 2/3 - \tau^2,\\
(\psi \circ \Psi_{2/3})(x,\tau) &= 1/3 + \tau^2,\\
(\psi \circ \Psi_{1/2})(x,\tau) &= 1 - \tau.
\end{aligned} \tag{3.10}$$

Let $B\colon \mathbb{R} \to [0,1]$ be a C^∞ function such that $B(t) = 0$ for $|t| \geqslant 5\mu/3$ and $B(t) = 1$ for $|t| \leqslant 4\mu/3$. Let $B_1 : \mathbb{R} \to \mathbb{R}^+$ be a C^∞ function such that

$$\operatorname{supp} B_1 \subset]\mu, 2\mu[, \qquad \int_0^\infty B_1(t)dt = C.$$

Let z_0 be a C^∞ vector field on V_0 such that $\Phi(z_0, 1)(\mathbb{L}_0(u_1)) \pitchfork \mathbb{X}$. Let z_1 be a C^∞ vector field on V_1 such that $\Phi(z_1, 1)(\mathbb{L}_1(-u_2)) \pitchfork \mathbb{Y}$. Let $z_{1/2}$ be a C^∞ vector field on $V_{1/2}$ such that $\Phi(z_{1/2}, 1)(\mathbb{L}_{1/2}(u_1)) \pitchfork \mathbb{L}_{1/2}(-u_2)$. We shall assume that the z_i (where $i = 0,\ 1/2,\ 1$) are chosen so small that $\sup_\tau |B_1(\tau)| \cdot \|z_i\| < C/9$.

3.2.2. *The Morse function F and its gradient u.*

1. *The Morse function F.*

Let $a_1, a_2 > 0$. Set

$$F(y) = \psi(y) \quad \text{for } y \in W \setminus (\mathrm{Tb}_{1/3}(2\mu) \cup \mathrm{Tb}_{2/3}(2\mu)),$$

and

$$(F \circ \Psi_{1/3})(x,\tau) = a_1 B(\tau) F_1(x) + 2/3 - \tau^2 \quad \text{for } (x,\tau) \in V_{1/3} \times [-2\mu, 2\mu],$$
$$(F \circ \Psi_{2/3})(x,\tau) = a_2 B(\tau) F_2(x) + 1/3 + \tau^2 \quad \text{for } (x,\tau) \in V_{2/3} \times [-2\mu, 2\mu].$$

These formulas correctly define a smooth function $F\colon W \to \mathbb{R}$ which is equal to f near ∂W. To find critical points of F, note that $S(F)$ is obviously contained in $\mathrm{Tb}_{1/3}(2\mu) \cup \mathrm{Tb}_{2/3}(2\mu)$. For $(x,\tau) \in V_{1/3} \times [-2\mu, 2\mu]$ we have

$$d(F \circ \Psi_1)(x,\tau) = (a_1 B'(\tau) F_1(x) - 2\tau)\, d\tau + a_1 B(\tau)\, dF_1(x).$$

For a_1 small enough, this expression can vanish only for $\tau = 0$. We therefore conclude (applying the same reasoning to $\mathrm{Tb}_{2/3}(2\mu)$) that $S(F) = S(F_1) \cup S(F_2)$ provided the numbers a_i are small enough, and we make this assumption from now on. To prove that F is a Morse function, we shall make explicit the standard charts for F. Let $p \in S(F_1)$ and write

$$(F_1 \circ \Phi_{1p}^{-1})(x) = F_1(p) + \sum_i \alpha_i x_i^2.$$

Consider the chart

$$\Phi_{1p} \times \mathrm{id} : U_{1p} \times]-\mu, \mu[\to B^n(0, r_{1p}) \times]-\mu, \mu[$$

of the manifold $V_{1/3} \times]-\mu, \mu[$ around the point $(p, 0)$. We have

$$F \circ \Psi_{1/3} \circ (\Phi_{1p} \times \mathrm{id})^{-1}(x,\tau) = a_1 F_1(p) + \frac{2}{3} + a_1 \sum_i \alpha_i x_i^2 - \tau^2.$$

Therefore the chart

$$(\Phi_{1p} \times \mathrm{id}) \mid (\Phi_{1p} \times \mathrm{id})^{-1}(B^{n+1}(0,\mu)) \circ (\Psi_{1/3} \mid \mathrm{Tb}_{1/3}(\mu))^{-1}$$

is a standard chart of radius μ for F at $p \in S(F_1)$. These charts together with similar ones for $q \in S(F_2)$ give an F-chart system of radius μ. We shall denote this system by $\mathcal{U}$. Note that $\mathrm{ind}_F\, p = \mathrm{ind}_{F_1} p + 1$ for $p \in S(F_1)$ and that $\mathrm{ind}_F\, q = \mathrm{ind}_{F_2} q$ for $q \in S(F_2)$. Note also that if a_1 and a_2 are chosen sufficiently small, then

$$F(\mathrm{Tb}_{1/3}(2\mu)) \subset [2/3 - \varepsilon, 2/3 + \varepsilon], \qquad F(\mathrm{Tb}_{2/3}(2\mu)) \subset [1/3 - \varepsilon, 1/3 + \varepsilon].$$

In this case we also have $F^{-1}(1/2) = V_{1/2} \cup V_\theta \cup V_{\theta'}$, where $\varepsilon < \theta < 1/3 - \varepsilon$, $2/3 + \varepsilon < \theta' < 1 - \varepsilon$.

2. *The F-gradient u.*

Set $\mathrm{Tub} = \mathrm{Tb}_0(2\mu) \cup \mathrm{Tb}_{1/3}(2\mu) \cup \mathrm{Tb}_{1/2}(2\mu) \cup \mathrm{Tb}_{2/3}(2\mu) \cup \mathrm{Tb}_1(2\mu)$

Define a vector field u on W as follows:

$$u(y) = v(y) \quad \text{for } y \in (W_0 \setminus \mathrm{Tub}) \cup (W_2 \setminus \mathrm{Tub}),$$
$$u(y) = -v(y) \quad \text{for } y \in (W_{1/2} \setminus \mathrm{Tub}),$$
$$((\Psi_0^{-1})_*(u))(x,\tau) = (-B_1(\tau)z_0(x), C),$$
$$((\Psi_1^{-1})_*(u))(x,\tau) = (B_1(-\tau)z_1(x), C),$$
$$((\Psi_{1/3}^{-1})_*(u))(x,\tau) = (B(\tau)u_1(x), -B(\tau)\tau - (1-B(\tau))\cdot C\cdot \operatorname{sgn}\tau),$$
$$((\Psi_{1/2}^{-1})_*(u))(x,\tau) = (-B_1(-\tau)z_{1/2}(x), -C),$$
$$((\Psi_{2/3}^{-1})_*(u))(x,\tau) = (B(\tau)u_2(x), B(\tau)\tau + (1-B(\tau))\cdot C\cdot \operatorname{sgn}\tau).$$

An easy computation using the definition of F and u shows that these formulas correctly define a C^∞ vector field on W, and that $(F,u,\mathcal{U})$ is an $\mathcal{M}$-flow on W of radius μ, provided z_0, z_1, $z_{1/2}$ are small enough (this assumption we make from now on). It is also easy to see that u is an f-gradient in $W_0 \setminus V_{1/3}$ and $W_1 \setminus V_{2/3}$; $(-u)$ is an f-gradient in $W_{1/2} \setminus (V_{1/3} \cup V_{2/3})$.

We claim that $\mathcal{V} = (F,u,\mathcal{U})$ satisfies all the conclusions of 3.15. Items (1) and (2) follow immediately from the construction. To prove (3), note that $V_{1/3}$ is a closed submanifold of W° and $u|V_{1/3}$ is tangent to $V_{1/3}$; therefore $V_{1/3}$ is $(\pm u)$-invariant (the same is true for $V_{2/3}$). To prove that $W_{1/2}$ is u-invariant, let $x \in W^\circ_{1/2}$. The trajectory $\gamma(x,\,\cdot\,;u)$ cannot intersect $V_{1/3}$ or $V_{2/3}$, since they are $(\pm u)$-invariant. Therefore $\gamma(x,\,\cdot\,;u)$ must stay in $W^\circ_{1/2}$ forever.

3.2.3. *Properties of F and u.*

1. *Estimate of the quickness of $\mathcal{V}$.*

To obtain an estimate of $\|u\|$, note that the inequality

$$|u(x)| \leqslant \frac{3}{2}(C + C_1 + C_2)$$

is to be checked only for $x \in \bigcup_i \mathrm{Tb}_i(2\mu)$, where it is a simple computation; we leave it to the reader. To estimate the time that a u-trajectory spends outside $\mathcal{U}(\mu)$, note first that for a trajectory starting at a point of $V_{1/3}$ (resp. $V_{2/3}$), this time is not more than β_1 (resp. β_2), since it is actually a u_1- (resp. u_2-) trajectory. Now let γ be a u-trajectory passing through a point of $W^\circ_{1/2}$. At the end of the preceding subsection, we proved that γ stays in $W^\circ_{1/2}$ forever. Since u is a $(-f)$-gradient in $W^\circ_{1/2}$, the function $t \mapsto f(\gamma(t))$ is strictly decreasing and

$$\lim_{t\to-\infty} f(\gamma(t)) = 2/3, \qquad \lim_{t\to\infty} f(\gamma(t)) = 1/3.$$

This implies

$$\lim_{t\to-\infty} \gamma(t) = q \in S(F_2), \qquad \lim_{t\to\infty} \gamma(t) = p \in S(F_1).$$

We shall estimate the time that γ spends outside $\mathcal{U}(\mu)$ between the various level surfaces of f.

1) $f(\gamma(t)) \in [2/3 - \mu/2, 2/3]$. By (3.9) and (3.10), this condition is equivalent to the inclusion $\gamma(t) \in \Psi_{2/3}(V_{2/3} \times [-\mu/2, 0[)$. The curve $\Psi_{2/3}^{-1}(\gamma(t))$ is a product of a u_2-trajectory and the curve $\tau \mapsto \alpha e^\tau$ with $\alpha < 0$. Since $\Psi_{2/3}^{-1}(\mathcal{U}(\mu))$ contains $U_p(\mu/2) \times]-\mu/2, \mu/2[$ for every $p \in S(F_2)$, it follows that the time that $\gamma(t)$ spends in $\Psi_{2/3}(V_{2/3} \times [-\mu/2, 0[)$ outside of $\Psi_{2/3}^{-1}(\mathcal{U}(\mu))$ is not more than β_2.

2) $f(\gamma(t)) \in [2/3 - 2\mu, 2/3 - \mu/2]$, or, in other words,

$$\gamma(t) \in \Psi_{2/3}(V_{2/3} \times [-2\mu, -\mu/2]).$$

The vector field $(\Psi_{2/3}^{-1})_*(u)$ equals $(B(\tau)u_2(x), \varkappa(\tau))$, where $\varkappa(\tau) \leqslant \tau$. Therefore the total time which γ can spend in this domain is not more than $\ln(2\mu/(\mu/2)) < 2$.

3) $f(\gamma(t)) \in [1/2, 2/3 - 2\mu]$. In the domain $f^{-1}([1/2, 2/3 - 2\mu])$ we have $u = v$. Therefore the time is not more than $2T/C$.

4) $f(\gamma(t)) \in [1/2 - 2\mu, 1/2]$. Here $\gamma(t) \in \Psi_{1/2}(V_{1/2} \times [-2\mu, 0])$. The second coordinate of $(\Psi_{1/2}^{-1})_*(u)$ is equal to $(-C)$. Therefore the total time spent here is not more than $2\mu/C \leqslant 1$.

As in 1)–3), one shows that the time γ spends in $f^{-1}(]1/3, 1/2 - 2\mu]) \setminus \mathcal{U}(\mu)$ is not more than $2T/C + 2 + \beta_1$. Summing up, we obtain that the time γ can spend in $W^\circ_{1/2} \setminus \mathcal{U}(\mu)$ is not more than $\beta_1 + \beta_2 + 4T/C + 5$.

Similar analysis of the behavior of u-trajectories in W_0 and W_1 shows that the same estimate holds in these cases also.

2. *Estimate of* $N(\gamma, \mu)$.

Let $\gamma(\cdot)$ be a u-trajectory in $W_{1/2}$. If γ is in $V_{1/3}$ (resp. in $V_{2/3}$), then obviously $N(\gamma, \mu) \leqslant N(\mathcal{V}_1, \mu)$ (resp. $N(\gamma, \mu) \leqslant N(\mathcal{V}_2, \mu)$). Assume that we have $\operatorname{Im}\gamma \subset W^\circ_{1/2}$. Let $\alpha \in \mathbb{R}$ (resp. $\beta \in \mathbb{R}$) be the unique number such that $f(\gamma(\alpha)) = 2/3 - \mu$ (resp. $f(\gamma(\beta)) = 1/3 + \mu$). We have

$$\bigcup_{p \in S(F_i)} U_p(\mu) \subset \mathrm{Tb}_i(\mu) \qquad (i = 1, 2).$$

Therefore $\gamma(t) \in \mathcal{U}(\mu)$ can occur only if $t \geqslant \beta$ or $t \leqslant \alpha$. For $t \leqslant \alpha$ (resp. $t \geqslant \beta$) the curve $\Psi_{2/3}^{-1}(\gamma(t))$ (resp. $\Psi_{1/3}^{-1}(\gamma(t))$) is an integral curve of the vector field $(u_2(x), \tau)$ (resp. $(u_1(x), -\tau)$). Since an integral curve of u_2 (resp. u_1) can intersect no more than $N(\mathcal{V}_2, \mu)$ (resp. $N(\mathcal{V}_1, \mu)$) standard coordinate neighborhoods of radius μ, the first part of (5) follows. The case of curves in W_0 and W_1 is treated similarly.

3. *Transversality properties.*

The next lemma implies that $\mathcal{V}$ satisfies the almost transversality condition. It also implies item 6) of our conclusions.

LEMMA 3.17. (1) *The family* $\{\bigcup_{p \in S_i(F_1)} D_p(-u)\}_{0 \leqslant i \leqslant n}$ *is equal to* $\mathbb{D}(-u_1)$. *The family* $\{\bigcup_{q \in S_i(F_2)} D_q(u)\}_{0 \leqslant i \leqslant n}$ *is equal to* $\mathbb{D}(u_2)$.

(2) *For each* $\lambda \in [0, 1]$ *the family* $\{(\bigcup_{q \in S_i(F_2)} D_q(-u)) \cap V_\lambda\}_{0 \leqslant i \leqslant n}$ *is an s-submanifold of* V_λ, *which is equal to*

(a) $\Phi(z_1, 1)(\mathbb{L}_0(-u_2))$ *if* $\lambda = 1$,

(b) $\mathbb{L}_\lambda(-u_2)$ *if* $\lambda \in [1/2, 1 - \varepsilon]$,

(c) $\varnothing$ *if* $\lambda \leqslant 1/3$.

(3) *For each* $\lambda \in [0, 1]$ *the family* $\{(\bigcup_{p \in S_i(F_1)} D_p(u)) \cap V_\lambda\}_{0 \leqslant i \leqslant n}$ *is an s-submanifold of* V_λ, *which is equal to*

(a) $\Phi(z_0, 1)(\mathbb{L}_0(u_1))$ *for* $\lambda = 0$,

(b) $\mathbb{L}_\lambda(u_1)$ *for* $\lambda \in [\varepsilon, 1/2 - 2\mu]$,

(c) $\Phi(z_{1/2}, 1)(\mathbb{L}_{1/2}(u_1))$ *for* $\lambda = 1/2$,

(d) $\varnothing$ *for* $\lambda \geqslant 2/3$.

(4) *The vector field* u *satisfies the almost transversality condition, and we have*

$$\mathbb{D}(u) \cap V_0 \pitchfork \mathbb{X}, \qquad \mathbb{D}(-u) \cap V_1 \pitchfork \mathbb{Y}.$$

PROOF. (1) For every $p \in S_i(F_1)$ the positive disc of F in p belongs to $V_{1/3}$, which immediately implies the first assertion; the second is proved similarly.

(2) It is not difficult to see that it suffices to prove the assertion for $\lambda \in [-2\mu + 2/3, 2\mu + 2/3]$. For these values of λ, this follows from the analysis of the behavior of u-trajectories in $\mathrm{Tb}_{2/3}(2\mu)$ carried out in 3.2.3 1) above.

Item (3) is proved similarly.

(4) To prove that u satisfies the almost transversality condition, let $p, q \in S(F)$, $\operatorname{ind} p \leqslant \operatorname{ind} q$. Let γ be a $(-u)$-trajectory joining p with q. The case $p, q \in S(F_1)$ or $p, q \in S(F_2)$ follows easily from (1). Let $p \in S(F_1)$, $q \in S(F_2)$. Denote by z the (unique) point of intersection of γ with $V_{1/2}$, and denote $\operatorname{ind}_F p$ by k and $\operatorname{ind}_F q$ by r. Then $z \in (\Phi(z_{1/2}, 1)(\mathbb{L}_{1/2}(u_1)))_{k-1}$ and $z \in \mathbb{L}_{1/2}(-u_2)_{n-r}$, which is impossible by the choice of $z_{1/2}$. The last item is already proved. □

3.3. Proof of the theorem on the existence of quick flows.

3.3.1. *Introduction.* Let W be a Riemannian cobordism of dimension $n+1$. From 3.12 we shall deduce Theorem 3.18 below, which is somehow stronger than 3.13. Indeed, we have included in the conclusions of Theorem 3.18 not only the restrictions (1)–(3) from 3.13 on the resulting $\mathcal{M}$-flow $\mathcal{V}$, but also the restrictions $(\partial 1)$–$(\partial 3)$ which concern the behavior of $\mathcal{V}$ near ∂W (conditions $(\partial 1)$–$(\partial 3)$ are empty when $\partial W = \varnothing$).

THEOREM 3.18. *Let $g\colon W \to [a, b]$ be a Morse function on W. Let $B > 0$, $\mu_0 > 0$. Let u be a g-gradient with $\|u\| \leqslant B/2$. Let $\mathbb{X}$ be an s-submanifold of $\partial_0 W$ and $\mathbb{Y}$ an s-submanifold of $\partial_1 W$. There is an $\mathcal{AM}$-flow $\mathcal{V} = (f, v, \mathcal{U})$ on W such that*

(1) *$\mathcal{G}(\mathcal{V}) \leqslant 2$ and $D(\mathcal{V}) \leqslant \mu_0$;*
(2) *$N(\mathcal{V}) \leqslant (n+1)2^{n+1}$;*
(3) *$\mathcal{V}$ is $(B, S(n))$-quick, where $S(n) = 30 + n2^{n+5}$.*

Moreover,

$(\partial 1)$ *$f(W) = [a, b]$, and in a neighborhood of ∂W we have $f = g$ and $u = v$;*
$(\partial 2)$ *$\mathbb{D}_a(v) \pitchfork \mathbb{X}$ and $\mathbb{D}_b(-v) \pitchfork \mathbb{Y}$;*
$(\partial 3)$ *if a v-trajectory γ starts at a point of $\partial_0 W$, then $\gamma(t)$ converges to a critical point of f as $t \to \infty$, and if a $(-v)$-trajectory θ starts at a point of $\partial_1 W$, then $\theta(t)$ converges to a critical point of f as $t \to \infty$; moreover, for such trajectories γ we have $N(\gamma) \leqslant n2^n$ and $T(\gamma) \leqslant 8 + n2^{n+4}$.*

We shall prove the theorem by induction on the number of critical points of g. Lemma 3.20 below shows how to do it: if 3.18 is true for the two "halves" of the cobordism W, then it is true also for the whole cobordism W. To make our notation shorter, we shall introduce some terminological conventions. In the next definition and in Lemma 3.20, $g\colon W \to [a, b]$ is a Morse function.

DEFINITION 3.19. (1) An *input data* is a quintuple $(B, \mu_0, u, \mathbb{X}, \mathbb{Y})$, where $B > 0$ and $\mu_0 > 0$ are real numbers, u is a g-gradient with $\|u\| \leqslant B/2$, and $\mathbb{X}$, resp. $\mathbb{Y}$, are s-submanifolds of $\partial_0 W$, resp. of $\partial_1 W$.

(2) The pair (g, W) will be called *good* if Theorem 3.18 is true for (g, W) and any input data $(B, \mu_0, u, \mathbb{X}, \mathbb{Y})$.

(3) A v-trajectory starting at a point of $\partial_0 W$ or finishing at a point of $\partial_1 W$ will be called a *boundary trajectory.*

Let λ be a regular value of g. Set

$$W_0 = g^{-1}([a,\lambda]), \quad W_1 = g^{-1}([\lambda,b]), \quad g_0 = g|W_0, \quad g_1 = g|W_1.$$

LEMMA 3.20. *If (g_0, W_0) and (g_1, W_1) are good, then so is (g, W).*

PROOF. Let $(B, \mu_0, u, \mathbb{X}, \mathbb{Y})$ be input data. Set $u_0 = u|W_0$, $u_1 = u|W_1$. Apply 3.18 to (g_0, W_0) and to the input data $(B, \mu_0, u_0, \mathbb{X}, \varnothing)$. We obtain an $\mathcal{AM}$-flow $\mathcal{V}_0 = (f_0, v_0, \mathcal{U}_0)$ on W_0. Next apply Theorem 3.18 to the function $g_1 \colon W_1 \to [\lambda, b]$ and to the input data $(B, \mu_0, u_1, \mathbb{D}_\lambda(-v_0), \mathbb{Y})$. We obtain an $\mathcal{AM}$-flow $\mathcal{V}_1 = (f_1, v_1, \mathcal{U}_1)$ on W_1. Note that the condition $(\partial 1)$ from 3.18 allows us to glue together the $\mathcal{M}$-flows $\mathcal{V}_0$ on W_0 and $\mathcal{V}_1$ on W_1 to obtain an $\mathcal{M}$-flow $\mathcal{V} = (f, v, \mathcal{U})$ on W. I claim that $\mathcal{V}$ satisfies the conclusions of 3.18. We must check first that $\mathcal{V}$ is an $\mathcal{M}$-flow. This follows since $\mathcal{V}_0$ and $\mathcal{V}_1$ are $\mathcal{AM}$-flows and $\mathbb{D}_\lambda(v_1) \pitchfork \mathbb{D}_\lambda(-v_0)$. Next we check that $\mathcal{V}$ satisfies (1)–(3) of 3.13. Item (1) follows from the properties of $\mathcal{V}_0$ and $\mathcal{V}_1$. To prove (2), let γ be a v-trajectory. If it stays forever in one of the sets W_0, W_1, then we are done since the corresponding condition is true for $\mathcal{V}_0$, $\mathcal{V}_1$. If γ intersects $V_\lambda = f^{-1}(\lambda)$, say, $\gamma(t_0) \in V_\lambda$, then $\gamma(t) \in W_1$ for $t \geqslant t_0$ and $\gamma(t) \in W_0$ for $t \leqslant t_0$. Now the inequality $N(\gamma) \leqslant (n+1)2^{n+1}$ follows from condition $(\partial 3)$, which holds for $\mathcal{V}_0$ and for $\mathcal{V}_1$. The proof of (3) is similar.

Now we proceed to check $(\partial 1)$–$(\partial 3)$ for $\mathcal{V}$. The properties $(\partial 1)$ and $(\partial 3)$ follow immediately from the corresponding properties of $\mathcal{V}_0$, $\mathcal{V}_1$. To check $(\partial 2)$, note that no v-trajectory starting at a point of W_0 reaches $\partial_1 W$ (this follows since the property $(\partial 3)$ holds for $\mathcal{V}_1$). Therefore the condition $\mathbb{D}(-v) \pitchfork \mathbb{Y}$ is equivalent to $\mathbb{D}(-v \mid W_1) \pitchfork \mathbb{Y}$, which holds by assumption. □

Therefore it suffices to deduce $3.18(n+1)$ from $3.12(n)$ for Morse functions $g \colon W \to [a,b]$ without critical points and for those with one critical point. This is the aim of 3.3.2 and 3.3.3. We shall not use the full force of $3.12(n)$, but only the following corollary of $3.12(n)$.

LEMMA 3.21. *Let N be a closed Riemannian manifold of dimension n. Suppose that $D > 0$, $\beta > 0$, $A > 1$. Then there is a $\nu_0 > 0$ such that for every $\mu < \nu_0$ there exists an $\mathcal{AM}$-flow $\mathcal{W}$ satisfying*

(1) *$\mathcal{W}$ is of radius μ and is (D, β)-quick;*
(2) *$N(\mathcal{W}) \leqslant n2^n$;*
(3) *$\mathcal{G}(\mathcal{W}) \leqslant A$.*

3.3.2. *Functions without critical points.* Here we prove $3.12(n+1)$ for Morse functions without critical points. Theorem 3.15 applies in this situation, and we shall first of all state a corollary of (3.15 + 3.21).

COROLLARY 3.22. *Let $g \colon W \to [a,b]$ be a Morse function without critical points. Let $(B, \mu_0, w, \mathbb{L}, \mathbb{K})$ be input data for g. Denote by T the maximal length of the interval of definition of the* $\mathrm{grd}(g)$*-trajectory. Then there is an $\mathcal{M}$-flow $\mathcal{V} = (F, u, \mathcal{U})$ of radius μ on W such that*

(1) *in a neighborhood of ∂W we have $F = g$ and $u = w$;*
(2) *$\mathcal{G}(\mathcal{V}) \leqslant 2$ and $\mu \leqslant \mu_0$;*
(3) *$N(\mathcal{V}) \leqslant n2^{n+1}$;*
(4) *$\mathcal{V}$ is $(B, 7 + n2^{n+4} + 8(T/B))$-quick;*
(5) *every boundary trajectory γ of v converges to a critical point of F, and for such trajectories we have $N(\gamma) \leqslant n2^n$;*

(6) $\mathbb{D}_a(u) \nmid \mathbb{L}$ *and* $\mathbb{D}_b(-u) \nmid \mathbb{K}$.

PROOF. As in the proof of 3.15, set

$$g^{-1}\left(\frac{2a+b}{3}\right) = V_{1/3}, \qquad g^{-1}\left(\frac{2b+a}{3}\right) = V_{2/3}.$$

Apply Lemma 3.21 to $V_{1/3}$ and to $V_{2/3}$, obtaining two $\mathcal{AM}$-flows $\mathcal{V}_1$, $\mathcal{V}_2$ of some radius $\mu < \mu_0$ on $V_{1/3}$, $V_{2/3}$ such that $\mathcal{G}(\mathcal{V}_i) \leqslant 2$, $N(\mathcal{V}_i) \leqslant n2^n$, and the $\mathcal{V}_i$ are $(B/12, 1)$-quick. Note that the $\mathcal{V}_i$ are $(B/12, 1 + n2^{n+3}, \mu/2)$-quick (by 3.11). Now just apply 3.15, obtaining an $\mathcal{AM}$-flow $\mathcal{V} = (F, u, \mathcal{U})$ on W of radius μ, which is $(B, 7 + n2^{n+4} + 8(T/B))$-quick. We also have $N(\mathcal{V}) \leqslant n2^{n+1}$. The other items of our conclusion also follow from the properties of $\mathcal{V}$ listed in 3.15. □

Of course this corollary does not yet give an $\mathcal{AM}$-flow satisfying the conclusions of 3.18, since T can be arbitrarily large. So what we shall do is to cut W into thin slices so that for each slice W_s the corresponding time T_s is small, then apply 3.22 to each of the W_s, and then glue the resulting flows together.

Proceeding to the precise constructions, recall that we are given the input data $(B, \mu_0, u, \mathbb{X}, \mathbb{Y})$. Choose a sequence $a = a_0 < a_1 < a_2 < \cdots < a_N = b$ of regular values of g so close to each other that for each s the time that a grd(g)-trajectory spends in $W_s = g^{-1}([a_s, a_{s+1}])$ is not more than $B/8$. We shall define by induction on s a sequence of $\mathcal{AM}$-flows $\mathcal{V}_i = (F_i, u_i, \mathcal{U}_i)$ on W_s. Assume that $\mathcal{V}_i$ is already constructed for $i \leqslant s-1$. Apply 3.22 to W_s, choosing the following input data: $(B, \mu_0, u, \mathbb{L}_s, \mathbb{K}_s)$, where $\mathbb{L}_s$ and $\mathbb{K}_s$ are defined as follows:

$$\begin{aligned} s = 0: &\quad \mathbb{L}_s = \mathbb{X},\ \mathbb{K}_s = \varnothing, \\ 0 < s < N-1: &\quad \mathbb{L}_s = \mathbb{D}_{a_s}(-u_{s-1}),\ \mathbb{K}_s = \varnothing, \\ s = N-1: &\quad \mathbb{L}_s = \mathbb{D}_{a_s}(-u_{s-1}),\ \mathbb{K}_s = \mathbb{Y}. \end{aligned}$$

Thus we obtain a sequence of $\mathcal{AM}$-flows $\mathcal{V}_i$ of radius $\mu_i \leqslant \mu_0$, and each of the flows $\mathcal{V}_i$ is $(B, 8 + n2^{n+4})$-quick and satisfies the inequality $N(\mathcal{V}_i) \leqslant n2^{n+1}$. Now glue all the $\mathcal{V}_i$ together (this is possible, since near every $g^{-1}(a_i)$ the functions f_i and f_{i+1} coincide with g and the vector fields u_i and u_{i+1} coincide with u). We obtain an $\mathcal{M}$-flow $\mathcal{V} = (F, u, \mathcal{U})$ on W, and I claim that it satisfies the conclusions of 3.18. Indeed, note first that $\mathcal{V}$ is an $\mathcal{AM}$-flow (this follows from the restrictions $\mathbb{D}_{a_s}(u_s) \nmid \mathbb{D}_{a_s}(-u_{s-1})$ imposed on $\mathcal{V}_s$ in the process of construction). Further, (1) and (2) follow immediately (actually, here we obtain a slightly stronger estimate $N(\mathcal{V}) \leqslant n2^{n+1}$). As to the quickness estimate, note that no v-trajectory can intersect more than two of the slices W_s. Thus $\mathcal{V}$ is $(B, 2(8 + n2^{n+4}))$-quick. The other conditions follow immediately.

3.3.3. *Functions with one critical point.* Here we prove 3.18($n+1$) for functions $g\colon W \to [a, b]$ with one critical point. Let p be this critical point, let k be its index, and let $(B, \mu_0, u, \mathbb{X}, \mathbb{Y})$ be our input data for the construction of the flow $\mathcal{V}$. Changing if necessary the g-gradient u in a neighborhood of p, we can assume that there is a Morse chart $\Phi\colon U \to B^n(0, r) \subset \mathbb{R}^{n+1}$ for g such that $\Phi(p) = 0$ and $\Phi_*(v) = \mathfrak{n}_k$.

LEMMA 3.23. *There is a δ_0, $0 < \delta_0 < r$, such that for every $\delta < \delta_0$ the time that a v-trajectory can spend in $g^{-1}([g(p) - \delta^2, g(p) + \delta^2]) \setminus U_p(\delta/2)$ is not more than* 6.

PROOF. Denote $g^{-1}([g(p)-\delta^2, g(p)+\delta^2])$ by S_δ. Let γ be a w-trajectory. The function $dg(w)$ is bounded from below in $S_\delta \setminus U_p$; therefore for $\delta > 0$ sufficiently small the time that γ spends in $S_\delta \setminus U_p$ is less than 1. The time that γ spends in $U_p \cap (S_\delta \setminus U_p(\delta))$ is not more than 2 (by 3.10), and the time that γ spends in $U_p(\delta) \setminus U_p(\delta/2)$ is not more than $\ln 8 \leqslant 3$ (by 3.9). □

Now let $\delta > 0$ be so small that

(1) $\delta < \mu_0$, $\delta < \delta_0$;

(2) $\mathcal{G}(U(\delta), \Phi|U(\delta)) \leqslant 2$.

Set

$$W_+ = g^{-1}([g(p)+\delta^2, b]), \qquad g_+ = g|W_+,$$
$$W_- = g^{-1}([a, g(p)-\delta^2]), \qquad g_- = g|W_-,$$
$$W_0 = g^{-1}([g(p)-\delta^2, g(p)+\delta^2]), \qquad g_0 = g|W_0,$$
$$V_- = g^{-1}(g(p)-\delta^2), \qquad V_+ = g^{-1}(g(p)+\delta^2),$$
$$S_+ = D(p, -v) \cap g^{-1}(g(p)+\delta^2).$$

By abuse of notation we keep the symbol u to denote all the restrictions of u to $W_+, W_-, \ldots$. Apply Theorem 3.18 to g_+ (proved in 3.3.2). Choose as the input data the following string: $(B, \mu_0, u, S_+, \mathbb{Y})$.

Denote the resulting $\mathcal{AM}$-flow by $\mathcal{V}_+ = (f_+, v_+, \mathcal{U}_+)$. Glue it to the $\mathcal{AM}$-flow $\mathcal{V}_0 = (g|W_0, u, \{\Phi|U(\delta/2)\})$, obtaining an $\mathcal{AM}$-flow on $W_+ \cup W_0$, which will be denoted by $\mathcal{V}_1 = (F_1, v_1, \mathcal{U}_1)$. Now apply Theorem 3.18 to g_-, choosing as the input data the following string: $(B, \mu_0, u, \mathbb{X}, \mathbb{D}_{f(p)-\delta^2}(v_1))$. We obtain an $\mathcal{AM}$-flow. Denote it by $\mathcal{V}_- = (f_-, v_-, \mathcal{U}_-)$ and glue it to $\mathcal{V}_1$.

Denote the resulting $\mathcal{M}$-flow by $\mathcal{V} = (f, v, \mathcal{U})$. I claim that it satisfies the conclusions of 3.18. Indeed, condition (1) holds by construction, as well as the condition $\|v\| \leqslant B$. To check (2), note that it is already proved for trajectories of v which stay forever in W_- or W_+. Assume that a v-trajectory γ intersects V_+. If $\lim_{t\to-\infty}\gamma(t) = p$, then $N(\gamma) \leqslant 1 + n2^n$. If γ intersects V_-, then $N(\gamma) \leqslant 1 + n2^{n+1}$. The time that a v-trajectory γ can spend outside $\bigcup_p U(p)$ is estimated from above using Lemma 3.23 and the estimates of Corollary 3.22.

The properties $(\partial 1)$–$(\partial 3)$ of Theorem 3.18 are verified without difficulty.

3.4. Final remark. The results of this section are essentially contained in [**35**]. In particular, Theorem 3.12 is a generalization of Theorem 1.12 from [**35**] to the case of cobordisms (in [**35**] it was stated only for closed manifolds). Theorem 3.13 is a generalization of Theorem 1.13 from [**35**] to the case of cobordisms (also we reduced the corresponding constant from $100 + n2^{n+7}$ in [**35**] to $30 + n2^{n+5}$; we think that it could be diminished still more). Theorems 1.12 and 1.13 of [**35**] are in turn exactly Theorems 1.17 and 1.18 of the preprint [**31**].

4. Regularizing the gradient descent map

Let W be a Riemannian cobordism of dimension n, $f\colon W \to [a, b]$ a Morse function and v an f-gradient. Consider the open set $U_1 \subset \partial_1 W$ of all points x such that the trajectory $\gamma(x, \cdot\,; -v)$ reaches $\partial_0 W$ and intersects it at some point y. The correspondence $x \mapsto y$ is a diffeomorphism of U_1 onto an open subset $U_0 \subset \partial_0 W$, which we denote by $(-v)^{\leadsto}$. The aim of this section is to prove an analog of the cellular approximation theorem for the map $(-v)^{\leadsto}$. The problem is that $(-v)^{\leadsto}$ in general cannot be extended to a continuous map on the whole of $\partial_1 W$.

Recall, however, from 2.5 that $(-v)^{\leadsto}$ is well defined on the category of s-submanifolds (under a mild transversality condition; see 2.5.2). The collection of descending discs of a gradient of a Morse function forms an s-submanifold (if the gradient satisfies the almost transversality condition); we choose a Morse function ϕ_1 on $\partial_1 W$ and a ϕ_1-gradient u_1 satisfying the almost transversality condition. Consider the corresponding s-submanifold $\mathbb{D}(u_1)$ formed by the descending discs of u_1. Denote $\mathbb{D}(u_1)$ by $\mathbb{A}$ and $\bigcup_{i\leqslant j} D(\mathrm{ind}{=}i\,; u_1)$ by $A_{\leqslant j}$. Proceed similarly with $\partial_0 W$ and obtain the s-submanifold $\mathbb{B} = \mathbb{D}(u_0)$, where u_0 is a gradient of a Morse function on $\partial_0 W$, and u_0 satisfies the almost transversality condition. The decomposition of a manifold into the union of descending discs of a gradient strongly resembles a cellular decomposition, so one can try to reproduce the cellular approximation theorem in this situation. Of course, we cannot expect that by perturbing v we shall obtain a direct analog of the cellular approximation condition for the new gradient u: $(-u)^{\leadsto}(A_{\leqslant j}) \subset B_{\leqslant j}$. We can, though, deform v so that for the new gradient u the following will hold:

$$(-u)^{\leadsto}(A_{\leqslant j}) \subset D_\delta(\mathrm{ind}{\leqslant}j\,; u_0). \tag{4.1}$$

Indeed, first perturb v so as to make the submanifold $(-u)^{\leadsto}(\mathbb{A})$ transversal to the s-submanifold $\widehat{\mathbb{B}} = \mathbb{D}(-u_0)$. Then $(-u)^{\leadsto}(A_{\leqslant j})$ does not intersect $\mathbb{D}_{\leqslant n-2-j}(-u_0)$ and for T sufficiently large the diffeomorphism $\Phi(T, -u_0)$ pushes $(-u)^{\leadsto}(A_{\leqslant j})$ into a small neighborhood of $\mathbb{D}_{\leqslant j}(-u_0)$. We can modify v in a small neighborhood of $\partial_0 W$, so that for the new vector field $\tilde{v}$ we have

$$(-\tilde{v})^{\leadsto} = \Phi(T, -u_0) \circ (-v)^{\leadsto}.$$

This is done by adding to v a horizontal component proportional to u_0 (see 2.1.6 for the definition of the construction of adding a horizontal component). Thus for $\tilde{v}$ the inclusion (4.1) holds.

It is essential for us that the property (4.1) (and actually a somewhat stronger property) can be achieved by a C^0-small perturbation $\tilde{v}$ of v. The proof of this result (Theorem 4.3) is based on the main theorem of the previous section.

4.1. Condition ($\mathfrak{C}$).

4.1.1. *Condition ($\mathfrak{C}$): Definition and statement of the density theorem.*

Terminology. Let $g\colon W \to [a, b]$ be a Morse function on a cobordism W of dimension n, and let v be a g-gradient.

Let $\lambda \in [a, b]$ be a regular value of f. Let $A \subset \partial_1 W$. We say that v *descends A to $f^{-1}(\lambda)$* if every $(-v)$-trajectory starting in A reaches $f^{-1}(\lambda)$. Let $X \subset W$. We say that v *descends A to $f^{-1}(\lambda)$ without intersecting* X if v descends A to $f^{-1}(\lambda)$ and $T(A, -v) \cap X = \varnothing$.

We leave to the reader the definition of the expressions "*v lifts B to $f^{-1}(\lambda)$*" and "*v lifts B to $f^{-1}(\lambda)$ without intersecting* Y", where $B \subset \partial_0 W$, $Y \subset f^{-1}([a, \lambda])$.

We recall the sets $B_\delta(\mathrm{ind}{\leqslant}s\,; v)$ and $D_\delta(\mathrm{ind}{\leqslant}s\,; v)$ from Definition 2.16, and introduce one more set with similar properties, the set

$$C_\delta(\mathrm{ind}{\leqslant}s\,; v) = W \setminus B_\delta(\mathrm{ind}{\geqslant}n-s-1\,; -v).$$

The following lemma is proved by standard Morse-theoretical arguments.

LEMMA 4.1. *Assume that v is δ-separated with respect to ϕ and to the ordering sequence $(a_0, \ldots, a_{n+1})$. Then for every s we have*

$$\partial_0 W \cup B_\delta(\text{ind}{\leqslant s}\,; v) \subset \phi^{-1}([a_0, a_{s+1}]) \subset C_\delta(\text{ind}{\leqslant s}\,; v) \cup \partial_0 W,$$

and these inclusions are homotopy equivalences.

Now we can formulate condition $(\mathfrak{C})$.

DEFINITION 4.2. We say that v *satisfies condition* $(\mathfrak{C})$ if there are objects 1)–4), listed below, with the properties (A), (B0), (B1) below.

Objects.

1) An ordered Morse function ϕ_1 on $\partial_1 W$ with ordering sequence $(\alpha_0, \ldots, \alpha_n)$, and a ϕ_1-gradient u_1.
2) An ordered Morse function ϕ_0 on $\partial_0 W$ with ordering sequence $(\beta_0, \ldots, \beta_n)$, and a ϕ_0-gradient u_0.
3) An ordered Morse function ϕ on W with ordering sequence $(a_0, \ldots, a_{n+1})$, adjusted to (f, v).
4) A number $\delta > 0$.

Properties.

(A) u_0 is δ-separated with respect to ϕ_0, u_1 is δ-separated with respect to ϕ_1, and v is δ-separated with respect to ϕ.

(B1) $(-v)^{\rightsquigarrow}(C_\delta(\text{ind}{\leqslant j}\,; u_1)) \cup (D_\delta(\text{ind}{\leqslant j+1}\,; v) \cap \partial_0 W) \subset B_\delta(\text{ind}{\leqslant j}\,, u_0)$ for every j.

(B0) $\overset{\rightsquigarrow}{v}(C_\delta(\text{ind}{\leqslant j}\,; -u_0)) \cup (D_\delta(\text{ind}{\leqslant j+1}\,; -v) \cap \partial_1 W) \subset B_\delta(\text{ind}{\leqslant j}\,; -u_1)$ for every j.

The set of all f-gradients satisfying $(\mathfrak{C})$ will be denoted by $\mathcal{G}\mathfrak{C}(f)$.

COMMENTS. 1. As we have already mentioned in the introduction, conditions (B0), (B1) together form an analog of the cellular approximation condition. Indeed, the sets $C_\delta(\text{ind}{\leqslant j}\,; u_1)$ and $B_\delta(\text{ind}{\leqslant j}\,; u_0)$ are "thickenings" of $D(\text{ind}{\leqslant j}\,; u_1)$, resp. of $D(\text{ind}{\leqslant j}\,; u_0)$ (the first one being "thicker" than the second). Thus condition (B1) requires that the map $(-v)^{\rightsquigarrow}$ sends a certain thickening of a j-skeleton of $\partial_1 W$ to a certain thickening of a j-skeleton of $\partial_0 W$. Warning: (B1) actually requires more than that: for every j the soles of δ-handles of W of indices $\leqslant j+1$ must belong to the set $B_\delta(\text{ind}{\leqslant j}\,; u_0)$.

2. Condition $(\mathfrak{C})$ is stronger than condition $(\mathcal{C})$ from [**28**, §1.2]. This is obvious, since $(\mathfrak{C})$ is formulated similarly to $(\mathcal{C})$, except that in $(\mathcal{C})$ we required that

$$(-v)^{\rightsquigarrow}(\phi^{-1}([a_0, a_{j+1}])) \subset B_\delta(\text{ind}{\leqslant j}\,; u_0).$$

Condition $(\mathcal{C})$ is in turn stronger than condition (RP) from [**35**, §4]; this is a bit less obvious, and we shall not prove it here.

3. If v satisfies condition $(\mathfrak{C})$, then v obviously satisfies the almost transversality condition. Note that if we are given a ϕ_0-gradient u_0, a ϕ_1-gradient u_1, and a ϕ-gradient v that satisfy the almost transversality condition, then condition (A) is always true if δ is sufficiently small.

THEOREM 4.3. *$\mathcal{G}\mathfrak{C}(f)$ is open and dense in $\mathcal{G}(f)$ with respect to the C^0 topology. Moreover, if v_0 is any f-gradient, then one can choose a C^0-small perturbation v of v_0 such that $v \in \mathcal{G}\mathfrak{C}(f)$, $v = v_0$ in a neighborhood of ∂W, and $v = v_0$ in a neighborhood of every critical point of f.*

PROOF. The proof occupies 4.1.2–4.1.5.

Convention. We shall say that ($\mathfrak{C}$) is *open*, resp. *dense*, if the set of all f-gradients satisfying ($\mathfrak{C}$) is open, resp. dense, in $\mathcal{G}(f)$.

4.1.2. *The C^0-openness of $\mathfrak{C}$.*

Let v be an f-gradient satisfying ($\mathfrak{C}$). Choose the corresponding functions ϕ_0, ϕ_1, ϕ, their gradients u_0, u_1, and a number $\delta > 0$ satisfying (A) and (B0), (B1). Now fix the string $\mathcal{S} = (\delta, u_0, u_1, \phi_0, \phi_1, \phi)$. For a ϕ-gradient w, we shall denote condition (A) with respect to w and $\mathcal{S}$ by (A)(w). Similarly, we shall denote conditions (B1), (B0) with respect to w and $\mathcal{S}$ by (B1)(w), resp. (B0)(w). We know that (A)(v) & (B1)(v) & (B0)(v) holds; we shall prove that (A)(w) & (B1)(w) & (B0)(w) holds for every w sufficiently C^0 close to v.

Let w be another f-gradient. Corollary 2.75 implies that if $\|w-v\|$ is sufficiently small, then w is a ϕ-gradient and is also δ-separated, so condition (A)(w) is also satisfied for w.

It is convenient to reformulate conditions (B1)(w) and (B0)(w). Introduce three new conditions:

$\beta_1(w)$: For every j, w descends $C_\delta(\operatorname{ind}\leqslant j\,; u_1)$ to $\phi^{-1}(a_{j+1})$ without intersecting $\bigcup_{\operatorname{ind} p \geqslant j+1} D_\delta(p)$.

$\beta_2(w)$: For every j, w lifts $C_\delta(\operatorname{ind}\leqslant n-j-2\,; -u_0)$ to $\phi^{-1}(a_{j+2})$ without intersecting $\bigcup_{\operatorname{ind} p \leqslant j+1} D_\delta(p)$.

$\beta_3(w)$: For every j the sets

$$R_j^+(w) = (-v)^{\leadsto}_{[b,a_{j+2}]}(C_\delta(\operatorname{ind}\leqslant j\,; u_1)), \qquad R_j^-(w) = v^{\leadsto}_{[a,a_{j+2}]}(C_\delta(\operatorname{ind}\leqslant n-j-2\,; -u_0))$$

are disjoint.

Note that $\beta_1(w)$ and $\beta_2(w)$ are in a sense dual to each other.

LEMMA 4.4. *Assume that* (A)(w) *holds. Then* (B1)(w) & (B0)(w) $\Longleftrightarrow$ $\beta_1(w)$ & $\beta_1(w)$ & $\beta_3(w)$.

PROOF. $\Longrightarrow$. Condition β_1 follows from (B1), and β_2 follows from (B0). To prove β_3, note that if $R_j^+(w) \cap R_j^-(w) \neq \varnothing$, then there is a $(-v)$-trajectory joining $x \in C_\delta(\operatorname{ind}\leqslant j\,; u_1)$ with $y \in C_\delta(\operatorname{ind}\leqslant n-j-2\,; -u_0) = \partial_0 W \setminus B_\delta(\operatorname{ind}\leqslant j\,; u_0)$. But this contradicts (B1).

$\Longleftarrow$. β_1 says that for each j we have

$$\Bigg(\bigcup_{\operatorname{ind} p \geqslant j+1} D_\delta(p, -v)\Bigg) \cap \partial_1 W \subset \partial_1 W \setminus C_\delta(\operatorname{ind}\leqslant j\,; u_1) = B_\delta(\operatorname{ind}\leqslant n-2-j\,; -u_1).$$

In other words,

$$D_\delta(\operatorname{ind}\leqslant n-1-j\,; -v) \cap \partial_1 W \subset B_\delta(\operatorname{ind}\leqslant n-2-j\,; -u_1).$$

Further, β_3 implies that for each j no v-trajectory starting in $C_\delta(\operatorname{ind}\leqslant n-j-2\,; -u_0)$ and reaching $\partial_1 W$ can intersect $\partial_1 W$ at a point of $C_\delta(\operatorname{ind}\leqslant j\,; u_1)$. Therefore it intersects $\partial_1 W$ at a point of $B_\delta(\operatorname{ind}\leqslant n-2-j\,; -u_1)$. Therefore (B0)($w$) holds. The proof of (B1)(w) is similar. $\square$

Returning to the proof of the C^0-openness of ($\mathfrak{C}$), note that conditions $\beta_1(w)$ and $\beta_2(w)$ are obviously open. We know that $\beta_1(v)$ and $\beta_2(v)$ hold, and so for every j the sets $R_j^+(v)$ and $R_j^-(v)$ are compact. Since they are disjoint (by $\beta_3(v)$), we can choose two disjoint open subsets U_j, V_j of $\phi^{-1}(a_{j+2})$ such that $R_j^+(v) \subset U_j$,

$R_j^-(v) \subset V_j$. Then for every w sufficiently close to v and for every j we have (by Corollary 2.72)

$$(-w)^{\leadsto}_{[b,a_{j+2}]}(C_\delta(\mathrm{ind}{\leqslant}j\,;u_1)) \subset U_j, \qquad w^{\leadsto}_{[a,a_{j+2}]}(C_\delta(\mathrm{ind}{\leqslant}n-2-j\,;-u_0)) \subset V_j,$$

and $\beta_3(w)$ holds. Therefore $\beta_1(w)\&\beta_2(w)\&\beta_3(w)$ is C^0 open, and we have proved the C^0 openness of $(\mathfrak{C})$.

4.1.3. *C^0-density of $(\mathfrak{C})$: Idea of the proof.* In this subsection I present the idea of the proof on the intuitive level.

Choose some pair (ϕ_0, u_0), where ϕ_0 is an ordered Morse function on $\partial_0 W$, and u_0 is an ϕ_0-gradient. Let us try to perturb a given f-gradient v_0 in order to obtain (B1). The set

$$\mathfrak{N}_j(v,\delta) = (-v)^{\leadsto}(C_\delta(\mathrm{ind}{\leqslant}j\,;u_1)) \cup (D_\delta(\mathrm{ind}{\leqslant}j+1\,;v) \cap \partial_0 W)$$

contains in particular the set

$$\mathfrak{N}_j(v) = (-v)^{\leadsto}(D(\mathrm{ind}{\leqslant}j\,;u_1)) \cup (D(\mathrm{ind}{\leqslant}j+1\,;v) \cap \partial_0 W).$$

We know from Proposition 2.52 and Lemma 2.51 that, perturbing v if necessary, we can assume that $\{\mathfrak{N}_j(v)\}$ forms an s-submanifold transversal to $\mathbb{D}(-u_0)$. In particular, $\mathfrak{N}_j(v)$ is disjoint from $D(\mathrm{ind}{\leqslant}i\,, -u_0)$ for $i+j < n-1$. Using Proposition 2.54, we conclude that for $\varepsilon > 0$ small enough the sets $\mathfrak{N}_j(v,\varepsilon)$ and $D_\varepsilon(\mathrm{ind}{\leqslant}i\,; -u_0)$ are disjoint if $i+j < n-1$.

Now we shall modify v by adding a horizontal component near $\partial_0 W$. Let ξ be a vector field equivalent to u_0 (see Definition 3.3). Then for some $T > 0$ we have

$$\Phi(T, -\xi)(\partial_0 W \setminus B_\varepsilon(\mathrm{ind}{\leqslant}i\,, -u_0)) \subset B_\varepsilon(\mathrm{ind}{\leqslant}n-2-i\,; u_0).$$

Hence if we apply this to the AHC-construction with respect to ξ and to a function h satisfying $\int_0^l h(\tau)\,d\tau = T$, for the resulting vector field w we shall have

$$\mathfrak{N}_j(w) \subset B_\varepsilon(\mathrm{ind}{\leqslant}n-2-i\,, u_0)$$

(see AHC2, page 52; actually we should also take some care to ensure that w be an f-gradient).

Of course if the vector field u_0 and ξ are chosen arbitrarily, then T may be large and therefore $\|v-w\|$ may also be large.

Now I shall explain why it is possible to choose T and $\|v-w\|$ small. This step uses Theorem 3.4 in an essential way. We shall choose the pair (ϕ_0, u_0) in a special way. Let $C > 0$ and $t > 0$ be real numbers, and let (ϕ_0, u_0) be a pair satisfying the conclusions of 3.4 with respect to C, t. Choose $\tilde{u}_0$, equivalent to u_0 with $\|\tilde{u}_0\| \leqslant C$ and having the property that the flow generated by $(-\tilde{u}_0)$ pushes $\partial_0 W \setminus B_\varepsilon(\mathrm{ind}{\leqslant}i\,; -u_0)$ inside $B_\varepsilon(\mathrm{ind}{\leqslant}n-2-i\,; u_0)$, the time necessary for this procedure being $\leqslant t$. Now set $\xi = \tilde{u}_0$ and to v add a horizontal component proportional to ξ near $\partial_0 W$. The input data (a, ξ, h) for the AHC-construction will be chosen so that $\int_0^a h(\tau)\,d\tau = t$. The estimates from AHC1 (page 51) show that $\|v-w\|$ will be small enough provided C and t are chosen sufficiently small.

4.1.4. *C^0-density of $(\mathfrak{C})$: terminology.* Let v_0 be an f-gradient. Choose regular values $\alpha, \beta \in \,]a, b[$ of f such that $a < \alpha < \beta < b$ and the intervals $[a, \alpha]$ and $[\beta, b]$ are regular. Denote $f^{-1}(\lambda)$ by V_λ, so that $\partial_1 W = V_b$, $\partial_0 W = V_a$. Denote $f^{-1}([\alpha,\beta])$ by W'. The maps $(-v_0)^{\leadsto}_{[b,\beta]}\colon V_b \to V_\beta$ and $(v_0)^{\leadsto}_{[a,\alpha]}\colon V_a \to V_\alpha$ are diffeomorphisms. We shall denote these maps by Ψ_1, resp. Ψ_0. Set

$$m = \inf_{x \in W \setminus W'} \|df(v)(x)\|, \qquad M = \sup_{x \in W \setminus W'} \|df(x)\|.$$

Let E_0, E_1 be the expansion constants for Ψ_0, resp. Ψ_1. Set $E = \max(E_0, E_1)$. Now we shall define the horizontal components that will be added to v. Let $\varepsilon > 0$. Our aim is to construct a vector field v such that (B1) and (B0) hold and $\|v - v_0\| < \varepsilon$. Choose closed collars

$$C_0 = \gamma(\partial_0 W, [0, l]; v), \qquad C_1 = \gamma(\partial_1 W, [0, l]; -v)$$

so that $C_0 \subset \phi^{-1}([a, \alpha[)$, $C_1 \subset \phi^{-1}(]\beta, b])$. Choose any positive C^∞ function h on $[0, l]$ such that $h(x) = 0$ for x in a neighborhood of 0 and in a neighborhood of l. Denote $\sup_{\tau \in [0,l]} h(\tau)$ by H. Set $t_0 = \int_0^l h(\tau)\, d\tau$. Choose a real number $A > 0$ such that

$$A < \frac{m}{EMH}, \qquad A < \frac{\varepsilon}{2EH}. \tag{4.2}$$

REMARK 4.5. It is easy to deduce from AHC1 and AHC3 that with this choice of A, t_0, h, H we have the following statement: Let ξ be a vector field $\partial_0 W$ with $\|\xi\| \leqslant A$. Let v be an f-gradient. Let w be the result of the AHC-construction with input data (l, ξ, h) applied to v. Then w is also an f-gradient and $\|w - v\| \leqslant \varepsilon/2$. (A similar assertion holds for adding a horizontal component near $\partial_1 W$.)

Let $\phi_0 \colon \partial_0 W \to \mathbb{R}$ be a Morse function and u_0 a ϕ_0-gradient satisfying the conclusions of Theorem 3.4 with respect to $C = A$, $t = t_0$. We can obviously assume that ϕ_0 is ordered; let $(\alpha_0, \dots, \alpha_n)$ be the corresponding ordering sequence.

Similarly, let $\phi_1 \colon \partial_1 W \to \mathbb{R}$ be a Morse function and u_1 a ϕ_1-gradient satisfying the conclusions of Theorem 3.4 with respect to $C = A$, $t = t_0$. We assume that ϕ_1 is ordered; let $(\beta_0, \dots, \beta_n)$ be the corresponding ordering sequence.

Assume that u_0 and u_1 are δ-separated. We obtain the corresponding gfn-systems $\mathbf{D}(u_1), \mathbf{D}(u_0), \mathbf{D}(-u_1), \mathbf{D}(-u_0)$ with the cores $\mathbb{D}(u_0), \mathbb{D}(u_1), \dots$. We fix $]0, \delta[$ as the interval of definition for these systems.

The diffeomorphism Ψ_1 identifies V_b with V_β. Therefore we can transfer to V_β the function ϕ_1 and its gradient u_1. The resulting function and gradient will be denoted by ϕ_1' and u_1'. We also obtain the gfn-systems: $\mathbf{A}^- = \Psi_1(\mathbf{D}(u_1))$ and $\mathbf{A}^+ = \Psi_1(\mathbf{D}(-u_1))$ in V_β with cores $\mathbb{D}(u_1')$ and $\mathbb{D}(-u_1')$. (*Warning*: Ψ_1 is not in general an isometry, so $\mathbf{A}^-$ need not be the same as $\mathbf{D}(u_1')$.) A similar construction gives gfn-systems $\mathbf{B}^-$ and $\mathbf{B}^+$ on V_α with cores $\mathbb{D}(u_0)$ and $\mathbb{D}(-u_0)$.

4.1.5. *C^0-density of $\mathcal{GC}(f)$: perturbation of a given gradient.* In this subsection we finish the proof of Theorem 4.3. Namely, we show how to perturb v_0 in order to obtain a gradient v satisfying condition ($\mathfrak{C}$).

A. *First modification of the gradient.*

Applying Proposition 2.52, we obtain an $(f|W')$-gradient v_1 having the following properties:

(i) v_1 satisfies the almost transversality condition.
(ii) $v_1 = v$ in a neighborhood of $V_\alpha \cup V_\beta$.
(iii) The s-submanifolds $\mathbb{T}(\mathbb{D}(u_1'), -v_1)$ and $\mathbb{T}(\mathbb{D}(-u_0'), v_1)$ are defined.
(iv) $(-v_1)^{\leadsto}_{[\beta,\alpha]}(\mathbb{D}(u_1')) \nmid \mathbb{D}(-u_0')$ and $(v_1)^{\leadsto}_{[\alpha,\beta]}(\mathbb{D}(-u_0')) \nmid \mathbb{D}(u_1')$.

Moreover, v_1 can be chosen arbitrarily C^∞ close to v_0; thus we can assume that $\|v_1 - v_0\| < \varepsilon/2$. Recall from 2.5.3 that there is a $\lambda > 0$ such that the pairs $(\mathbf{A}, v_1)$ and $(\mathbf{B}, -v_1)$ are λ-ordered. Therefore we can regard $]0, \lambda[$ as the interval of definition of the gfn-systems $\mathbf{T}(\mathbf{A}^-, -v_1)$ and $\mathbf{T}(\mathbf{B}^+, v_1)$.

Applying Proposition 2.54, we deduce from (iv) above that there is a $\nu < \lambda$ such that for every s

$$(-v_1)^{\leadsto}_{[\beta,\alpha]}(\mathbf{A}^-_{(\leqslant s)}(\nu)) \cup (D_\nu(\text{ind}\leqslant s+1\,;v_1)\cap V_\alpha) \text{ does not intersect } \mathbf{B}^+_{(\leqslant n-2-s)}(\nu), \tag{4.3}$$

$$(v_1)^{\leadsto}_{[\alpha,\beta]}(\mathbf{B}^+_{(\leqslant s)}(\nu)) \cup (D_\nu(\text{ind}\leqslant s+1\,;-v_1)\cap V_\beta) \text{ does not intersect } \mathbf{A}^-_{(\leqslant n-2-s)}(\nu). \tag{4.4}$$

We shall choose ν so small that v_1 is ν-separated. Extend the vector field v_1 to the whole of W by setting $v_1(x) = v_0(x)$ for $x \in W \setminus W'$. Then v_1 is a ν-separated f-gradient. Choose an ordered Morse function ϕ on W', adjusted to (f, v_1), and extend ϕ to the whole of W by setting $\phi(x) = f(x)$ for $x \in W \setminus W'$.

B. *Second modification of the gradient.*

Here we start with the gradient v_1 constructed in the previous modification. The second modification consists in adding to v_1 horizontal components near $\partial_1 W$ and $\partial_0 W$. We have chosen a number $\nu > 0$. Now choose (by Theorem 3.4) a ϕ_0-gradient $\tilde{u}_0$ and a ϕ_1-gradient $\tilde{u}_1$, equivalent to u_0, resp. to u_1, and such that $\tilde{u}_0$, $\tilde{u}_1$ are (t_0,ν)-rapid and satisfy $\|\tilde{u}_0\| \leqslant A$, $\|\tilde{u}_1\| \leqslant A$. That is, for $i = 0, 1$ and for every s, we have

$$\Phi(\tilde{u}_i, t_0)(C_\nu(\text{ind}\leqslant s\,;\tilde{u}_i)) \subset B_\nu(\text{ind}\leqslant s\,;\tilde{u}_i),$$
$$\Phi(-\tilde{u}_i, t_0)(C_\nu(\text{ind}\leqslant s\,;-\tilde{u}_i)) \subset B_\nu(\text{ind}\leqslant s\,;-\tilde{u}_1).$$

Apply to v_1 the AHC-construction near $\partial_0 W$ with input data $(l, \tilde{u}_0, h)$ and near $\partial_1 W$ with input data $(l, \tilde{u}_1, h)$ (recall that l and h were chosen at the beginning of 4.1.4).

Denote the resulting vector field by v. Then $\|v - v_0\| \leqslant \varepsilon$, and that v is an f-gradient (see Remark 4.5). Since

$$\operatorname{supp}(v_1 - v) \subset W \setminus W' \quad \text{and} \quad f|W \setminus W' = \phi|W \setminus W',$$

the vector field v is also a ϕ-gradient. To compute $\overset{\leadsto}{v}$ and $(-v)^{\leadsto}$, denote the diffeomorphism $(-v)^{\leadsto}_{[b,\beta]}$ by $\widetilde{\Psi}_1$ and $v^{\leadsto}_{[a,\alpha]}$ by $\widetilde{\Psi}_0$.

From the construction of v it follows (see AHC2, page 52) that for every s we have

$$\widetilde{\Psi}_1(C_\nu(\text{ind}\leqslant s\,;u_1)) \subset \mathbf{A}^-_{(\leqslant s)}(\nu), \tag{4.5}$$

$$\widetilde{\Psi}_1^{-1}(V_\beta \setminus \mathbf{A}^+_{(\leqslant s)}(\nu)) \subset B_\nu(\text{ind}\leqslant n-2-s\,;-u_1). \tag{4.6}$$

Similarly,

$$\widetilde{\Psi}_0(C_\nu(\text{ind}\leqslant s\,;-u_0)) \subset \mathbf{B}^+_{(\leqslant s)}(\nu), \tag{4.7}$$

$$\widetilde{\Psi}_0^{-1}(V_\alpha \setminus \mathbf{B}^+_{(\leqslant n-s-2)}(\nu)) \subset B_\nu(\text{ind}\leqslant s\,;u_0). \tag{4.8}$$

Now let us write

$$(-v)^{\leadsto}_{[b,a]} = \widetilde{\Psi}_0^{-1} \circ (-v_1)^{\leadsto}_{[\beta,\alpha]} \circ \widetilde{\Psi}_1, \qquad v^{\leadsto}_{[b,a]} = \widetilde{\Psi}_1^{-1} \circ (v_1)^{\leadsto}_{[\alpha,\beta]} \circ \widetilde{\Psi}_0,$$

and combine (4.5)–(4.8) with (4.3), (4.4) to conclude that (B1) and (B0) hold for v with $\delta = \nu$ and for every s. The proof of Theorem 4.3 is completed. □

4.1.6. *Cyclic cobordisms and condition* $(\mathfrak{C}\mathcal{Y})$. Condition $(\mathfrak{C})$ and related techniques developed in this section are the basic ingredients in the proof of the main results of our work (Theorems 5.5 and 5.14). In these theorems we apply the results of the present section in the special situation where the parts $\partial_1 W$ and $\partial_0 W$ of the cobordism are diffeomorphic (this cobordism is obtained as the result of cutting a manifold M along a regular value of a Morse map $M \to S^1$). The present subsection contains a version of Theorem 4.3 adapted to this situation.

DEFINITION 4.6. *A cyclic cobordism* W is a Riemannian cobordism together with an isometry $\Phi \colon \partial_0 W \to \partial_1 W$.

Let $f \colon W \to [a, b]$ be a Morse function on a cyclic cobordism W of dimension n, and let v be an f-gradient.

DEFINITION 4.7. We say that v satisfies $(\mathfrak{C}\mathcal{Y})$ if v satisfies condition $(\mathfrak{C})$ (see 4.1.1), and, moreover, the Morse functions ϕ_1, ϕ_0 and their gradients u_1, u_0 from Definition 4.2 can be chosen so that $\phi_1 \circ \Phi = \phi_0$, $u_1 = \Phi_*(u_0)$. The set of all f-gradients satisfying $(\mathfrak{C}\mathcal{Y})$ will be denoted by $\mathcal{G}\mathfrak{C}\mathcal{Y}(f)$.

THEOREM 4.8. *$\mathcal{G}\mathfrak{C}\mathcal{Y}(f)$ is open and dense in $\mathcal{G}(f)$ with respect to the C^0 topology. Moreover, if v_0 is any f-gradient, then one can choose a C^0 small perturbation v of v_0 such that $v \in \mathcal{G}\mathfrak{C}\mathcal{Y}(f)$ and $v = v_0$ in a neighborhood of ∂W.*

PROOF. The proof is carried out in exactly the same way as that of Theorem 4.3. To prove the C^0-openness of the set $\mathcal{G}\mathfrak{C}\mathcal{Y}(f)$, we do not need to change anything in the first part of the proof of Theorem 4.3.

For the proof of the C^0 density, just note that in the proof of the C^0 density of $(\mathfrak{C})$ (see 4.1.4) we can choose $\phi_0 = \phi_1 \circ \Phi$ and $u_1 = \Phi_*(u_0)$. □

4.2. Homological gradient descent. Now we shall deduce consequences from condition $(\mathfrak{C})$. In this section $f \colon W \to [a, b]$ is a Morse function on a Riemannian cobordism W of dimension n, and v is an f-gradient satisfying condition $(\mathfrak{C})$. Recall that condition $(\mathfrak{C})$ requires the existence of Morse functions $\phi_0 \colon \partial_0 W \to \mathbb{R}$, $\phi_1 \colon \partial_1 W \to \mathbb{R}$ and their gradients u_0, u_1, satisfying conditions (B1) and (B0) from 4.1.1. There must be also an ordered Morse function $\phi \colon W \to [a, b]$, adjusted to (f, v) with an ordering sequence $a = a_0 < a_1 < \cdots < a_{n+1} = b$. We shall denote $\phi^{-1}(\lambda)$ by V_λ, so that $\partial_1 W = V_b$, $\partial_0 W = V_a$. From the gradients u_0, u_1 we obtain the filtrations of V_b and of V_a, namely $V_b^{\{\leqslant s\}}$ and $V_a^{\{\leqslant s\}}$ (s ranges over integers between 0 and $n-1$). We have proved that every f-gradient w sufficiently C^0 close to v satisfies condition $(\mathfrak{C})$ with respect to the same objects ϕ_0, ϕ_1, u_0, u_1, ϕ (see 4.1.2).

THEOREM 4.9. *For every integer s, $0 \leqslant s \leqslant n$, there is a homomorphism*

$$\mathcal{H}_s(-v) \colon H_*(V_b^{\{\leqslant s\}}, V_b^{\{\leqslant s-1\}}) \to H_*(V_a^{\{\leqslant s\}}, V_a^{\{\leqslant s-1\}})$$

with the following properties:

(1) *Let N be an oriented submanifold of V_b, contained in $V_b^{\{\leqslant s\}}$, such that the set $N \setminus \operatorname{Int} V_b^{\{\leqslant s-1\}}$ is compact. Then the manifold $N' = (-v)^{\leadsto}_{[b,a]}(N)$ is in $V_a^{\{\leqslant s\}}$, the set $N' \setminus \operatorname{Int} V_a^{\{\leqslant s-1\}}$ is compact, and the fundamental class of N' modulo $V_a^{\{\leqslant s-1\}}$ equals $\mathcal{H}_s(-v)([N])$.*

(2) *There is an $\varepsilon > 0$ such that for every f-gradient w with $\|w - v\| \leqslant \varepsilon$ and every s we have $\mathcal{H}_s(-v) = \mathcal{H}_s(-w)$.*

The proof of the theorem occupies 4.2.1, 4.2.2, and 4.2.3.

4.2.1. *Main idea of the proof and definition of* $\mathcal{H}_s(-v)$. We shall first give an informal description of $\mathcal{H}_s(-v)$. Let $\alpha \in H_m(V_b^{\{\leqslant s\}}, V_b^{\{\leqslant s-1\}})$. Assume for simplicity that α is represented by a closed oriented submanifold A of V_b, $\dim A = m$, contained in $V_b^{\{\leqslant s\}}$. Denote $V_{a_{s+1}}$ by V' for brevity. We descend A along the $(-v)$-trajectories from V_b to V'. Condition $(\mathfrak{C})$ implies that the whole of $V_b^{\{\leqslant s\}}$ descends to V' without hanging on critical points of f; that is, every $(-v)$-trajectory starting in $V_b^{\{\leqslant s\}}$ reaches $V_{a_{s+1}}$. Therefore $A' = (-v)^{\leadsto}_{[b,a_{s+1}]}(A)$ is a closed oriented submanifold of V'. Now let us try to apply the same procedure in order to descend A' to V_a. In general A' intersects the upper soles of the ascending discs $D(p,-v)$ with $\operatorname{ind} p \leqslant s$, and so $(-v)^{\leadsto}_{[a_{s+1},a]}(A')$ is a submanifold of V_a, which is not closed. But note that the intersection of the upper soles of these ascending discs with A' is contained in the neighborhood $\bigcup_{\operatorname{ind} p \leqslant s} B_\delta(p,-v)$ of $\bigcup_{\operatorname{ind} p\leqslant s} D(p,-v)$, and this neighborhood is carried to $V_a^{\{\leqslant s-1\}}$ by $(-v)^{\leadsto}_{[a_{s+1},a]}$. So the fundamental class of $(-v)^{\leadsto}(A')$ can still be defined modulo $V_a^{\{\leqslant s-1\}}$, and this was our aim.

We proceed now to the precise definitions. The homomorphism $\mathcal{H}_s(-v)$ will be defined as the composition of two homomorphisms: $\mathcal{H}1_s(-v)$ and $\mathcal{H}0_s(-v)$. (Informally, $\mathcal{H}1$ corresponds to the descent from the level b to the level a_{s+1}, and $\mathcal{H}0$ corresponds to the descent from a_{s+1} to a.) We shall denote the maps $(-v)^{\leadsto}_{[b,a_{s+1}]}$ and $(-v)^{\leadsto}_{[a_{s+1},a]}$ by $(-v1)^{\leadsto}$ and $(-v0)^{\leadsto}$ respectively. Every $(-v)$-trajectory starting in $V_b^{\{\leqslant s\}}$ reaches $V_{a_{s+1}}$. Therefore $V_b^{\{\leqslant s\}}$ is in the domain of $(-v1)^{\leadsto}$ and $(-v1)^{\leadsto}$ defines a homeomorphism of the pair $(V_b^{\{\leqslant s\}}, V_b^{\{\leqslant s-1\}})$ to its image, which is a pair of compact subsets of V'.

Now let A, B, C be subsets of V' satisfying the following conditions:

$$A \supset B \supset C, \qquad \overline{C} \subset \operatorname{Int} B, \tag{R1}$$

$$(-v1)^{\leadsto}(V_b^{\{\leqslant s\}}) \subset A, \qquad (-v1)^{\leadsto}(V_b^{\{\leqslant s-1\}}) \subset B, \tag{R2}$$

$$\Big(\bigcup_{\operatorname{ind} p\leqslant s} D(p,-v)\Big) \cap V' \subset C, \tag{R3}$$

$$(-v0)^{\leadsto}(A) \subset V_a^{\{\leqslant s\}}, \qquad (-v0)^{\leadsto}(B) \subset V_a^{\{\leqslant s-1\}}. \tag{R4}$$

Later we shall check that such subsets exist. Now we construct $\mathcal{H}_s(-v)$ assuming that they do, and show that $\mathcal{H}_s(-v)$ does not depend on the choice of A, B, C satisfying (R1)–(R4).

The composition of $(-v1)^{\leadsto}$ with the inclusion defines a continuous map

$$\phi\colon (V_b^{\{\leqslant s\}}, V_b^{\{\leqslant s-1\}}) \to (A,B).$$

Set $\mathcal{H}1_s(-v) = \phi_*$. Consider the excision isomorphism

$$\operatorname{Exc}\colon H_*(A\setminus C, B\setminus C) \xrightarrow{\approx} H_*(A,B). \tag{4.9}$$

The conditions (R3) and (R4) imply that the gradient descent from a_{s+1} to a defines a continuous map of pairs

$$\rho\colon (A\setminus C, B\setminus C) \to (V_a^{\{\leqslant s\}}, V_a^{\{\leqslant s-1\}}). \tag{4.10}$$

Set $\mathcal{H}0_s(-v) = \rho_* \circ (\operatorname{Exc})^{-1}$ and $\mathcal{H}_s(-v) = \mathcal{H}0_s(-v) \circ \mathcal{H}1_s(-v)$.

Now we check that the homomorphism $\mathcal{H}_s(-v)$ does not depend on the choice of A, B, C. Let (A, B, C), (A', B', C') be two triples satisfying (R1)–(R4); denote the corresponding homomorphisms by $\mathcal{H}$, $\mathcal{H}'$. It is easy to see that the triple $A \cap A'$, $B \cap B'$, $C \cap C'$ again satisfies conditions (R1)–(R4). Therefore it suffices to verify $\mathcal{H} = \mathcal{H}'$ assuming $A \supset A'$, $B \supset B'$, $C \supset C'$. This follows from an easy functoriality argument, which we leave to the reader.

In the following subsection we give two examples of triples (A, B, C) satisfying the conditions (R1)–(R4).

4.2.2. *The sets $T_k(v)$ and $t_k(v)$.* Let k be an integer, $0 \leqslant k \leqslant n$.

DEFINITION 4.10. Denote by $T_k(v)$ the set of all $y \in \phi^{-1}([a, a_{k+1}])$ such that either

i) $\gamma(y, \cdot\,; -v)$ reaches V_a and intersects it at a point $z \in V_a^{\{\leqslant k-1\}}$, or
ii) $\lim_{t\to\infty} \gamma(y, t; -v) = p$, where $p \in S(f)$.

One can reformulate this definition slightly. Consider the cobordism $W' = \phi^{-1}([a, a_{k+1}])$. We have the restriction $\phi|W' \colon W' \to [a, a_{k+1}]$ of the Morse function ϕ and the restriction $v|W'$ of its gradient v. Then

$$T_k(v) = T(V_a^{\{\leqslant k-1\}}, v|W') \cup D(-v|W')$$

(see the definition of $T(\,\cdot\,,\,\cdot\,)$ in 2.5.1). This implies immediately that $T_k(v)$ is compact. Before proceeding to the following definition, recall that we have fixed $\delta > 0$ satisfying $(\mathfrak{C})$.

DEFINITION 4.11. Denote by $t_k(v)$ the set of all $y \in \phi^{-1}([a, a_{k+1}])$ such that either

i) $\gamma(y, \cdot\,; -v)$ reaches V_a and intersects it at a point $z \in V_a^{[\leqslant k-1]}(\delta)$, or
ii) $\lim_{t\to\infty} \gamma(y, t; -v) = p$, where $p \in S(f)$.

We obviously have $t_k(v) \subset T_k(v)$. The same argument that we used for $T_k(v)$ shows that $t_k(v)$ is compact.

LEMMA 4.12. *There is $\delta' > \delta$ such that v is δ'-separated and*

$$\bigcup_{\operatorname{ind} p \leqslant k} D_{\delta'}(p, -v) \cap V_{a_{k+1}} \subset t_k(v). \tag{4.11}$$

PROOF. Condition (B1) implies that

$$D_\delta(\mathrm{ind}{\leqslant}k\,; v) \cap \partial_0 W \subset B_\delta(\mathrm{ind}{\leqslant}k{-}1\,; u_0).$$

Since v is δ-separated, it is δ''-separated for some $\delta'' > \delta$, and the interval of definition of the gfn-system $\mathbf{D}(v)$ contains $]0, \delta''[$. Therefore $\{B_\theta(\mathrm{ind}{\leqslant}k\,; v) \cap \partial_0 W\}_{\theta>\delta}$ is a fundamental system of neighborhoods for $D_\delta(\mathrm{ind}{\leqslant}k\,; v) \cap \partial_0 W$. Thus for some $\delta' : \delta < \delta' < \delta''$ we have

$$D_{\delta'}(\mathrm{ind}{\leqslant}k\,; v) \cap \partial_0 W \subset B_\delta(\mathrm{ind}{\leqslant}k{-}1\,; u_0).$$

The lemma follows. □

REMARK 4.13. The previous lemma implies that

$$t_k(v) \cap V_{a_{k+1}} \subset \mathrm{Int}(T_k(v) \cap V_{a_{k+1}}).$$

Indeed, the inclusion $t_k(v) \subset T_k(v)$ is obvious. Further, let $z \in t_k(v) \cap V_{a_{k+1}}$. If $z \in \bigcup_{\operatorname{ind} p \leqslant k} D(p, -v)$, then z is contained in

$$\Big(\bigcup_{\operatorname{ind} p \leqslant k} B_\delta(p, -v) \Big) \cap V_{a_{k+1}},$$

which is in $t_k(v)$. If $\gamma(z, \cdot\,; -v)$ reaches V_a, then it intersects V_a at a point $z' \in V_a^{[\leqslant k-1]}(\delta)$, which is in the interior of $V_a^{\{\leqslant k-1\}}$, and this implies our assertion. A similar argument proves that $t_k(v) \cap \phi^{-1}([a, a_{k+1}[)$ is in $\operatorname{Int}(T_k(v) \cap \phi^{-1}([a, a_{k+1}[))$.

Returning to the definition of the sets A, B, C from 4.2.1, set

$$\begin{cases} A = T_{s+1}(v) \cap V', \\ B = T_s(v) \cap V', \\ C = (\bigcup_{\operatorname{ind} p \leqslant s} D(p, -v)) \cap V'. \end{cases} \tag{4.12}$$

It is now obvious that this triple satisfies (R1)–(R4). Another triple satisfying the conditions is

$$\begin{cases} A = t_{s+1}(v) \cap V', \\ B = t_s(v) \cap V'; \\ C = (\bigcup_{\operatorname{ind} p \leqslant s} D(p, -v)) \cap V'. \end{cases} \tag{4.13}$$

4.2.3. *Properties of $\mathcal{H}_s$. Proof of* (1).

Consider the submanifold $N'' = (-v1)^{\leadsto}(N)$ of $V' = V_{a_{s+1}}$. Since $(-v1)^{\leadsto}$ is a diffeomorphism of the open subset $\operatorname{Int} V_b^{\{\leqslant s-1\}}$ of V_b onto its image in V', the manifold $N'' \setminus (-v1)^{\leadsto}(\operatorname{Int} V_b^{\{\leqslant s-1\}})$ is compact.

Thus the compactness of $N' \setminus \operatorname{Int} V_a^{\{\leqslant s-1\}}$ follows from the next lemma.

LEMMA 4.14. *Let L be a submanifold of V' such that the set*

$$L \setminus (-v1)^{\leadsto}(\operatorname{Int} V_b^{\{\leqslant s-1\}})$$

is compact. Set $L' = (-v0)^{\leadsto}(L)$. Then $L' \setminus \operatorname{Int} V_a^{\{\leqslant s-1\}}$ is compact.

PROOF. The set $L \setminus \operatorname{Int} T_s$ is compact. Since $(-v0)^{\leadsto}$ is defined on $V' \setminus \operatorname{Int} T_s$, the set

$$(-v0)^{\leadsto}(L' \setminus \operatorname{Int} T_s) = (-v0)^{\leadsto}(L') \setminus (-v0)^{\leadsto}(\operatorname{Int} T_s)$$

is a compact subset in V_a. Note that $\operatorname{Int} V_a^{\{\leqslant s-1\}}$ is an open set containing the set $(-v0)^{\leadsto}(\operatorname{Int} T_s)$, and this implies finally that $L' \setminus \operatorname{Int} V_a^{\{\leqslant s-1\}}$ is compact. □

To verify that $[N'] = \mathcal{H}_s(-v)([N])$, use the definition of $\mathcal{H}_s(-v)$ with the help of A, B, C as in (4.12) and check through the definitions of N' and of $\mathcal{H}_s(-v)$. We leave the details to the reader.

Proof of (2).

Here for the definition of $\mathcal{H}_s(-v)$ it is convenient to use the triple (A, B, C), where

$$\begin{cases} A = T_{s+1}(v) \cap V', \\ B = T_s(v) \cap V', \\ C = (\bigcup_{\operatorname{ind} p \leqslant s} B_\delta(p, -v)) \cap V'. \end{cases} \tag{4.14}$$

In the rest of this subsection we shall abbreviate $V_b^{\{\leqslant s\}}$ to R_b, and $V_b^{\{\leqslant s-1\}}$ to Q_b, $V_a^{\{\leqslant s\}}$ to R_a, $V_a^{\{\leqslant s-1\}}$ to Q_a. First we shall show that the triple (A, B, C)

defined in (4.14) satisfies the conditions (R1)–(R4) not only for v but also for every ϕ-gradient w sufficiently C^0 close to v. Of course this is true for (R1), since it does not depend on the gradient.

Proceeding to (R2), recall that (by Remark 4.13) $t_{s+1}(v) \cap V' \subset \operatorname{Int} A$ and $t_s(v) \cap V' \subset \operatorname{Int} B$. Therefore, the pair of compact sets

$$((-v1)^{\leadsto}(R_b), (-v1)^{\leadsto}(Q_b))$$

is in $(\operatorname{Int} A, \operatorname{Int} B)$. Now it follows from Corollary 2.72 that there is an $\varepsilon > 0$ such that, for every vector field w on $\phi^{-1}([a_{s+1}, b])$ with $\|w - v\| < \varepsilon$, each $(-w)$-trajectory starting in $V_b^{\{\leqslant s\}}$ reaches V' and

$$((-w)^{\leadsto}(R_b), (-w)^{\leadsto}(Q_b)) \subset (\operatorname{Int} A, \operatorname{Int} B). \tag{4.15}$$

Therefore for every ϕ-gradient w with $\|w - v\| < \varepsilon$ condition (R2) is true with respect to w.

Since C is open, the property (R3) with respect to w is true by Proposition 2.76 if w is sufficiently close to v. Proceeding to (R4), note that

$$(-v0)^{\leadsto}(A) \subset V_a^{[\leqslant s]}(\delta) \subset \operatorname{Int} V_a^{\{\leqslant s\}}.$$

So each v-trajectory starting in $V_a^{\{\geqslant s\}}$ reaches V' and intersects it at a point outside A. Then the same is true for every w-trajectory starting in $V_a^{\{\geqslant s\}}$, provided w is sufficiently C^0 close to v. This implies the first half of property (R4) with respect to w. The second half can be verified similarly.

Thus we can use the triple (A, B, C) defined in (4.14) for the definition of $\mathcal{H}_s(-w)$. It remains to show that $\mathcal{H}_s(-w) = \mathcal{H}_s(-v)$ for all w sufficiently C^0 close to v.

LEMMA 4.15. *For every ϕ-gradient w sufficiently C^0-close to v we have*

$$\mathcal{H}1_s(-w) = \mathcal{H}1_s(-v).$$

PROOF. Apply Corollary 2.73 to the cobordism $\phi^{-1}([a_{s+1}, b])$, and the vector field $(-v)$. □

LEMMA 4.16. *For every ϕ-gradient w sufficiently C^0-close to v we have*

$$\mathcal{H}0_s(-w) = \mathcal{H}0_s(-v).$$

PROOF. Recall that $A \setminus C$ is in the domain of definition of $(-v)^{\leadsto}_{[a_{s+1}, a]}$. Moreover, the condition $(\mathfrak{C})$ implies that $(-v)^{\leadsto}_{[a_{s+1}, a]}$ sends the pair of compact sets $(A \setminus C, B \setminus C)$ to the pair $(V_a^{[\leqslant s]}(\delta), V_a^{[\leqslant s-1]}(\delta))$, which is in the pair of open sets $(\operatorname{Int} R_a, \operatorname{Int} Q_a)$. Therefore (by Corollary 2.73) for every w sufficiently C^0 close to v, the map $(-w)^{\leadsto}_{[a_{s+1}, a]}$ is defined on $A \setminus C$, sends the pair $(A \setminus C, B \setminus C)$ to $(\operatorname{Int} R_a, \operatorname{Int} Q_a)$, and is homotopic to $(-v)^{\leadsto}_{[a_{s+1}, a]}$ as a map of these pairs. □

4.2.4. *A pairing between $C_s(u_1)$ and $C_s(v)$.* Here we present a construction related to the homological gradient descent homomorphism. This construction will be used in Section 5 for the computation of Novikov incidence coefficients.

Let $0 \leqslant s \leqslant n - 1$. As in the previous subsection, we denote $\phi^{-1}(a_{s+1})$ by V'. Every $(-v)$-trajectory starting at $C_\delta(\operatorname{ind}_{\leqslant s}; u_1)$ reaches V'. Therefore the map $(-v)^{\leadsto}_{[b, a_{s+1}]}$ defines a homeomorphism of the pair

$$(\operatorname{Int} C_\delta(\operatorname{ind}_{\leqslant s}; u_1), \operatorname{Int} C_\delta(\operatorname{ind}_{\leqslant s-1}; u_1))$$

to the pair (N_s, N_{s-1}) of open subsets of V'. Let τ be the restriction of $(-v)^{\leadsto}_{[b,a_{s+1}]}$ to the pair $((\partial_1 W)^{\{\leqslant s\}}, (\partial_1 W)^{\{\leqslant s-1\}})$. Then $\operatorname{Im}\tau$ is a compact subpair $(n_s, n_{s-1}) \subset (N_s, N_{s-1})$. Let $q \in S_s(f)$. Consider the submanifold $S(q) = D(q, -v) \cap V'$ of V'; it is an $(n-s-1)$-dimensional cooriented embedded sphere. Denote by $]S_q[$ the cohomology class in $H^s(V', V' \setminus S(q))$ dual to $S(q)$. We have $S(q) \cap n_{s-1} = \varnothing$ (this follows from $(\mathfrak{C})$). Therefore (N_s, N_{s-1}) is a subpair of $(V', V' \setminus S(q))$, and we can restrict the class $]S_q[$, obtaining a cohomology class of (N_s, N_{s-1}). The resulting cohomology class will again be denoted by $]S_q[$. For

$$x \in C_s(u_1) = H_s((\partial_1 W)^{\{\leqslant s\}}, (\partial_1 W)^{\{\leqslant s-1\}})$$

set

$$\langle q, x\rangle_v = \langle]S_q[, \tau_*(x)\rangle.$$

(The index v reminds us that the pairing depends on the vector field v.) This pairing has some natural properties gathered in Proposition 4.17.

PROPOSITION 4.17. (1) *Let L be an s-dimensional oriented submanifold of $\partial_1 W$ such that $L \subset (\partial_1 W)^{\{\leqslant s\}}$ and $L \setminus \operatorname{Int}(\partial_1 W)^{\{\leqslant s-1\}}$ is compact. Then $L' = (-v)^{\leadsto}_{[b,a_{s+1}]}(L)$ is an s-dimensional oriented submanifold of V' such that $L' \subset n_s$ and $L' \setminus \operatorname{Int} n_{s-1}$ is compact. If $L' \pitchfork S(q)$, then*

$$L' \# S(q) = \langle q, [L]\rangle_v.$$

(2) *Let U be a neighborhood of $S(f)$. There is a $\delta > 0$ such that for every ϕ-gradient w with $v|U = w|U$ and $\|v - w\| < \delta$ we have $\langle q, x\rangle_v = \langle q, x\rangle_w$.*

PROOF. Part (1) is obvious. Proceeding to (2), note first that the objects N_s, N_{s-1}, n_s, n_{s-1}, τ, S_q, etc., introduced at the beginning of this subsection depend on v, and we shall now show it in the notation, writing $N_s(v)$, $N_{s-1}(v)$, $n_s(v)$, $n_{s-1}(v)$, $\tau(v)$, $S(q,v)$, etc., in order to make the dependence on the gradient explicit.

Recall that if w is sufficiently C^0 close to v, then w still satisfies $(\mathfrak{C})$ with respect to the same string $\mathcal{S} = (\delta, u_0, u_1, \phi_0, \phi_1, \phi)$. In particular, every $(-w)$-trajectory starting in $(\partial_1 W)^{\{\leqslant s\}}$ reaches V'. It follows easily that if w is sufficiently close to v, then $n_s(w) \subset N_s(v)$ and $n_{s-1}(w) \subset N_{s-1}(v)$.

The compact sets $n_{s-1}(v)$ and $S(q,v)$ are disjoint subsets of V'. Therefore there are disjoint open subsets Λ, Γ of V' with $n_{s-1}(v) \subset \Lambda$, $S(q,v) \subset \Gamma$. We have the following inclusions:

$$(n_s(v), n_{s-1}(v)) \xhookrightarrow{i(v)} (V', \Lambda) \xhookrightarrow{j} (V', V' \setminus \Gamma) \xhookrightarrow{k(v)} (V', V' \setminus S(q,v)). \quad (4.16)$$

Note that, obviously,

$$\langle q, x\rangle_v = \langle (k(v))^*(]S_q(v)[), (j \circ i(v))_*(\tau_*(v)(x))\rangle. \quad (4.17)$$

If w is sufficiently C^0 close to v, then we still have $n_{s-1}(w) \subset \Lambda$, $S(q,w) \subset \Gamma$, and formula (4.17) still holds if we replace v by w. Therefore it suffices to show that for w sufficiently C^0 close to v we have

$$i(v) \circ \tau(v) \sim i(w) \circ \tau(w), \qquad (k(v))^*(]S_q(v)[) = (k(w))^*(]S_q(w)[). \quad (4.18)$$

The first follows from Corollary 2.73. To prove the second, note that the class

$$(k(v))^*(]S_q(v)[) \in H^s(V', V' \setminus \Gamma)$$

is Lefschetz-dual to the fundamental class $[S_q(v)] \in H_{n-s-1}(\Gamma)$. Therefore it suffices to prove that if $v|U = w|U$ and w is sufficiently close to v in the C^0 topology, then the fundamental classes of $S_q(v)$ and of $S_q(w)$ in $H_*(\Gamma)$ are equal. Now just apply Proposition 2.77. □

4.2.5. *Homological gradient descent: equivariant version.* Let $p\colon \widehat{W} \to W$ be a regular covering of W with structure group H. For a subset $A \subset W$ we denote by $\widehat{A}$ the set $p^{-1}(A)$. We thus obtain an H-invariant filtration $\widehat{V}_b^{\{\leqslant s\}}$ of $\widehat{V}_b$ and an H-invariant filtration $\widehat{V}_a^{\{\leqslant s\}}$ of $\widehat{V}_a$. Recall from subsection 2.1 the notation

$$K^+ = \bigcup_{x\in S(f)} D(x,-v), \qquad K^- = \bigcup_{x\in S(f)} D(x,v).$$

For $x \in \widehat{\partial_1 W} \setminus \widehat{K}^+$ consider the $(-v)$-trajectory $\gamma(q(x),t;-v)$. It reaches $\partial_0 W$, and the lifting $\widehat{\gamma}$ of γ to $\widehat{W}$ is an integral curve of $-p^*v$. This integral curve reaches $\widehat{\partial}_0 W$, and the intersection point of $\widehat{\gamma}$ with $\widehat{\partial}_0 W$ will be denoted by $(-\widehat{v})^{\leadsto}_{[b,a]}(x)$. Thus we obtain a diffeomorphism

$$(-\widehat{v})^{\leadsto}_{[b,a]}(x)\colon \widehat{\partial_1 W} \setminus \widehat{K}^+ \longrightarrow \widehat{\partial_0 W} \setminus \widehat{K}^-$$

which commutes with the action of H.

The homology $H_*((\widehat{\partial_1 W})^{\{\leqslant s\}}, (\widehat{\partial_1 W})^{\{\leqslant s-1\}})$ is a right $\mathbb{Z}H$-module, and Proposition 2.42 implies that it is a free $\mathbb{Z}H$-module with the base formed by liftings of descending discs of the critical points of ϕ_1 of index s. The following theorem is an "equivariant version" of Theorem 4.9. The proof is a direct generalization of the proof of Theorem 4.9, and we leave it to the reader.

THEOREM 4.18. *For every integer s, $0 \leqslant s \leqslant n$, there is a homomorphism*

$$\widehat{\mathcal{H}}_s(-v)\colon H_*(\widehat{V}_b^{\{\leqslant s\}}, \widehat{V}_b^{\{\leqslant s-1\}}) \to H_*(\widehat{V}_a^{\{\leqslant s\}}, \widehat{V}_a^{\{\leqslant s-1\}})$$

with the following properties:

(1) *Let N be an oriented submanifold of $\widehat{V}_b$, contained in $\widehat{V}_b^{\{\leqslant s\}}$ and such that $N \setminus \operatorname{Int} \widehat{V}_b^{\{\leqslant s-1\}}$ is compact. Then the manifold $N' = (-\widehat{v})^{\leadsto}_{[b,a]}(N)$ is in $\widehat{V}_a^{\{\leqslant s\}}$ and $N' \setminus \operatorname{Int} \widehat{V}_a^{\{\leqslant s-1\}}$ is compact, and the fundamental class of N' modulo $\widehat{V}_a^{\{\leqslant s-1\}}$ is equal to $\widehat{\mathcal{H}}_s(-v)([N])$.*
(2) *There is an $\varepsilon > 0$ such that for every f-gradient w with $\|w-v\| \leqslant \varepsilon$ and every s we have $\widehat{\mathcal{H}}_s(-v) = \widehat{\mathcal{H}}_s(-w)$.*

We proceed to the equivariant analogs of the results of 4.2.4. Let $0 \leqslant s \leqslant n-1$. We assume the terminology of 4.2.4. We denote

$$\widehat{C}_s(u_1) = H_s((\widehat{\partial_1 W})^{\{\leqslant s\}}, (\widehat{\partial_1 W})^{\{\leqslant s-1\}}).$$

This is a free right $\mathbb{Z}H$-module. There is an H-equivariant homeomorphism

$$\widehat{\tau}(v)\colon ((\widehat{\partial_1 W})^{\{\leqslant s\}}, (\widehat{\partial_1 W})^{\{\leqslant s-1\}}) \to (\hat{n}_s(v), \hat{n}_{s-1}(v)).$$

Let $q \in S_s(f)$ and choose a lifting $\hat{q}$ of q to $\widehat{W}$. Then we obtain a lifting $S(\hat{q},v)$ of $S(q,v)$ to $\widehat{W}$, and a cohomology class $]S_{\hat{q}}[\in H^s(\widehat{N}_s(v), \widehat{N}_{s-1}(v))$.

For $x \in H_s((\widehat{\partial_1 W})^{\{\leqslant s\}}, (\widehat{\partial_1 W})^{\{\leqslant s-1\}})$, set

$$\langle \hat{q}, x\rangle_v = \langle]S_{\hat{q}}[, (\widehat{\tau}(v))_*(x)\rangle.$$

Thus $\langle \hat{q}, x\rangle_v$ is an integer. If we fix $\hat{q}$ and consider x as the variable, we obtain a homomorphism $\widehat{C}_s(u_1) \to \mathbb{Z}$.

Next we give an equivariant version of Proposition 4.17. The first two claims of this proposition are direct generalization of the corresponding points of Proposition 4.17. Part (3) is obvious, and (4) follows from the fact that for a given singular chain $y \in H_*(\hat{n}_s(v), \hat{n}_{s-1}(v))$ there are only finitely many $h \in H$ such that $\operatorname{supp} y \cap S(\hat{q}, v) \neq 0$.

PROPOSITION 4.19. (1) *Let L be an s-dimensional oriented submanifold of $\widehat{\partial_1 W}$ such that $L \subset (\widehat{\partial_1 W})^{\{\leqslant s\}}$ and $L \setminus \operatorname{Int}(\widehat{\partial_1 W})^{\{\leqslant s\}}$ is compact. Then $L' = (-\widehat{v})^{\leadsto}_{[b,a_{s+1}]}(L)$ is an s-dimensional oriented submanifold of $\widehat{V}'$ such that $L' \subset \hat{n}_s(v)$ and $L' \setminus \operatorname{Int} \hat{n}_{s-1}(v)$ is compact. If $L' \pitchfork S(\hat{q}, v)$, then*

$$L' \# S(\hat{q}, v) = \langle \hat{q}, [L] \rangle_v.$$

(2) *Let U be a neighborhood of $S(f)$. There is a $\delta > 0$ such that for every ϕ-gradient w with $v|U = w|U$ and $\|v - w\| < \delta$ we have $\langle \hat{q}, x \rangle_v = \langle \hat{q}, x \rangle_w$.*

(3) *Let $h \in H$. Then $\langle \hat{q}h, xh \rangle_v = \langle \hat{q}, x \rangle_v$.*

(4) *For every x the set of $h \in H$ such that $\langle \hat{q}h, x \rangle_v \neq 0$ is finite.*

5. C^0-generic rationality of the boundary operators in the Novikov complex

In this section we prove the theorems that are the aim of this paper. We refer to the corresponding parts of the Introduction (1.2.3 and 1.2.4) for the exposition of the main ideas of the proof, and we proceed directly to the precise definitions and statements of results.

5.1. Basic terminology. In 5.1.1, 5.1.2 we recall the basic definitions concerning the algebraic (5.1.1) and geometric (5.1.2) framework of Novikov theory, so that our exposition is self-contained, at least with regard to terminology.

5.1.1. *Novikov rings.* We shall denote the integer Laurent polynomial ring $\mathbb{Z}[t, t^{-1}]$ by L. The very first example of a Novikov ring is the ring $\mathbb{Z}((t))$, which consists by definition of all the Laurent series $\lambda = \sum_{-\infty}^{\infty} a_n t^n$ with integer coefficients and finite negative part (so that $\lambda \in \mathbb{Z}((t))$ if for all n we have $a_n \in \mathbb{Z}$ and there is an $N = N(\lambda)$ such that $a_n = 0$ for $n < N(\lambda)$). We shall often denote $\mathbb{Z}((t))$ by $\widehat{L}$. The localization of L with respect to the set of all polynomials of the form $1 + tP(t)$, where P is a polynomial, is denoted by $\widetilde{L}$; we have $L \subset \widetilde{L} \subset \widehat{L}$.

To consider "equivariant versions" of the Novikov complex, one needs to introduce some more complicated rings. Let G be a group, and $\xi \colon G \to \mathbb{R}$ a group homomorphism. Set $\Lambda = \mathbb{Z}G$, and denote by $\widehat{\widehat{\Lambda}}$ the abelian group of all formal linear combinations of the form $\sum_{g \in G} n_g g$ (infinite in general). For $\lambda \in \widehat{\widehat{\Lambda}}$, $\lambda = \sum_{g \in G} n_g g$, set as usual $\operatorname{supp} \lambda = \{g \in G \mid n_g \neq 0\}$. Consider the subset $\widehat{\Lambda}_\xi \subset \widehat{\widehat{\Lambda}}$ defined by

$$\widehat{\Lambda}_\xi = \{\lambda \in \widehat{\widehat{L}} \mid \forall c \in \mathbb{R} \text{ the set } \operatorname{supp} \lambda \cap \xi^{-1}([c, \infty[) \text{ is finite}\}.$$

It is not difficult to see that $\widehat{\Lambda}_\xi$ has the natural structure of a ring for which the embedding $\Lambda \hookrightarrow \widehat{\Lambda}_\xi$ is a homomorphism.[9] The most important case for us is the case

[9]In all my earlier papers on this subject I used the notation Λ_ξ^- for this ring in order to stress that the power series $\lambda \in \Lambda_\xi^-$ are allowed to have terms with $\xi(g)$ an arbitrary large *negative* number. Now it seems to me more important to stress that the definition of the ring uses the completion procedure. The symbol $\widehat{\ }$ is included in the notation of the ring in the original paper [**20**], and in many other papers, see e.g. [**41**].

of a group epimorphism $\xi\colon G \to \mathbb{Z}$. In this case set $H = \operatorname{Ker}\xi$. For $n \in \mathbb{Z}$ denote $\xi^{-1}(n)$ by $G_{(n)}$ and $\{x \in \mathbb{Z}G \mid \operatorname{supp} x \subset G_{(n)}\}$ by $\mathbb{Z}G_{(n)}$. Denote $\xi^{-1}(]-\infty,-1])$ by G_- and $\{x \in \mathbb{Z}G \mid \operatorname{supp} x \subset G_-\}$ by $\mathbb{Z}G_-$. Choose $\theta \in G_{(-1)}$. It is easy to see that $\widehat{\mathbb{Z}G_\xi}$ is identified with the ring of power series of the form

$$\{a_{-n}\theta^n + \cdots + a_1\theta + \cdots \mid a_i \in \mathbb{Z}H\}.$$

5.1.2. *Morse maps $M \to S^1$.* The circle S^1 is for us the quotient $\mathbb{R}/\mathbb{Z}$. Let M be a closed connected manifold. A *Morse map $f\colon M \to S^1$* is a C^∞ map such that all critical points of f are nondegenerate. As in 2.1.1 the set of all critical points of f is denoted by $S(f)$, and the set of all critical points of index k is denoted by $S_k(f)$.

We leave to the reader the formulation of the definition of the f-gradient of a Morse map (following Definition 2.3 and Remark 2.4). The set of all gradients of f is denoted by $\mathcal{G}(f)$. Since M is closed, the integral curves of vector fields on M are defined on the whole of $\mathbb{R}$. For $p \in S(f)$ the stable manifold $D(p,v)$ and the unstable manifold $D(p,-v)$ are defined. These are immersed smooth manifolds in M. We have the corresponding notions of gradients satisfying the transversality condition and, respectively, the almost transversality condition. The set of all f-gradients satisfying the transversality condition, resp. the almost transversality condition, will be denoted by $\mathcal{GT}(f)$, resp. by $\mathcal{GA}(f)$. (Warning: In our present setting there are no analogs of Lemma 2.19 or Proposition 2.30, mainly because the v-trajectories can pass infinitely many times more and more closely to the initial position.)

The next lemma implies that $\mathcal{GT}(f)$ is C^0 dense in $\mathcal{G}(f)$. This is a version of the Kupka–Smale theorem. (It would be more precise to say "a version of a result weaker than the Kupka–Smale theorem", since we make no assertions concerning the closed orbits of the vector field considered.) The proof of the Kupka–Smale theorem given in [**36**] works in our case with only minor modifications. We leave the details to the reader.

LEMMA 5.1. *Let $v \in \mathcal{G}(f)$. Let U be a neighborhood of $S(f)$, and $\mathcal{V}$ a neighborhood of v in $\mathcal{G}(f)$ with respect to the C^∞ topology. Then there is an f-gradient v_0 such that*

$$v_0 \in \mathcal{GT}(f), \quad v_0 \in \mathcal{V}, \quad \operatorname{supp}(v - v_0) \subset U,$$

and there is a neighborhood U_0 of $S(f)$ such that $v_0|U_0 = v|U_0$.

We shall assume that f is not homotopic to 0 (otherwise f lifts to a Morse function $M \to \mathbb{R}$ and the situation is the same as considered previously in Section 1). There is a unique connected infinite cyclic covering $\mathcal{C}\colon \bar{M} \to M$ such that $f \circ \mathcal{C}$ is homotopic to zero. The map f lifts to a Morse function $F\colon \bar{M} \to \mathbb{R}$. Let t be the generator of the structure group ($\approx \mathbb{Z}$) of $\mathcal{C}$, such that for every x we have $F(xt) < F(x)$. (We invite the reader to look at Figure 1, back on page 34, where one can visualize most of the constructions of this subsection.) We shall assume that f is primitive; that is, $f_*\colon H_1(M) \to H_1(S^1)$ is epimorphic (one can always reduce the situation to this case by considering the map F/n with some $n \in \mathbb{N}$). In this case we have $F(xt) = F(x) - 1$. To simplify the notation, we shall assume that $1 \in S^1$ is a regular value of f, and therefore every $n \in \mathbb{N}$ is a regular value of F. Denote $f^{-1}(1)$ by V. Set

$$V_\alpha = F^{-1}(\alpha), \quad W = F^{-1}([0,1]), \quad V^- = F^{-1}(]-\infty,1]).$$

The cobordism W can be thought of as the result of cutting M along V. Note that $V_\alpha t = V_{\alpha-1}$. Denote Wt^s by W_s; then $\bar{M}$ is the union $\bigcup_{s\in\mathbb{Z}} W_s$, the neighboring copies W_s and W_{s+1} intersecting in V_{-s}. For any $n \in \mathbb{Z}$, the restriction of $\mathcal{C}$ to V_n is a diffeomorphism $V_n \to V$. Endow M with an arbitrary Riemannian metric and lift it to a t-invariant Riemannian metric on $\bar{M}$. Now W is a Riemannian cobordism, and actually it is a cyclic cobordism with respect to the isometry $t^{-1}\colon \partial_0 W = V_0 \to \partial_1 W = V_1$. The $\mathcal{C}$-preimage of a subset $A \subset M$ will be denoted by $\bar{A}$. On the other hand, if $x \in M$, we shall reserve the symbol $\bar{x}$ for liftings of x to $\bar{M}$ (that is, $\bar{x} \in \bar{M}$ and $\mathcal{C}(\bar{x}) = x$). Let $v \in \mathcal{G}(f)$. Lift v to $\bar{M}$; we obtain a complete vector field on $\bar{M}$, which is obviously an F-gradient. The restriction of this F-gradient to any cobordism $W_{[\lambda,\mu]} = F^{-1}([\lambda,\mu])$ (where λ, μ are regular values of F) is an $(F|W_{[\lambda,\mu]})$-gradient. We shall denote this F-gradient by the same symbol v, since there is no possibility of confusion. Note that if v satisfies the transversality condition, so does the F-gradient $v|W_{[\lambda,\mu]}$.

We shall consider "equivariant" versions of the Novikov complex. This requires considering one more covering (which can be, for example, the universal covering). Let $\mathcal{P}\colon \widehat{M} \to M$ be a regular covering with structure group G. Assume that $f \circ \mathcal{P}$ is homotopic to zero; then the covering $\mathcal{P}$ is factored through $\mathcal{C}$ as follows:

$$\begin{array}{ccc} \widehat{M} & \xrightarrow{\;p\;} & \bar{M} \\ & \searrow^{\mathcal{P}} & \big\downarrow \mathcal{C} \\ & & M \end{array} \tag{5.1}$$

Therefore there is a lifting of $f\colon M \to S^1$ to a Morse function $\widehat{F} = \bar{F} \circ p\colon \widehat{M} \to \mathbb{R}$. Further, the homomorphism $f_*\colon \pi_1 M \to \mathbb{Z}$ can be factored through a homomorphism $\xi\colon G \to \mathbb{Z}$. Recall that we assume ξ is epimorphic. So, applying the terminology of 5.1.1, $H = \operatorname{Ker}\xi$ and $G_- = \xi^{-1}(]-\infty, 0])$.

5.1.3. *Novikov complex: the definition.* We use the terminology of 5.1.2. Let $v \in \mathcal{GT}(f)$. For each point $x \in S(f)$ choose a lifting $\bar{x}$ of x to $\bar{M}$. Moreover, for each point $x \in S(f)$ choose an orientation of the local stable manifold $D(x,v)$, and lift it to an orientation of the corresponding local stable manifold of $\bar{x} \in S(F)$.

These choices made, we can define the Novikov complex. The rth term of this complex is defined to be the free $\mathbb{Z}((t))$-module generated by $S_r(f)$. For $x, y \in S(f)$ satisfying $\operatorname{ind} x = \operatorname{ind} y + 1$, define an element $n(x,y) \in \mathbb{Z}((t))$, called the *Novikov incidence coefficient*, as follows: For $k \in \mathbb{Z}$ consider the set $\Gamma_k(\bar{x},\bar{y})$ of all v-orbits joining $\bar{x}$ with $\bar{y}t^k$. This is a finite set, and for every $\gamma \in \Gamma_k(\bar{x},\bar{y})$ the orientation sign $\varepsilon(\gamma) = \pm 1$ is defined. (Indeed, choose $\lambda, \mu \in \mathbb{R}$ so that $\bar{x}, \bar{y}t^k \in F^{-1}([\lambda,\mu])$; then all the trajectories of $\Gamma_k(\bar{x},\bar{y})$ are in the cobordism $F^{-1}([\lambda,\mu])$ and we can apply the argument from page 60). Set

$$n_k(\bar{x},\bar{y}) = \sum_{\gamma\in\Gamma_k(x,y)} \varepsilon(\gamma), \qquad n(x,y) = \sum_k n_k(\bar{x},\bar{y}) \cdot t^k.$$

Now the boundary operator is defined on the generators of C_r as follows:

$$\partial_r x = \sum_{y\in S_{r-1}(f)} y \cdot n(x,y).$$

Since all the coefficients $n_k(x,y)$ depend on v, we shall sometimes include v in the notation, writing $n_k(\bar{x},\bar{y};v)$, $n(\bar{x},\bar{y};v)$, and $C_*(v)$. The dependence on the particular choice of liftings of critical points to $\bar{M}$ is of lesser importance, and we shall often write for example $n(x,y)$ instead of $n(\bar{x},\bar{y})$. Note also that the resulting complex does not depend on the particular choice of the function f for which v is the gradient.

The next lemma says that $C_*(v)$ is indeed a chain complex.

LEMMA 5.2. *For every r we have $\partial_r \circ \partial_{r+1} = 0$.*

PROOF. We shall assume that for every $x \in S(f)$ the lifting $\bar{x}$ is in W_0. Then for every $x, y \in S(f)$ the Novikov incidence coefficient $n(\bar{x},\bar{y};v)$ is actually an element of $\mathbb{Z}[[t]]$. Denote by $\partial_r^{(n)}$ the quotient homomorphism

$$C_r/t^n C_r \to C_{r-1}/t^n C_{r-1}.$$

It suffices to prove that $\partial_r^{(n)} \circ \partial_{r+1}^{(n)} = 0$ for every n. But $\partial_r^{(n)}$ is easily identified with the boundary operator in the Morse complex corresponding to the cobordism $F^{-1}([-n,0])$. □

The main aim of this paper is to investigate the properties of the incidence coefficients $n(x,y;v)$, but not the chain homotopy properties of $C_*(v)$. We hope to give a systematic treatment of these properties in the second part of this work. So we shall leave the next proposition without proof. (This proposition is proved in [**25**] for the case when the f-gradient v is an f-Gradient. The proof can be carried over to the general case without difficulty.)

PROPOSITION 5.3. $H_*(C_*(v)) \approx H_*(\bar{M}) \otimes_{\mathbb{Z}[t,t^{-1}]} \mathbb{Z}((t))$.

5.2. Condition $(\mathfrak{C}\mathcal{C})$, and the C^0-generic rationality of the Novikov incidence coefficients. We shall use the terminology of 5.1.2.

We say that v satisfies condition $(\mathfrak{C}\mathcal{C})$ ($\mathfrak{C}$ for *cellular* and $\mathcal{C}$ for *circle*) if the $(F|W)$-gradient v satisfies the condition $(\mathfrak{C}\mathcal{Y})$ from 4.1.6. The set of f-gradients v satisfying $(\mathfrak{C}\mathcal{C})$ will be denoted by $\mathcal{G}\mathfrak{C}\mathcal{C}(f)$. First of all we establish that the properties $(\mathfrak{C}\mathcal{C})$ and $(\mathfrak{C}\mathcal{C}\mathcal{T})$ are C^0-generic.

PROPOSITION 5.4. (1) *The set $\mathcal{G}\mathfrak{C}\mathcal{C}(f)$ is open and dense in $\mathcal{G}(f)$ with respect to the C^0-topology. The set $\mathcal{G}\mathfrak{C}\mathcal{C}\mathcal{T}(f)$ is open and dense in $\mathcal{G}\mathcal{T}(f)$ with respect to the C^0-topology.*

(2) *Let $v \in \mathcal{G}(f)$, and let $\mathcal{V}$ be a C^0-neighborhood of v in $\mathcal{G}(f)$. Then there is a $v_0 \in \mathcal{G}\mathfrak{C}\mathcal{C}(f)$ such that $v_0 \in \mathcal{V}$ and $v = v_0$ in some neighborhood of $S(f)$.*

(3) *Let $v \in \mathcal{G}\mathcal{T}(f)$, and let $\mathcal{V}$ be a C^0-neighborhood of v in $\mathcal{G}(f)$. Then there is a $v_0 \in \mathcal{G}\mathfrak{C}\mathcal{C}(f)$ such that $v_0 \in \mathcal{V}$ and $v = v_0$ in some neighborhood of $S(f)$.*

PROOF. It suffices to prove (2) and (3). Assertion (2) follows immediately from Theorem 4.8. To prove (3), let v_0 be an f-gradient such that $v_0 \in \mathcal{G}\mathfrak{C}\mathcal{C}(f) \cap \mathcal{V}$ and $v_0 = v$ in a neighborhood of $S(f)$ (such a v_0 exists by (2)). Now to obtain an f-gradient satisfying assertion (3), apply Lemma 5.1 and recall that $\mathcal{G}\mathfrak{C}\mathcal{C}(f) \cap \mathcal{V}$ is open. □

Therefore $\mathcal{G}\mathfrak{C}\mathcal{C}\mathcal{T}(f) = \mathcal{G}\mathfrak{C}\mathcal{C}(f) \cap \mathcal{G}\mathcal{T}(f)$ is C^0-open-and-dense in $\mathcal{G}\mathcal{T}(f)$.

THEOREM 5.5. *Let $v \in \mathcal{GCCT}(f)$. Let $x, y \in S(f)$, $\operatorname{ind} x = \operatorname{ind} y + 1$. Then $n(x, y; v) \in \widetilde{L}$; that is,*

$$n(x, y; v) = \frac{P(t)}{t^m Q(t)},$$

where $m \in \mathbb{N}$ while P, Q are polynomials with integer coefficients and $Q(0) = 1$.

PROOF. Let $v \in \mathcal{GCC}(f)$. Condition $(\mathfrak{CY})$ provides a Morse function ϕ_1 on V_1 together with its gradient u_1, and a Morse function $\phi_0 = \phi_1 \circ t^{-1}$ on V_0 together with its gradient $u_0 = (t)_*(u_1)$. For every $k \in \mathbb{Z}$ we also obtain a Morse function $\phi_0 \circ t^{-k}$ on V_{-k} and a $\phi_0 \circ t^{-k}$-gradient $u_k = (t^k)_*(u_0)$. We also have an ordered Morse function $\phi \colon W \to [0, 1]$, adjusted to (f, v), with an ordered sequence $(a_0, \dots, a_{n+1})$. Note that $v | W_k$ satisfies $(\mathfrak{C})$ with respect to the string $\mathcal{S}_k = (\delta, (t^k)_*(u_0), (t^k)_*(u_1), \phi_0 \circ t^{-k}, \phi_1 \circ t^{-k})$.

Let $x, y \in S(f)$, $\operatorname{ind} x = s + 1$, $\operatorname{ind} y = s$. Set $V' = \phi^{-1}(a_{s+1})$. We can assume that $\bar{x} \in W$ and $\bar{y} \in W_1$. Then $n_k(x, y; v) = 0$ for $k \leqslant 0$. Consider the oriented submanifold $X = D(\bar{x}, v) \cap V_0$ of V_0.

Note that $X \subset V_0^{\{\leqslant s\}}$ and that $X \setminus \operatorname{Int} V_0^{\{\leqslant s-1\}}$ is compact. (Indeed, let $\Sigma = D(\bar{x}, v) \cap V'$. Then Σ is an embedded s-dimensional sphere in V'. Using the terminology of 4.2.1, write $X = (-v0)^{\leadsto}(\Sigma)$. Then the compactness of $X \setminus \operatorname{Int} V_0^{\{\leqslant s-1\}}$ follows from Lemma 4.14.) Denote by $[X]$ the homology class of X modulo $V_0^{\{\leqslant s-1\}}$.

Applying Theorem 4.9(1) several times, we deduce by induction that for every k the submanifold $X_k = (-v)^{\leadsto}_{[0,-k]}(X)$ of V_{-k} is in $V_{-k}^{\{\leqslant s\}}$ and $X_k \setminus \operatorname{Int}(V_{-k})^{\{\leqslant s-1\}}$ is compact. Set

$$V'_{-(k+1)} = V' \cdot t^{k+1}, \qquad X'_k = (-v)^{\leadsto}_{[-k,-k-1+a_{s+1}]}(X_k).$$

Set $Y_k = D(\bar{y}t^k, -v) \cap V'_{-k}$; this is a cooriented embedded $(n - s - 1)$-sphere. Since v satisfies the transversality condition, we have $X'_k \pitchfork Y_k$. Proposition 4.17 implies that

$$n_k(x, y; v) = X'_k \# Y_k = \langle \bar{y}t^k, [X_k] \rangle_v. \tag{5.2}$$

Recall the homomorphism from subsection 4.2:

$$\mathcal{H}_s(-v) \colon H_s(V_1^{\{\leqslant s\}}, V_1^{\{\leqslant s-1\}}) \to H_s(V_0^{\{\leqslant s\}}, V_0^{\{\leqslant s-1\}})$$

associated with the cobordism W, and set $h = \mathcal{H}_s(-v) \circ (t_*^{-1})$. Then h is an endomorphism of $H_s(V_0^{\{\leqslant s\}}, V_0^{\{\leqslant s-1\}})$.

Apply Theorem 4.9 to obtain

$$n_k(x, y; v) = \langle \bar{y}, h^k([X]) \rangle_v.$$

Now the proof is finished by the application of the next lemma.

LEMMA 5.6. *Let G be a finitely generated abelian group, A an endomorphism of G, and $\lambda \colon G \to \mathbb{Z}$ a homomorphism. Then for every $p \in G$ the series*

$$\sum_{k \geqslant 0} \lambda(A^k p) t^k \in \mathbb{Z}[[t]]$$

is a rational function of t of the form $P(t)/Q(t)$, where P, Q are polynomials with integral coefficients and $Q(0) = 1$.

PROOF. It is sufficient to consider the case of a free finitely generated abelian group G. Consider the free finitely generated $\mathbb{Z}[[t]]$-module $R = G[[t]]$ and the homomorphism $\phi: R \to R$ given by $\phi = 1 - At$. Then ϕ is invertible, and the inverse homomorphism is given by the formula $\phi^{-1} = \sum_{k \geqslant 0} A^k t^k$. On the other hand, the inverse of ϕ is given by Cramer formulas, which imply that each matrix entry of the homomorphism inverse to ϕ is the ratio of two integer polynomials. Moreover, the denominator is equal to $\det(1 - At)$. □

This also completes the proof of Theorem 5.5. □

The next result deals with the continuity properties of $n(x, y; v)$ under small perturbations of v.

THEOREM 5.7. *Let $v \in \mathcal{GCCT}(f)$. Let U be an open neighborhood of $S(f)$. Then there is a $\delta > 0$ such that for every $w \in \mathcal{GCCT}(f)$ with $\|w - v\| < \delta$ and $v|U = w|U$, and for every $x, y \in S(f)$ with $\operatorname{ind} x = \operatorname{ind} y + 1$, we have $n(x, y; v) = n(x, y; w)$*

PROOF. We shall use the terminology of the proof of Theorem 5.5, but we modify it a bit. Namely, we denote $D(\bar{x}, v) \cap V_0$ by $X(v)$ in order to stress the dependence on the vector field v. We assume a similar convention for the sets X_k, Y_k, the endomorphism h, the sphere Σ, etc. Thus we have

$$n_k(x, y; v) = \langle \bar{y}, (h(v))^k([X(v)]) \rangle_v.$$

Recall from Theorem 4.8 that if $w \in \mathcal{GT}(f)$ is sufficiently C^0 close to v, then w satisfies ($\mathfrak{CC}$) with respect to the same string $S = (\delta, u_0, u_1, \phi_0, \phi_1)$. Therefore the constructions in the proof of Theorem 5.5 also apply to w; thus we obtain the corresponding objects $X(w)$, $X_k(w)$, $Y_k(v)$, etc. Thus (5.2) holds with w replacing v.

The endomorphism $h(v) = (t^{-1})_* \circ \mathcal{H}_s(-v)$ is invariant with respect to C^0-small perturbations of v (see Theorem 4.9). Therefore we have only to show that if $v|U = w|U$ and $\|u - w\|$ is small enough,then $\langle \cdot, \cdot \rangle_v = \langle \cdot, \cdot \rangle_w$ and $[X(v)] = [X(w)]$. The first property follows from Proposition 4.17.

Proceed to the second property. We use the terminology of 4.2.3. In particular, for the definition of $\mathcal{H}$ and $\mathcal{H}0$ we use the triple (A, B, C) from (4.14). Note that $\Sigma(v) \subset \operatorname{Int} A$. Denote by $s(v)$ the fundamental class of $\Sigma(v)$ reduced modulo B; then $[X(v)] = \mathcal{H}0_s(-v)(s(v))$. If w is sufficiently C^0 close to v, then $\Sigma(w) \subset \operatorname{Int} A$ and $s(v) = s(w)$ by Propositions 2.76 and 2.77.

Lemma 4.16 applies to show that $\mathcal{H}0_s(-v) = \mathcal{H}0_s(-w)$, and that completes the proof. □

5.3. Incidence coefficients with values in completions of group rings.

5.3.1. *Definitions.* In this section we deal with "equivariant" analogs of the Novikov complex. Let $f: M \to S^1$ be a Morse map and $\mathcal{P}: \widehat{M} \to M$ a regular covering satisfying $f \circ \mathcal{P} = 0$. We assume here the terminology of subsection 5.1.

Let $v \in \mathcal{GT}(f)$. For every $x \in S(f)$ we choose a lifting $\hat{x}$ of x to $\widehat{M}$, and an orientation of the descending disc $D(x, v)$. Set $\bar{x} = p(\hat{x})$. Let $x, y \in S(f)$, $\operatorname{ind} x = \operatorname{ind} y + 1$. The set $\Gamma(x, y)$ of all $(-v)$-trajectories joining x to y splits naturally into disjoint parts defined as follows.

Let $g \in G$. We say that $\gamma \in \Gamma_g(\hat{x}, \hat{y})$ if the lifting of γ to $\widehat{M}$ starting at $\hat{x}$ finishes at $\hat{y} \cdot g$. Then $\{\Gamma_g(\hat{x}, \hat{y})\}_{g \in G}$ is a family of disjoint subsets of $\Gamma(x, y)$, and their union is equal to $\Gamma(x, y)$. Note also that the union of $\Gamma_g(\hat{x}, \hat{y})$ over all $g \in \xi^{-1}(k)$ is in a natural bijective correspondence with $\Gamma_k(\bar{x}, \bar{y})$. In particular, the set $\bigcup_{\xi(g)>C} \Gamma_g(\hat{x}, \hat{y})$ is finite for every $C \in \mathbb{R}$. Set

$$\hat{n}_g(\hat{x}, \hat{y}) = \sum_{\gamma \in \Gamma_g(\hat{x}, \hat{y})} \varepsilon(\gamma) \cdot g, \tag{5.3}$$

$$\hat{n}(\hat{x}, \hat{y}) = \sum_{g \in G} \hat{n}_g(\hat{x}, \hat{y}). \tag{5.4}$$

Then $\hat{n}_g(\hat{x}, \hat{y})$ is an element of $\mathbb{Z}G$ and $\hat{n}(\hat{x}, \hat{y})$ is a priori an element of $\widehat{\widehat{\Lambda}}$. Since for a given C there are only finitely many $(-v)$-trajectories joining $\hat{x}$ to $\hat{y}g$ with $\xi(g) > C$, we conclude that $\hat{n}(\hat{x}, \hat{y})$ is actually an element of $\widehat{\Lambda}_\xi$. Note that the Novikov incidence coefficient depends on the particular choice of the liftings of the critical points to $\widehat{M}$: it is easy to obtain the formula

$$\hat{n}(\hat{x}g_1, \hat{y}g_2) = g_2^{-1} \hat{n}(\hat{x}, \hat{y}) g_1. \tag{5.5}$$

REMARK 5.8. Let $\widehat{C}_r(f)$ be the free right $\widehat{\Lambda}_\xi$-module generated by $S_r(f)$. As in 5.1.3, one can prove that, setting

$$\widehat{\partial}_r(x) = \sum_{y \in S_{r-1}(f)} x \cdot \hat{n}(\hat{x}, \hat{y})$$

for $x \in S_r(f)$, we obtain a homomorphism of $\widehat{C}_r(v)$ to $\widehat{C}_{r-1}(v)$ satisfying $\widehat{\partial}_{r+1} \circ \widehat{\partial}_r = 0$. The homology $H_*(\widehat{C}_*(v))$ of the resulting complex is isomorphic to the homology of the completed chain complex $C_*^{\Delta}(\widehat{M}) \otimes_\Lambda \widehat{\Lambda}_\xi$. We shall not give the proofs here. See [**25**], where these theorems are proved in the case when v is an f-Gradient.

5.3.2. *Some noncommutative algebra.* In order to formulate an "equivariant" analog of Theorem 5.5, we need first of all an analog of the localization ring $\widetilde{L}$. This analog will be introduced via the following noncommutative localization procedure.

Recollection ([**7**, p. 255]). Let R be a (noncommutative) ring. Assume that for every n a subset S_n of the ring $\mathrm{Mat}(n \times n, R)$ is given; set $S = \bigsqcup_n S_n$. For each matrix $\alpha \in S_n$ and each pair $1 \leqslant i, j \leqslant n$, add to the ring R the generators $\bar{\alpha}_{i,j}$ subject to the relations $A \cdot \bar{A} = \bar{A} \cdot A = 1$, where $\bar{A} = (\bar{\alpha}_{i,j})$. We obtain a ring, denoted by $S^{-1}R$, together with a homomorphism $\ell \colon R \to S^{-1}R$. If R is a subring of a ring L and every $\alpha \in S$ is invertible over L, then there is a commutative diagram

$$\begin{array}{ccc} R & \hookrightarrow & L \\ & \searrow & \uparrow \ell \\ & & S^{-1}R \end{array} \tag{5.6}$$

Apply this procedure to our ring Λ, setting

$$S_n = \{1 + A \mid A \in \mathrm{Mat}(n \times n, \mathbb{Z}G_{(-1)})\}.$$

We obtain a ring which will be denoted by $\widetilde{\Lambda}_\xi$. Since every matrix in S_n is invertible in $\widehat{\Lambda}_\xi$ (the inverse of $1 + A$ being given by the formula $\sum_{i \geqslant 0} (-1)^i A^i$), we have the

commutative diagram

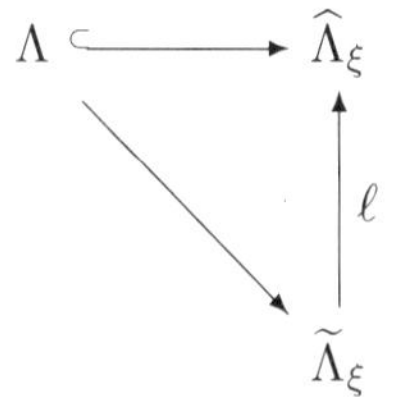

(5.7)

We shall see that the Novikov incidence coefficients of C^0-generic gradients of Morse maps $M \to S^1$ are among the simplest possible elements of $\widetilde{L}\Lambda_\xi$. They will be all of the form $g_1 Y(1+A)^{-1}Xg_2$, where

$$A \in \mathrm{Mat}(m \times m, \mathbb{Z}G_{(-1)}), \quad X, Y \in (\mathbb{Z}H)^m, \quad g_1, g_2 \in G.$$

We shall give a special name to this class of elements of $S^{-1}R$.

DEFINITION 5.9. An element $a \in \widehat{\Lambda}_\xi$ is *of type* $(\mathcal{L})$ if there are vectors

$$X = (X_i)_{1 \leqslant i \leqslant n}, \quad Y = (Y_i)_{1 \leqslant i \leqslant n}$$

with $X_i, Y_j \in \mathbb{Z}H$, elements $g_1, g_2 \in G$, and an $m \times m$ matrix $A = (a_{ij})$ with entries in $\mathbb{Z}G_{(-1)}$ such that

$$a = g_1 \sum_{s \geqslant 0} \Big(\sum_{i,j} Y_i a_{ij}^{(s)} X_i \Big) g_2,$$

where $a_{ij}^{(s)}$ are the matrix entries of A^s.

The elements of type $(\mathcal{L})$ are obviously in $\operatorname{Im} \ell$.

In the rest of this subsection we develop some basic noncommutative linear algebra over the ring $R = \mathbb{Z}H$. These noncommutative linear-algebraic constructions will be used later. Let F be a right R-module. Recall that we have chosen an element $\theta \in G_{(-1)}$. We say that an abelian group homomorphism $\xi \colon F \to F$ is *θ-semilinear* if $\xi(x\lambda) = \xi(x) \cdot \theta\lambda\theta^{-1}$ for every $\lambda \in R$. Assume that F is a free right R-module with free base $(e_i)_{1 \leqslant i \leqslant m}$. Associate to each θ-semilinear homomorphism $\xi \colon F \to F$ the matrix $\widehat{\xi} = (\xi_{ij})_{1 \leqslant i,j \leqslant m}$ as follows: $\xi(e_i) = \sum_{j=1}^m e_j \xi_{ji}$.

LEMMA 5.10. *Let $x = \sum_{i=1}^m e_i x_i$ be a vector in F, and let $n \in \mathbb{N}$. Then*

$$\xi^n(x) = \sum_{i,j} e_j \eta_{ji}^{(n)} x_i \theta^{-n}, \tag{5.8}$$

where $\eta_{ji}^{(n)}$ stands for the entries of the matrix $(\widehat{\xi}\theta)^n$.

PROOF. Standard computation by induction on n. □

COROLLARY 5.11. *Let $\lambda \colon F \to R$ be a homomorphism of right R-modules. Then the element*

$$\sum_{k \geqslant 0} \lambda(\xi^k(x)) \cdot \theta^k$$

of $\widehat{\Lambda}_\xi$ is of type $(\mathcal{L})$.

PROOF. Just apply λ to the right hand side of (5.8). □

The homomorphisms $F \to R$ to which we shall apply our corollary will be constructed with the help of the following procedure. Let $l\colon F \to \mathbb{Z}$ be a homomorphism of abelian groups such that for every $x \in F$ the set $\{h \in H \mid l(xh) \neq 0\}$ is finite. Define a homomorphism $\underline{l}\colon F \to R$ as follows:

$$\underline{l}(x) = \sum_{h \in H} l(xh^{-1}) \cdot h.$$

Then $\underline{l}$ is obviously a right $\mathbb{Z}H$-module homomorphism.

5.3.3. *Elements of exponential growth in Novikov rings.* Let

$$x \in \mathbb{Z}((t)), \qquad x = \sum_n x_n t^n.$$

We say that x is *of exponential growth* if there exist $A, B > 0$ such that $|x_n| \leqslant Ae^{nB}$ for every n. The elements of exponential growth obviously form a subring of $\mathbb{Z}((t))$. In this subsection, we suggest an analog of this notion for elements of the Novikov completion of group rings.

We use the terminology of 5.1.1. In particular, $\Lambda = \mathbb{Z}G$. For an element $\lambda \in \Lambda$, $\lambda = \sum_{g \in G} n_g g$, set $|\lambda| = \sum |n_g|$. Thus $|\cdot|$ is the restriction to $\mathbb{Z}G$ of the L^1-norm on the Banach algebra $L^1(G)$; see [**5**, Chapter 1, §2.2]. In particular, we have $|\lambda\mu| \leqslant |\lambda| \cdot |\mu|$. Recall that $\widehat{\Lambda}_\xi$ is the Novikov ring associated with a homomorphism $\xi\colon G \to \mathbb{R}$. For $\lambda \in \widehat{\Lambda}_\xi$, $\lambda = \sum_{g \in G} n_g g$, and $c \in \mathbb{R}$, set $\lambda_{\llcorner c \lrcorner} = \sum_{\xi(g) \geqslant c} n_g g$. It follows from the definition of $\widehat{\Lambda}_\xi$ that $\lambda_{\llcorner c \lrcorner} \in \mathbb{Z}G$ for every c.

DEFINITION 5.12. We say that $\lambda \in \widehat{\Lambda}_\xi$ is *of exponential growth* if there exist $A, B > 0$ such that for every $c \in \mathbb{R}$ we have $|\lambda_{\llcorner c \lrcorner}| \leqslant Ae^{-Bc}$.[10] The subset of all $\lambda \in \widehat{\Lambda}_\xi$ of exponential growth will be denoted by $\Lambda_\xi^{\text{e.g.}}$.

It is easy to see that $\Lambda_\xi^{\text{e.g.}}$ is a subring of $\widehat{\Lambda}_\xi$.

LEMMA 5.13. *Elements of type* $(\mathcal{L})$ *are of exponential growth.*

PROOF. Let $\lambda \in \widehat{\Lambda}_\xi$ be a element of type $(\mathcal{L})$, i.e.,

$$\lambda = g_1 \Big(\sum_{s \geqslant 0} X_i a_{ij}^{(s)} Y_j \Big) g_2, \quad A^s = (a_{ij}^s), \qquad a_{ij} \in \mathbb{Z}G_{(-1)},\ g_i \in G.$$

We can assume $g_1 = g_2 = 1$, $X_i = \delta_{ik}$, $Y_i = \delta_{js}$. Thus we must show that each matrix entry of $\sum_{s \geqslant 0} A^s$ is in $\Lambda_\xi^{\text{e.g.}}$. Set $\|A\| = \max_{i,j} |a_{ij}|$. Then for every $A, B \in \operatorname{Mat}(m \times m, \mathbb{Z}H)$ we have $\|AB\| \leqslant \|A\| \|B\| m$, and for every $s \geqslant 0$ we have $\|A^s\| \leqslant \|A\|^s m^{s-1}$. This implies easily that if N is a natural number greater than $\max(2, \|A\| m)$, then $\| \sum_{s=0}^k A^s \| \leqslant N e^{k \ln N}$, and our result is proved. □

5.3.4. *Rationality of Novikov incidence coefficients.*

THEOREM 5.14. *Let* $v \in \mathcal{GCCT}(f)$. *Then for every* $x, y \in S(f)$ *with* $\operatorname{ind} x = \operatorname{ind} y + 1$ *the Novikov incidence coefficient* $\hat{n}(\hat{x}, \hat{y}; v)$ *is of type* $(\mathcal{L})$. *Therefore* $\hat{n}(\hat{x}, \hat{y}; v)$ *is in* $\operatorname{Im} \ell$, *and is of exponential growth.*

---

[10]One should probably say *not more than exponential growth.* We chose our terminology for brevity.

PROOF. The proof follows along the lines of the proof of Theorem 5.5, and we shall use the terminology from there. Choose an element $\theta \in G_{(-1)}$. In view of (5.5), we can choose liftings $\hat{x}$ of x and $\hat{y}$ of y to $\widehat{M}$ as we like. Choose them so that their projections $p(\hat{x})$, $p(\hat{y})$ to $\bar{M}$ satisfy $p(\hat{x}) \subset W$, $p(\hat{y}) \subset W_1$. Set $\bar{x} = p(\hat{x})$, $\bar{y} = p(\hat{y})$.

Thus the descending disc $D(\bar{x}, v)$ lifts to $\widehat{M}$ in a natural way, and the manifolds $X_k \subset V_{-k}$ are also lifted to some manifolds $\widehat{X}_k \subset \widehat{V}_{-k}$. Note that for every $\rho \in H$ the point $\hat{y}\rho\theta^k$ is a lifting of $\bar{y}t^k$. For the Novikov incidence coefficient we have

$$\hat{n}(\hat{x}, \hat{y}; v) = \sum_{k \geqslant 0, \rho \in H} \hat{n}_{\rho\theta^k}(\hat{x}, \hat{y}; v).$$

To compute the generic term of this sum, we apply Proposition 4.19 to obtain a formula similar to (5.2):

$$\hat{n}_{\rho\theta^k}(\hat{x}, \hat{y}) = \langle \hat{y}\rho\theta^k, [\widehat{X}_k] \rangle_v \cdot \rho\theta^k. \tag{5.9}$$

Define a θ-semilinear endomorphism $\widehat{h}$ of $\widehat{C}_s(u_0)$ by setting $\widehat{h} = \theta_*^{-1} \circ \widehat{\mathcal{H}}_s(-v|W_1)$. Applying Theorem 4.18, it is not difficult to see that $[\widehat{X}_k] = \theta_*^k \circ \widehat{h}^k$. Therefore

$$\hat{n}_{\rho\theta^k}(\hat{x}, \hat{y}) = \langle \hat{y}\rho, \widehat{h}^k([\widehat{X}]) \rangle = \langle \hat{y}, \widehat{h}^k([\widehat{X}])\rho^{-1} \rangle \cdot \rho.$$

Denote by Υ the homomorphism $\langle \hat{y}, \cdot \rangle \colon \widehat{C}_l(u_0) \to \mathbb{Z}$, and set $\lambda = \underline{\Upsilon}$ (see 5.3.2 for the meaning of the underline). Then

$$\sum_{\rho} \hat{n}_{\rho\theta^k}(\hat{x}, \hat{y}) \cdot \rho\theta^k = \lambda(\widehat{h}^k([\widehat{X}]))\theta^k.$$

Thus

$$\hat{n}(\hat{x}, \hat{y}; v) = \sum_{k \geqslant 0} \lambda(\widehat{h}^k([\widehat{X}]))\theta^k,$$

and it suffices to apply Corollary 5.11 to complete the proof. □

References

[1] R. Abraham and J. Robbin, *Transversal mappings and flows*, Benjamin, New York, 1967.

[2] V. Arnold, *Dynamics of intersections*, Analysis, et cetera: Research Papers Published in Honor of Jürgen Moser's 60th Birthday (P. H. Rabinowitz and E. Zehnder, eds.), Academic Press, 1990, pp. 77–84.

[3] ———, *Sur quelques problèmes de la théorie des systèmes dynamiques*, Topological Methods in Nonlinear Analysis **4** (1994), no. 2, 209–225.

[4] G. Birkhoff and G.-C. Rota, *Ordinary differential equations*, Wiley, New York, 1984.

[5] N. Bourbaki, *Eléments de Mathématiques. Théories spectrales*, Hermann, Paris, 1967.

[6] J. Cerf, *La stratification naturelle des espaces de fonctions différentiables réelles et le théorème de la pseudo-isotopie*, Inst. Hautes Études Sci. Publ. Math. **39** (1970), 5–173.

[7] P. M. Cohn, *Free rings and their relations*, Academic Press, New York–London, 1971.

[8] A. Dold, *Lectures on algebraic topology*, Springer-Verlag, Berlin–Heidelberg, 1972.

[9] M. Sh. Farber, *Exactness of the Novikov inequalities*, Funktsional. Anal. i Prilozhen. **19** (1985), no. 1, 49–59; English transl. in Functional Anal. Appl. **19** (1985), no. 1.

[10] T. Farrell, *The obstruction to fibering a manifold over a circle*, Indiana Univ. J. **21** (1971), 315–346.

[11] M. Hirsch, *Differential topology*, Springer-Verlag, Berlin–Heidelberg, 1976.

[12] W. Klingenberg, *Lectures on closed geodesics*, Springer-Verlag, Berlin–Heidelberg, 1978.

[13] F. Latour, *Existence de 1-formes fermées non singulières dans une classe de cohomologie de de Rham*, Inst. Hautes Études Sci. Publ. Math. **80** (1994), 135–194.

[14] F. Laudenbach and J.-C. Sikorav, *Persistance d'intersection avec la section nulle au cours d'une isotopic hamiltonienne dans un fibré cotangent*, Invent. Math. **82** (1985), 349–357.

[15] J. Milnor, *Morse theory*, Princeton Univ. Press, Princeton, NJ, 1963.

[16] ———, *Lectures on the h-cobordism theorem*, Princeton Univ. Press, Princeton, NJ, 1965.

[17] ———, *Whitehead torsion*, Bull. Amer. Math. Soc. **72** (1966), 358–426.

[18] J. Milnor and J. Stasheff, *Characteristic classes*, Princeton Univ. Press, Princeton, NJ, 1974.

[19] M. Morse, *Calculus of Variations in the Large*, Colloq. Publ., vol. 18, Amer. Math. Soc. New York, 1934.

[20] S. P. Novikov, *Multivalued functions and functionals. An analogue of the Morse theory*, Dokl. Akad. Nauk SSSR **260** (1981), no. 1, 31–35; English transl. in Soviet Math. Dokl. **24** (1981), 222–226.

[21] ———, *Variational methods and periodic solutions of equations of Kirchhoff type.* II, Funktsional. Anal. i Prilozhen. **15** (1981), no. 4, 37–52; English transl. in Functional Anal. Appl. **15** (1981), no. 4.

[22] ———, *The Hamiltonian formalism and a multivalued analogue of Morse theory* Uspekhi Mat. Nauk **37** (1982), 3–49; English transl. in Russian Math. Surveys **37** (1982), 1–56.

[23] ———, *Quasiperiodic Structures in topology*, Topological Methods in Modern Mathematics (J. Milnor 60th Birthday Sympos., Stony Brook, NY, 1991; L. R. Goldberg and A. V. Phillips, eds.) Publish or Perish Inc., Houston, TX, 1993, pp. 223–235.

[24] S. P. Novikov and I. Schmeltser, *Periodic solutions of Kirchhoff equations for the free motion of a rigid body in a fluid and the extended Lyusternik–Shnirelman–Morse theory*, I, Funktsional. Anal. i Prilozhen., **15** (1981), no. 3, 54–66; English transl. Functional Anal. Appl. **15** (1981),, no. 3, 197–207.

[25] A. V. Pazhitnov (Pajitnov), *On the Novikov complex for rational Morse forms*, Ann. Fac. Sci. Toulouse Math. (6) **4** (1995), 297–338.

[26] ———, *Surgery on the Novikov complex*, K-Theory **10** (1996), 323–412.

[27] ———, *Rationality and exponential growth properties of the boundary operators in the Novikov complex*, Math. Res. Lett. **3** (1996), 541–548.

[28] ———, *Simple homotopy type of Novikov complex for closed 1-forms and Lefschetz ζ-functions of the gradient flow*, e-print dg-ga/9706014.

[29] ———, *On the Novikov complex for rational Morse forms*, Preprint of Odense University, Odense, 1991.

[30] ———, *Surgery on the Novikov complex*, Rapport de Recherche CNRS URA 758, Nantes, 1993.

[31] ———, *The incidence coefficients in the Novikov complex are generically rational functions*, e-print dg-ga/9603006.

[32] ———, *Incidence coefficients in the Novikov complex for Morse forms: rationality and exponential growth properties*, e-print dg-ga/9604004.

[33] ———, *Exactness of Novikov-type inequalities for the case $\pi_1(M) = \mathbb{Z}^m$ and for Morse forms whose cohomology classes are in general position*, Dokl. Akad. Nauk SSSR **306** (1989), 544-548; English transl., Soviet Math. Dokl. **39** (1989), 528–532.

[34] ———, *On the sharpness of inequalities of Novikov type for manifolds with a free abelian fundamental group*, Mat. Sb. **180** (1989), 1486–1523; English transl., Math. USSR–Sb. **68** (1991), 351–389.

[35] ———, *The incidence coefficients in the Novikov complex are generically rational functions*, Algebra i Analiz **9** (1997), no. 5, 92–139; Englsih transl., St. Petersburg Math. J. **9** (1998), 969–1006

[36] M. M. Peixoto, *On an approximation theorem of Kupka and Smale*, J. Differential Equations **3** (1966) 214–227.

[37] M. Pozniak, *Triangulation of compact smooth manifolds and Morse theory*, Warwick preprint: 11/1990, November 1990.

[38] A. Ranicki, *Finite domination and Novikov rings*, Preprint, 1993.

[39] ———, *Finite domination and Novikov rings*, Topology **34** (1995), 619–632.

[40] J.-Cl. Sikorav, *Points fixes de diffeomorphismes symplectiques, intersections de sous-variétés lagrangiennes, et singularites de un-formes fermées*, Thèse d'Etat, Orsay, 1987.

[41] ———, *Un problème de disjonction par isotopie symplectique dans un fibré cotangent*, Ann. Sci. École Norm. Sup. **19** (1986), 543–552.

[42] S. Smale, *On the structure of manifolds*, Amer. J. Math. **84** (1962), 387–399.

[43] ———, *Generalized Poincaré conjecture in dimensions greater than four*, Ann. Math. **74** (1961), 391–406.

[44] R. Thom, *Sur une partition en cellules associée à une fonction sur une variété*, C. R. Acad. Sci. Paris Sér. Gén. Vie Sci. **228** (1949), 973–975.
[45] E. Witten, *Supersymmetry and Morse theory*, J. Differential Goem. **17** (1982), no. 2, 661–682.

UMR 6629 CNRS, Université de Nantes, Faculté des Sciences 2, rue de la Houssinière, 44072, Nantes Cedex
E-mail address: pajitnov@math.univ-nantes.fr

Amer. Math. Soc. Transl.
(2) Vol. **197**, 1999

Pseudoperiodic Mappings

D. A. Panov

Abstract. This paper is a survey of pseudoperiodic topology. The main subject of the paper will be pseudoperiodic mappings and polyintegrable flows. Investigations in this direction were initiated by V. I. Arnold. The main results are due to Arnold himself, as well as to I. A. Dynnikov, K. M. Khanin, D. A. Panov, and Ya. G. Sinai.

1. Introduction

Fix an integer lattice $\mathbb{Z}^n$ in the space $\mathbb{R}^n$. A mapping $P\colon \mathbb{R}^n \to \mathbb{R}^m$ is called *periodic* if $P(x+k) = P(x)$ for any $x \in \mathbb{R}^n$ and any $k \in \mathbb{Z}^n$. A mapping $\mathbb{R}^n \to \mathbb{R}^m$ is called *pseudoperiodic* if it can be presented as the sum of a periodic and a linear mapping.

On a pseudoperiodic mapping we usually impose the following genericity condition: the restriction of its linear part to the lattice $\mathbb{Z}^n$ is a monomorphism, and the image of the lattice under this linear part is a dense subset in $\mathbb{R}^m$.

The papers [**1, 2, 3, 4, 5**] treat pseudoperiodic mappings for the two extremal cases: mappings $\mathbb{R}^n \to \mathbb{R}^{n-1}$ and mappings $\mathbb{R}^n \to \mathbb{R}^1$.

To each pseudoperiodic mapping $\mathbb{R}^n \to \mathbb{R}^{n-1}$ one can associate a *polyintegrable* flow on the space $\mathbb{R}^n$. Trajectories of this flow are contained in the level curves of the mapping. Let μ, ν be volume forms on $\mathbb{R}^n$ and $\mathbb{R}^{n-1}$, respectively, and $H\colon \mathbb{R}^n \to \mathbb{R}^{n-1}$ a pseudoperiodic mapping. Then the corresponding polyintegrable flow v on the space $\mathbb{R}^n$ is defined by the requirement $i_v\mu = H^*\nu$. This flow descends naturally to the quotient $T^n = \mathbb{R}^n/\mathbb{Z}^n$. Polyintegrable flows on the two-dimensional torus were considered in [**1**]. It was shown that the torus splits into a number of domains consisting of periodic trajectories of the flow and an ergodic domain of positive measure. In the ergodic domain, the flow is isomorphic to the so-called special flow over a circle rotation (see Theorem 7). It was shown in [**4**] that the flow in the ergodic domain possesses the mixing property (although the mixing is slow, of logarithmic growth, so that the Lyapunov exponent is zero).

The ergodicity of polyintegrable flows for $n > 2$ is known only for flows without singular points. It is known [**2**] that a polyintegrable vector field v on the torus T^n can be deformed, by a smooth diffeomorphism, to the form fv_0, where v_0 is a constant vector field on the torus and f is a positive function (Theorem 9).

1991 *Mathematics Subject Classification.* Primary 58C27, 58Fxx.

The trajectories of polyintegrable flows, i.e., the level curves of pseudoperiodic mappings, were studied in [**3**], but only regular level curves of pseudoperiodic mappings are studied there. A regular level curve of a pseudoperiodic mapping from $\mathbb{R}^n$ to $\mathbb{R}^{n-1}$ splits into components of two types: unbounded components diffeomorphic to the line, and bounded ones diffeomorphic to the circle. It was shown in [**3**] that unbounded components are finite perturbations of the line $L = \text{const}$, where L is the linear part of the corresponding pseudoperiodic mapping (Theorem 1). In the 2-dimensional case (modulo the lattice $\mathbb{Z}^2$) unbounded components fill in the ergodic domain of the torus T^2 described above.

All regular level curves of a pseudoperiodic mapping $\mathbb{R}^n \to \mathbb{R}^{n-1}$ contain the same number of unbounded components (and this number is odd). In the 2-dimensional case the unbounded component is unique (since the ergodic domain on the two-dimensional torus is connected), but already in the 3-dimensional case [**5**] the number of unbounded components of a regular level curve can be arbitrarily large (Theorem 8).

Pseudoperiodic mappings $\mathbb{R}^n \to \mathbb{R}^1$ were studied in [**1**]. It was shown that a level surface of a generic pseudoperiodic mapping $\mathbb{R}^n \to \mathbb{R}^1$ divides $\mathbb{R}^n$ into two unbounded domains (and a number of bounded domains). For a compact manifold, the natural analog of a pseudoperiodic mapping to $\mathbb{R}^1$ is a closed 1-form with incommensurable periods. The study of the family of hypersurfaces on which the form vanishes is equivalent to the study of hyperplane sections of periodic manifolds in Euclidean spaces. It is shown in [**1**] that a generic hyperplane section divides an analytic periodic manifold into precisely two unbounded components (Theorem 10′).

2. Properties of level curves of pseudoperiodic mappings $\mathbb{R}^n \to \mathbb{R}^{n-1}$

DEFINITION 1. Let γ_1, γ_2 be two unbounded curves in $\mathbb{R}^n$. The curve γ_2 is called a *d-deformation* of γ_1 if there exist parametrizations of γ_1 and γ_2 such that $|\gamma_1(t) - \gamma_2(t)| < d$ for all $t \in \mathbb{R}^1$.

Consider a pseudoperiodic mapping $H\colon \mathbb{R}^n \to \mathbb{R}^{n-1}$ represented as the sum $H = L + P$ of a linear mapping L and a periodic mapping P (from now on, the mapping L is assumed generic: the restriction of L to the lattice $\mathbb{Z}^n$ must be monomorphic, and it must map $\mathbb{Z}^n$ to a dense subset of $\mathbb{R}^{n-1}$). Let γ_a denote the a-level curve of the mapping H, $\gamma_a = H^{-1}(a)$, and let l_a denote the a-level curve of the mapping L.

THEOREM 1 ([**3**]). *Let $H\colon \mathbb{R}^n \to \mathbb{R}^{n-1}$ be a pseudoperiodic mapping. There exists a positive number d such that for any regular value a of H the following statements are true. Each unbounded component of γ_a is a d-deformation of the line, and the diameter of each bounded component of l_a is not greater than d. The number of unbounded components is the same for all regular level curves of the mapping H, and this number is odd.*

In our proof of Theorem 1 we use the following statement.

PROPOSITION 1. *Consider a pseudoperiodic mapping $H\colon \mathbb{R}^n \to \mathbb{R}^{n-1}$, and let a be a regular value of this mapping. There exists a positive number $d(a)$ such that each unbounded component of the level curve γ_a is a $d(a)$-deformation of the line l_a, and the diameter of each bounded component of γ_a is not greater than $d(a)$.*

PROOF OF PROPOSITION 1. Let a be a regular value of H. We are going to construct a family of hypersurfaces in $\mathbb{R}^n$ satisfying the following conditions.

1) Each hypersurface meets only unbounded components of the curve γ_a, and not more than once.

2) Each hypersurface of the family can be obtained from any of the others by a shift along a vector of the integer lattice.

3) Outside a cylinder with axis l_a, the hypersurfaces coincide with hyperplanes orthogonal to the line l_a.

4) Each of the hypersurfaces divides the space $\mathbb{R}^n$ into precisely two parts.

5) The hypersurfaces do not intersect each other and split the space $\mathbb{R}^n$ into layers whose thickness does not exceed some fixed number $d_1(a)$ (the *layer thickness* is the length of the orthogonal projection of the layer onto the line l_a).

The existence of such a family of hypersurfaces easily implies the statement of the proposition. Indeed, since the mapping H is pseudoperiodic, the distance from points of the curve γ_a to the line l_a does not exceed some fixed number $d_2(a)$. Therefore, the diameter of the intersection of γ_a with any layer does not exceed $\sqrt{d_1(a)^2 + (2d_2(a))^2}$. Hence, the diameter of a bounded component of γ_a also does not exceed $\sqrt{d_1(a)^2 + (2d_2(a))^2}$ (since, by assumption 1), bounded components of γ_a do not intersect the hypersurfaces of the family). This is also easy to handle for the unbounded components of γ_a. Each of these components intersects each hypersurface of the family precisely once, and it is, therefore, a $\sqrt{d_1(a)^2 + d_2(a)^2}$-deformation of the line l_a.

Now we can construct the required family of hypersurfaces.

Construction. We start by constructing one hypersurface satisfying conditions 3) and 4). In order to do this, we choose in $\mathbb{R}^n$ an arbitrary hyperplane intersecting the curve γ_a transversally and orthogonal to the line l_a. Choose the arcs of the curve γ_a with ends on the chosen hyperplane belonging to one of the half-spaces with respect to this hyperplane. The required hypersurface is obtained by attaching cylinders to the hyperplane along each of the chosen arcs (see Figure 1). Denote the constructed surface by M_a.

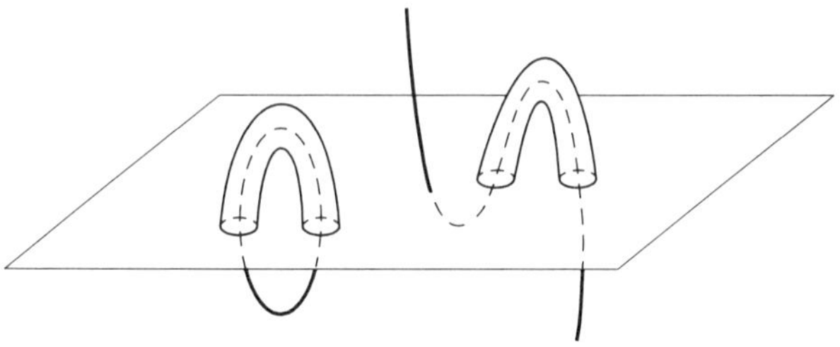

FIGURE 1

With a hypersurface satisfying conditions 3) and 4) in hand, we are able to construct a family of hypersurfaces satisfying all five conditions. Let us find a number ε such that for any value b, $|b - a| < \varepsilon$ ($a, b \in \mathbb{R}^{n-1}$), the curve γ_b is transversal to the surface M_a. Consider in $\mathbb{R}^n$ the cylinder $L^{-1}(U_{\varepsilon/2}(a))$, and shift the hypersurface M_a by all integer vectors of the lattice $\mathbb{Z}^n$ belonging to the cylinder. The number of intersections of the hypersurface M_a shifted by an integer vector k with the curve γ_a is equal to the number of intersections of the curve

$\gamma_{a-L(k)}$ with the surface M_a; by the choice of ε, this number coincides with the number of intersections of γ_a itself with M_a. Hence, the number of intersections with γ_a is the same for all shifted hypersurfaces. Now it follows easily that 1) is satisfied.

The shifted hypersurfaces divide the space $\mathbb{R}^n$ into layers of thickness not exceeding some fixed number, since integer points are distributed in the cylinder $L^{-1}(U_{\varepsilon/2}(a))$ fairly uniformly. Therefore, 5) is satisfied as well. □

PROOF OF THEOREM 1. Let a be a regular value of the mapping H; denote by d the least upper bound of diameters of the bounded components of the curve γ_a. The closure of the union of all integer shifts of γ_a coincides with $\mathbb{R}^n$; therefore, integer shifts of bounded components of γ_a are dense in the space of bounded components of all level curves of the space H. Thus, the diameters of bounded components of all regular level curves of H are also bounded from above by d. Hence, one of the statements of the theorem is proved. The other statements can be proved similarly. □

Theorem 1 justifies the following definition.

DEFINITION 2. A pseudoperiodic mapping $\mathbb{R}^n \to \mathbb{R}^{n-1}$ is said to be a *k-component* mapping if each of its regular curves contains precisely k unbounded components.

We are going to show below that all pseudoperiodic mappings $\mathbb{R}^2 \to \mathbb{R}^1$ are one-component, but pseudoperiodic mappings $\mathbb{R}^3 \to \mathbb{R}^2$ can be k-component for arbitrary large values of k.

The construction of the family of hypersurfaces in the proof of Theorem 1 implies the following fact.

THEOREM 2 ([**5**]). *A small perturbation of a pseudoperiodic mapping cannot lead to an increase in the number of unbounded components of a regular level curve of the mapping.*

REMARK 1. Theorem 2 implies that there are unstable multi-component mappings such that an arbitrary small perturbation can cause a decrease in the number of unbounded components of a regular level curve. The stability of mappings is apparently related to the following fact.

THEOREM 3 ([**5**]). *Let H be a k-component pseudoperiodic mapping. Consider the closure N of the union of all unbounded components of all regular level curves of H. Then the number of connected components of the set N is not greater than k.*

Presumably, a k-component mapping is stable if and only if the corresponding set N contains precisely k connected components.

3. Polyintegrable vector fields: examples and properties [2]

3.1. Properties of polyintegrable vector fields. Polyintegrable vector fields are a multidimensional generalization of Hamiltonian vector fields on the plane. This generalization is based on the strong integrability property of the phase flow (for example, a 3-dimensional polyintegrable system has two first integrals). The simplest example is provided by the vector product of gradients of two functions in three-dimensional Euclidean space, which determines the differential equation

$$x' = [\operatorname{grad}\mathcal{H}, \operatorname{grad}\mathcal{G}].$$

The functions $\mathcal{H}$ and $\mathcal{G}$ are first integrals of this vector field.

Let (M, μ) be a smooth m-fold with a volume form μ, and let (N, ν) be a smooth $(m-1)$-fold, $\dim N = m-1$, with a volume form ν. Let $H\colon M \to N$ be a smooth mapping.

This data determines a vector field v on M, which is called a *polyintegrable field* with poly-Hamiltonian H.

DEFINITION 3. The *polyintegrable field* v (specified by the mapping $H\colon M \to N$ and the volume forms μ and ν) is determined by the condition

$$i_v\mu = H^*\nu,$$

where the $(m-1)$-form in the left-hand side is given by the equality

$$(i_v\mu)\cdot(a_1, \dots, a_{m-1}) = \mu(v, a_1, \dots, a_{m-1}).$$

EXAMPLE 1. $(M, \mu) = (\mathbb{R}^2, dp \wedge dq)$, $(N, \nu) = (\mathbb{R}, dh)$, and $h = H(p, q)$ is a mapping. In this case v is the Hamiltonian vector field with Hamiltonian H.

EXAMPLE 2. $(M, \mu) = (\mathbb{R}^3, dx_1 \wedge dx_2 \wedge dx_3)$, $(N, \nu) = (\mathbb{R}^2, df \wedge dg)$, and the mapping has the form $(f = \mathcal{H}(x), g = \mathcal{G}(x))$. In this case $v = [\operatorname{grad}\mathcal{H}, \operatorname{grad}\mathcal{G}]$.

The following two properties of polyintegrable systems are almost obvious.

THEOREM 4 ([**2**]). *Each coordinate function of the mapping H is a first integral of the polyintegrable vector field specified by the mapping H.*

PROOF. We have $i_v(H^*\nu) = i_v i_v \mu = 0$, and so $i_{H_*v}\nu = 0$. Hence, $H_*v = 0$, since the form ν is nondegenerate. □

THEOREM 5 ([**2**]). *The field v is divergence-free.*

PROOF. We have

$$L_v\mu = d(i_v\mu) = d(H^*\nu) = H^*d\nu = 0,$$

since the form ν is closed. □

REMARK 2. The Gelfand–Leray differential 1-form

$$\frac{\mu}{H^*\nu} = dt$$

arises on the curves $H = \text{const}$. This form has singularities at the critical points of the mapping H. It is related to the polyintegrable field by the equality

$$i_v\, dt = 1$$

(which generalizes the well-known relation $dt = dp \wedge dq/dH$ from the theory of Hamiltonian systems).

3.2. Finite polyintegrable systems. Now let us consider the standard bundle $H\colon (\mathbb{R}^m, \mu) \to (\mathbb{R}^{m-1}, \nu)$. We call the fibers of this bundle *vertical lines.*

DEFINITION 4. A *compactly supported polyintegrable system* is a system that differs from the standard vertical line bundle only in a bounded domain in $\mathbb{R}^m$.

We discuss below the simplest topological properties of generic compactly supported polyintegrable systems.

The preimage of a generic point under the mapping H consists of a nonsingular curve going "from $-\infty$ to $+\infty$" along a vertical line (deviating from the line only in a bounded domain of the space) and of a finite set of closed components. The closed components together form a finite set of bundles over open $(m-1)$-dimensional manifolds with circle fibers. The closures of cells formed by these closed components are bounded by parts of singular curves.

EXAMPLE 3. For a generic compactly supported polyintegrable system in 3-space, a typical part of a singular curve belonging to the boundary of a cell either contains an angle point, or reduces to one point. Besides, the boundary of a cell in a generic system can contain (a finite set of parts of) exceptional singular curves having either two singular points or a cusp.

Critical points of a generic mapping $H\colon \mathbb{R}^3 \to \mathbb{R}^2$ form a finite set of closed curves in $\mathbb{R}^3$ that are contained in boundaries of cells. Generic critical points can be either *elliptic* or *hyperbolic.* The corresponding local normal forms are

$$(x, y, z) \to (x^2 + y^2, z) \qquad \text{elliptic point,}$$
$$(x, y, z) \to (x^2 - y^2, z) \qquad \text{hyperbolic point.}$$

Parabolic points with the local normal form

$$(x, y, z) \to (x^3 + zx, z)$$

separate arcs formed by elliptic and hyperbolic points for a generic mapping. A generic mapping has no other complex singularities.

EXAMPLE 4. There is a compactly supported polyintegrable system in $\mathbb{R}^3$ without elliptic points with the set of critical values forming the 8-curve. One part of the singular curves on the boundary of both cells is diffeomorphic to the union of a circle standardly embedded in 3-space with its diameter. The other parts are curves with one singular point.

The space of connected components of the preimages for a typical compactly supported system determined by a mapping $\mathbb{R}^3 \to \mathbb{R}^2$ is the union of a plane and a polyhedral surface; the singularities of the last can be easily described (see, e.g., [**6**]). It is not known, however, what global topological types of surfaces with such singularities can be realized by a generic mapping.

For mappings of more complicated manifolds, $H\colon M^3 \to N^2$, this problem is also unsolved. De Rham and Burlet [**7**] reformulated the Poincaré conjecture in the following way: for any simply connected closed three-dimensional manifold M^3 there is a generic mapping to the plane without elliptic singular points (i.e., without a boundary in the space of components).

REMARK 3. A similar problem concerning the structure of the space of level curves for a Morse function $H\colon M^2 \to \mathbb{R}$ admits a simple solution.

For the surface M^2 of genus g, the space Γ of connected components of the level curve $H = c$ is a graph with g independent cycles. For example, the space of connected components of level curves for a Morse function on the two-dimensional surface is a tree.

On manifolds of higher dimensions, the graph of components is known at present for spheres: it is always a tree (see the theory of Kronrod et al. [**8**]).

In the simplest examples of compactly supported three-dimensional systems, closed trajectories are linked with the curve of singular points (this is true, for example, in neighborhoods of elliptic points).

The reasons for this effect are not clear, since the linking number (which becomes well defined after the choice, generally speaking not canonical, of the orientation of the components) can be zero. Nevertheless, in all known examples, closed trajectories are linked with the curve of singular points in the following sense: no curve can be moved far from the curve of singular points by an isotopy preserving the curve of singular points.

4. Pseudoperiodic functions on $\mathbb{R}^2$

4.1. Unbounded components of regular level curves.

THEOREM 6 ([**1**]). *A regular curve of a pseudoperiodic function on the plane contains precisely one unbounded component.*

DEFINITION 5. Let h be a pseudoperiodic function on the plane. Consider an unbounded curve on the plane connecting $h^{-1}(-\infty)$ with $h^{-1}(+\infty)$. Such a curve will be called *transversal* if the restriction of h to it is monotone.

THEOREM ON TRANSVERSAL CURVES ([**1**]). *There exists a transversal curve for each pseudoperiodic function on the plane.*

Let us deduce Theorem 6 from the theorem on transversal curves. Each unbounded component of the level curve of a pseudoperiodic function is a d-deformation of some line. Hence, each such component must intersect the transversal curve. But the transversal curve meets each level curve only once. Theorem 6 is proved.

DEFINITION 6. A segment connecting two unbounded components of regular level curves of a pseudoperiodic function and transversal to the level curves of the function is called *elementary transversal.*

PROOF OF THE THEOREM ON TRANSVERSAL CURVES. Consider a pseudoperiodic function $h = l + p$ on the plane $\mathbb{R}^2$. Let a be a regular value of h. Consider all unbounded components of the curve γ_a (there may be only one such component). Let γ denote the component that can be reached from the point $h^{-1}(+\infty)$ without intersecting other unbounded components of γ_a.

Let us find an integer vector k such that the curve γ and its shift under the vector k can be connected by an elementary transversal segment. In order to do this, connect the curve γ with $h^{-1}(+\infty)$ by a path $\rho\colon [0, +\infty) \to \mathbb{R}^2$ meeting no components of the curve γ_a. Choose a number ε such that the function $h(\rho(t))$ takes each value from the segment $[a, h(\rho(\varepsilon))]$ only once on the ray $t \in [0, +\infty]$. Then any integer vector k satisfying the condition $0 < l(k) < h(\rho(\varepsilon)) - a$ possesses the required property, and the segment $\rho[0, \varepsilon]$ can serve as the elementary transversal segment. Indeed, the curve γ shifted by the vector k must intersect the path $\rho[0, +\infty)$ (since γ is a d-deformation of the line l_a), and, by the choice of ε, the shifted curve γ can intersect the path only at the point $\rho(\varepsilon)$.

Hence, the curve γ can be connected by an elementary transversal segment with some integer shift of it, and, therefore, with an arbitrary integer shift. The theorem is proved. □

REMARK 4. It is easy to show that the transversal curve constructed in the proof of the theorem can be chosen as a periodic curve that maps, under factorization of the plane by the lattice $\mathbb{Z}^2$, to a closed nonselfintersecting curve on the torus.

Theorem 6 has the following multidimensional generalization.

THEOREM 6′ ([**5**]). *Let $H = (h_1, \dots, h_{n-1})\colon \mathbb{R}^n \to \mathbb{R}^{n-1}$ be a pseudoperiodic mapping. Suppose that for some $a_2, \dots, a_{n-1}$ the surface*

$$M = h_2^{-1}(a_2) \cap \dots \cap h_{n-1}^{-1}(a_{n-1})$$

is regular and is the union of a connected component π (diffeomorphic to the plane) and of a family of bounded surfaces. Then each regular level curve of the mapping H contains precisely one unbounded component.

4.2. Special flows. In this section we study the properties of a Hamiltonian curve specified by a pseudoperiodic function on the plane. We restrict ourselves to Morse functions only, since in this case sets of bounded components of various level curves of the function are fairly simple.

Let h be a Morse pseudoperiodic function on the plane. An unbounded connected component of an irregular level curve of the function h can contain only one singular point. This singular point divides the component into a bounded component and an unbounded component (which is a d-deformation of a line, Figure 2). We call the domain bounded by the bounded part of a curve a *trap*. Each bounded component of an arbitrary level curve of the function h belongs to some trap. Each trap is a trap from some finite set shifted by some integer vector.

FIGURE 2

The following theorem describes the Hamiltonian flow in the complement to the union of all traps. For the sake of convenience, we consider the flow not on the plane, but on the torus $T^2 = \mathbb{R}^2/\mathbb{Z}^2$. The flow on the torus is uniquely determined by the differential dh of the function h, which is a well-defined closed (but not exact) 1-form. The properties of the flow will be described by means of the transversal curve (constructed in the previous section), which is a closed curve on the torus transversal to the set of zeros of the 1-form dh.

DEFINITION 7. The *special flow over a mapping* $A: X \to X$ is the dynamical system determined by the following data. The phase space of the special flow is the domain Ω under the graph of a positive integrable function $f: X \to \mathbb{R}$,

$$\Omega = \{x, y : x \in X,\ 0 \geqslant y \geqslant f(x)\},$$

where the points $(x, f(x))$ are identified with the points $(A(x), 0)$.

The flow is defined as the uniform motion along the y-axis with velocity 1, i.e., $g^t(x, y+t)$ while $y + t \geqslant f(x)$.

THEOREM 7 ([**1**]). *Let dH be a generic 1-form with incommensurable periods on the two-dimensional torus with a fixed area element, and let κ denote the ratio of two periods, one of which is taken over some transversal curve of the 1-form dH. Then the corresponding Hamiltonian flow over the complement to the union of all traps is isomorphic to the special flow over the rotation of the circle by the angle $2\kappa\pi$. The function determining the special flow is given by the time between two subsequent intersections of the transversal curve. This function is smooth everywhere outside a finite set of points, where it has a logarithmic singularity. The number of these points coincides with the number of traps on the torus.*

PROOF. Let γ be a closed smooth curve on the torus transversal to all curves $dH = 0$, and consider the corresponding covering curve Γ on the covering plane. Without loss of generality, we can suppose that the coordinate x_1 is chosen so that the curve $\Gamma = \Gamma_0$ and its integer shifts Γ_t are specified by the equations $x_1 = t \in \mathbb{Z}$. The coordinates x_2 can be chosen proportional to H on the line Γ, i.e., $H(0, x_2) = \omega_2 x_2$. Hence, $H(t, x_2) = \omega_1 t + \omega_2 x_2$ for all integers t and for all x.

LEMMA 1. *The function x_1 tends to ∞ as a point moves to either direction along an unbounded component of a regular level curve of the function H.*

PROOF. The unbounded component of a regular level curve of the function H must intersect all vertical lines Γ_t with t an integer, since this component is a d-deformation of the line $\omega_1 x_1 + \omega_2 x_2 = 0$. Therefore, it intersects each of these lines precisely once, and $x_1 \in (+\infty, -\infty)$ for this component. □

LEMMA 2. *The unbounded component of a regular level curve of the function H passing through the point $(0, x_2)$ of the vertical line Γ_0 intersects Γ_1 at the point $(1, X)$, where $X = x_2 - (\omega_1/\omega_2)$.*

PROOF. Since the values of the function H at the points $(0, x_2)$ and $(1, X)$ coincide, we have

$$\omega_2 x_2 = \omega_1 + \omega_2 X,$$

and the lemma is proved. □

The line Γ_0 does not intersect bounded components of level curves of the function H. Therefore, the value of the first return map $\Gamma_0 \to \Gamma_1$ is well defined for almost all points, and it is the shift $(0, x_2) \in (1, x_2 - 2\pi\omega_1/\omega_2)$. Hence, the restriction of the first return map to the meridian of the torus is the rotation by the angle $\alpha = -2\pi\omega_1/\omega_2$.

Let $T(x_2)$ denote the time it takes to move from the point $(0, x_2)$ to the point $(1, X)$ along the Hamiltonian flow:

$$T = \int_{x_1=0}^{x_1=1} \tau/dH,$$

where τ is the area element on the torus, and the integral is taken over the segment of the level curve $H = \omega_2 x_2$ passing through the point $(0, x_2)$.

The curve is assumed to be oriented so that $T > 0$.

Let us also define the function y on all unbounded components of regular level curves as the time it takes to move from Γ_0 to z along the level curve,

$$y(z) = \int_{x_1=0}^{z} \tau/dH.$$

Then the phase flow of the Hamiltonian field becomes $g^t(x_2, y) = (x_2, y + t)$, while $y+t$ is less than $T(x_2)$. In these coordinates, the identification of the points $(1, x_2)$ and $(0, x_2)$ is expressed as the identification of $(x_2, T(x_2))$ with $(x_2 - (\omega_1, \omega_2), 0)$.

Finally, the logarithmic behavior of the integral $T(x_2)$ at the point of a singular level of the function H corresponding to the critical point in the region $0 < x_1 < 1$ is due to the standard normal form of the Hamiltonian in the neighborhood of a saddle singular point. The proof is complete. □

5. Example of a multi-component mapping

In this section we construct pseudoperiodic mappings $\mathbb{R}^3 \to \mathbb{R}^2$ with regular level curves containing arbitrarily many unbounded components. We start by defining a foliation of the space $\mathbb{R}^3$. Let V_0 denote the one-dimensional skeleton of the integer lattice in $\mathbb{R}^3$, the union of all lines parallel to the coordinate axes and passing through integer points. The shift of V_0 by the vector $(1/2, 1/2, 1/2)$ will be denoted by V_1. Introduce a fiber bundle structure on the space $\mathbb{R}^3 \setminus (V_0 \cup V_1)$. The base of the bundle is the interval $(0, 1)$, and the fiber $V(a)$, $a \in (0, 1)$, is the set of points at distance $a/2$ from the skeleton V_0 in the metric $\rho(x, y) = \max_i |x_i - y_i|$. Clearly, all fibers are pairwise diffeomorphic and the bundle is trivial. Of course, the surfaces $V(a)$ are not smooth, but a small periodic deformation can make them smooth.

Below, $V(a)$ will denote these smooth fibers. For a small ε the typical fiber $V(\varepsilon)$ is shown on Figure 3. The projection mapping can be extended to a mapping $\mathbb{R}^3 \to [0, 1]$ by continuity if we set $V(0) = V_0$, $V(1) = V_1$. We also introduce the notation $V[a, b] = \bigcup_{t \in [a,b]} V(t)$.

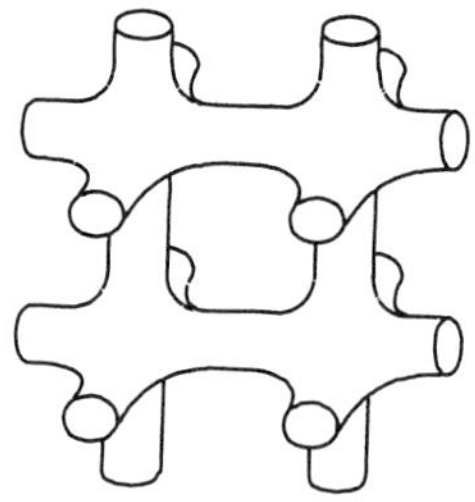

FIGURE 3

PROPOSITION 2. *Let $L = (l_1, l_2)\colon \mathbb{R}^3 \to \mathbb{R}^2$ be a linear mapping. Then there exist a number $\varepsilon > 0$ and a pseudoperiodic map $H = (x_1, x_2)$ with linear part L such that each regular level curve of H contains an unbounded component not intersecting the set $V[0, \varepsilon]$.*

PROOF OF PROPOSITION 2. We construct the required mapping H in the form $H = (l_1, l_2 + p_2)$, changing only the second coordinate of L. Of course, level curves of such a mapping belong to the planes $l_1 = \text{const}$. Let the perturbation p_2 be an arbitrary periodic function supported in $V[0,\varepsilon]$ and taking a very large positive value on the skeleton V_0.

Consider an arbitrary level curve of the perturbed mapping passing through a point of the skeleton V_0. Such curve cannot leave the intersection of the plane $l_1 = \text{const}$ with the domain $V[0,\varepsilon]$, since the value of h_2 at an arbitrary point of the skeleton is much greater than the value of h_2 at neighboring points of level $V(\varepsilon)$. Hence, for the perturbed mapping there is a domain containing the skeleton and consisting of bounded components. In order to complete the proof, it is sufficient to recall that any regular level curve of a pseudoperiodic mapping contains an unbounded component. □

The proposition can be strengthened, and unbounded components can be localized more precisely. as follows.

COROLLARY 1. *Let $L = (l_1, l_2)$ be a linear mapping, and let δ be a positive number. Then there exists a pseudoperiodic mapping H with linear part L that differs from L only inside the domain $V[1-2\delta, 1]$ and such that each regular level curve of H contains an unbounded component belonging to $V[1-\delta, 1]$.*

PROOF. Let us find a pseudoperiodic mapping $H = (h_1, h_2)$ with linear part L and a number ε satisfying the statement of Proposition 2.

Since the bundle V_s is trivial, there is a deformation of $\mathbb{R}^3$ contracting the domain $V[\varepsilon, 1]$ onto the domain $V[1-\delta, 1]$. Such a diffeomorphism of $\mathbb{R}^3$ can be chosen to be periodic, and this construction automatically leads to a pseudoperiodic mapping H' such that each regular level curve of H' contains an unbounded component in $V[1-\delta, 1]$. The linear part of H' coincides with L; therefore, there is a periodic mapping P' supported in $V[0, 1-\delta]$ such that $H' - P' = L$ in $V[0, 1-2\delta]$. The mapping $H = H' - P'$ is the required one. □

The result obtained above can be reformulated, omitting details, in the following brief form: any pseudoperiodic mapping can be perturbed in a neighborhood of the one-dimensional skeleton V_0 so that a regular level curve of the perturbed mapping contains an unbounded component close to V_0.

Now we directly start constructing a mapping with regular level curves containing many unbounded components. Let us shift the skeleton V_0 by the vectors $m_i = (i/k, i/k, i/k)$ $(0 \leqslant i < k)$ in $\mathbb{R}^3$, setting this time $V_i = V_0 + m_i$. Consider any pseudoperiodic mapping $\mathbb{R}^3 \to \mathbb{R}^2$. Using the corollary of Proposition 2, let us perturb this mapping in a neighborhood of each skeleton V_i so that any regular level curve contains an unbounded component close to V_i. The number of such components coincides with the number of skeletons V_i, and is therefore equal to k. The example is constructed.

The argument of the present section remains true in the multidimensional situation as well.

THEOREM 8 ([**5**]). *For $n \geqslant 3$ there exists a pseudoperiodic mapping $\mathbb{R}^n \to \mathbb{R}^{n-1}$ with regular level curves containing arbitrarily many unbounded components.*

The proof of this theorem is based on the fact that the complement to the $(n-2)$-dimensional skeleton of the integer lattice in $\mathbb{R}^n$ is contractible to the one-dimensional skeleton.

6. Polyintegrable flows without singular points

Consider a pseudoperiodic mapping $H\colon \mathbb{R}^m \to \mathbb{R}^{m-1}$,

$$H(x) = L(x) + p(x),$$

where $p(x+k) = p(x)$ for all $k \in \mathbb{Z}^m$. The mapping H generates a polyintegrable vector field v on the space $\mathbb{R}^m$. This field naturally descends to the quotient $T^m = \mathbb{R}^m/\mathbb{Z}^m$.

THEOREM 9 ([**2**]). *A smooth coordinate deformation of the torus T^m reduces a polyintegrable vector field v to the form fv_0, where v_0 is the constant vector field, and f is a positive function.*

In order to prove this theorem, we shall need the following statements.

PROPOSITION 3. *For any vector field without singular points on the torus T^m there exists an embedding of the torus T^{m-1} into T^m transversal to this field.*

LEMMA 3. *Let v be a polyintegrable vector field on T^m and let $i\colon T^{m-1} \to T^m$ be an embedding transversal to the field v. Then there exist coordinates on the torus $i(T^{m-1})$ such that the first return map in these coordinates is a shift by some vector.*

PROOF OF LEMMA 3. Let $H = L + p\colon \mathbb{R}^m \to \mathbb{R}^{m-1}$ be a pseudoperiodic mapping and let v be the corresponding vector field on the torus T^m. Also let $I\colon \Pi^{m-1} \to \mathbb{R}^m$ be the embedding covering the torus $i(T^{m-1})$; the hypersurface $I(\Pi)$ is periodic with respect to some subgroup $\mathbb{Z}^{m-1}$ of the integer lattice in $\mathbb{R}^m$. To be definite, suppose that $\mathbb{Z}^{m-1}$ consists of the vectors $(0, k_1, \dots, k_{m-1})$, $k_i \in \mathbb{Z}$.

Introduce coordinates in the space $\mathbb{R}^{m-1}$ so that the mapping L takes the group $\mathbb{Z}^{m-1}$ to the integer lattice in $\mathbb{R}^{m-1}$. Consider the coordinates $(x_1, \dots, x_{m-1})$ induced on the surface Π by the mapping H from the coordinates on $\mathbb{R}^{m-1}$. It can be easily shown that the first return map on the torus $i(T^{m-1})$ in the coordinates $(x_1 \bmod 1, \dots, x_{m-1} \bmod 1)$ looks like the shift by the vector $(-L(1,0,\dots,0) \bmod \mathbb{Z}^{m-1})$. The lemma is proved. □

Now we are ready to prove Theorem 9.

PROOF OF THEOREM 9. Let v be a polyintegrable vector field on the torus T^m corresponding to a pseudoperiodic mapping H. Let $i\colon T^{m-1} \to T^m$ be the embedding from Proposition 3. Introduce the coordinates $x_1, \dots, x_{m-1}$ from Lemma 3 on the embedded torus $i(T^{m-1})$.

Let f be a positive function on the torus T^m such that the mapping corresponding to time 1 of the flow generated by the vector field fv preserves the torus $i(T^{m-1})$, $g^1(i(T^{m-1})) = i(T^{m-1})$. Then it is possible to introduce a coordinate t on the torus T^m determined modulo 1, which is equal to the time of motion along the vector field fv until the torus $i(T^{m-1})$ is reached. Of course, the coordinate t vanishes on $i(T^{m-1})$. All other coordinates $(x_1, \dots, x_{m-1})$ on $i(T^{m-1})$ extend to T^m by the rule $x_i(g^t(x_1, \dots, x_i, \dots, x_{m-1})) = x_i + ta_i$. In these coordinates the vector field has the form

$$\frac{1}{f}\left(\frac{\partial}{\partial t}, a_1\frac{\partial}{\partial x_1}, \dots, a_{m-1}\frac{\partial}{\partial x_{m-1}}\right).$$

The theorem is proved. □

The proof of Proposition 3 requires the following two lemmas.

LEMMA 4. *Let f be a smooth function on the torus T^m with zero integral. Then, for any generic vector a and for any arbitrarily small ε, there exists a smooth function g such that $|f(x) - (g(x) - g(x-a))| < \varepsilon$ for any $x \in T^m$.*

PROOF OF LEMMA 4. Consider the operator D_a on the space $L^2(T^m)$ that takes a function $f(x)$ to the difference $f(x)-f(x-a)$. The operator D_a preserves the natural $\mathbb{Z}^m$-grading on the space of trigonometric polynomials on the torus T^m. For a generic value of a, all solutions of the equation $D_a(f) = 0$ are constants; therefore the equation $D_a(f) = p$ has a solution for any trigonometric polynomial p that is orthogonal to constants. The statement of the lemma follows automatically. □

LEMMA 5. *Let $H\colon \mathbb{R}^m \to \mathbb{R}^{m-1}$ be an arbitrary pseudoperiodic mapping without singular points. Choose a horizontal plane $\mathbb{Z}^{m-1}$ and a vertical line $\mathbb{Z}^1$ in the integer lattice $\mathbb{Z}^m$. Then there exists a hypersurface Π_0 in $\mathbb{R}^m$ periodic with respect to the shifts along vectors of $\mathbb{Z}^{m-1}$ and transversal to the level curves of the mapping H.*

PROOF OF LEMMA 5. Consider all embeddings of the $(m-1)$-ball D^{m-1} into $\mathbb{R}^m$ transversal to the level curves of H. For such an embedding i, denote by I the family of embeddings obtained by shifting i along the vectors of $\mathbb{Z}^{m-1}$. A set $i_1, \ldots, i_k$ of embeddings is called *complete* if the union of the images

$$H(I_1(D^{m-1})), \ldots, H(I_k(D^{m-1}))$$

of all embeddings covers the space $\mathbb{R}^{m-1}$. The existence of a complete set of embeddings is an easy corollary of the properties of H. Let us choose a complete set $i_1, \ldots, i_k$ of embeddings.

Now let us construct, for any level curve γ of H, the intersection point of this curve with the required hypersurface. The curve γ intersects the family of balls $I_1, \ldots, I_k$ at a finite set of points $x_1, \ldots, x_j$. Recall that $\mu/H^*\nu$ is the Gelfand–Leray 1-form on γ, which determines, up to an additive constant, a coordinate on it. Fix an arbitrary function Φ on the ball D^{m-1} which is positive in the interior of the ball and vanishes on its boundary together with all derivatives of arbitrary order. Choose the point on γ whose coordinate is equal to the weighted mean value of the coordinates of the points $x_1, \ldots, x_j$; the weight is proportional to the value of Φ at these points. We choose the resulting point for the point of intersection of the surface Π_0 with the curve γ. Repeating the construction above for all level curves of H, we obtain the required surface Π_0. Obviously, this surface is smooth and periodic. □

PROOF OF PROPOSITION 3. Let Π_0 be a periodic hypersurface transversal to the level curves of H constructed in Lemma 5. Denote by Π_1 the surface Π_0 shifted by the vector $(1, 0, \ldots, 0)$. In order to prove Proposition 3, let us deform the periodic hypersurface Π_0 so that Π_0 and Π_1 do not intersect. (In this case the quotient of Π_0 under the factorization $\mathbb{R}^m/\mathbb{Z}^m$ is a smooth hypersurface in the torus.)

Let $\varphi\colon \Pi_0 \to \mathbb{R}$ denote the function equal to the time of motion from Π_0 to Π_1 (this time equals zero at the intersection points of Π_0 and Π_1). Consider also the first return map $\Pi_0 \to \Pi_0$ constructed as follows: a point goes along the flow until it meets the surface Π_1 and then is shifted by the vector $(-1, 0, \ldots, 0)$. Introduce the coordinates $(x_1, \dots, x_{m-1})$ from Lemma 3 on Π_0. In these coordinates, the first return map is presented as a shift by some vector a, and the integral of φ is positive.

Consider a periodic function g on Π_0, and denote by Π_g the periodic surface in $\mathbb{R}^m$ obtained by moving each point of the surface along the trajectory for a time equal to the value of g at this point. Then the functions φ_g and φ satisfy the equation

$$\varphi(x) = \varphi_g(x) - (g(x) - g(x-a)).$$

Since the vector a is generic, and the integral of φ is positive, we can find, by Lemma 3, a periodic function g such that the function φ_g is positive. Then the surface Π_g does not intersect itself shifted by the vector $(1, 0, \dots, 0)$. The proposition is proved. □

7. 1-forms with incommensurable periods on manifolds [1]

THEOREM 10 ([**1**]). *Each set*

$$M_c = \{x \in \mathbb{R}^n : H(x) < c\}$$

of underlevel points of a pseudoperiodic function H has precisely one bounded connected component.

Let G be a connected component of the open set of underlevel points M_c of a continuous pseudoperiodic function H.

LEMMA 6. *If the function H is not bounded from below on G, then the images $G_t = G + t$ of the integer shifts of G are pairwise disjoint.*

PROOF. The domain G_t is a connected component of the open set $M_{c(t)}$, where $c(t) = c + l(t)$. All the numbers $c(t)$, $t \in \mathbb{Z}^n$, are different, since the periods are incommensurable.

If two connected components of two sets of underlevel points of the same continuous function intersect each other, then the component corresponding to the smaller level is a subset of the other one. Suppose that, contrary to the statement of the lemma, $G_t \supset G_s$. Shifting G_t by $r(s-t)$, we obtain a decreasing sequence of nonempty connected components $G_t = G_0 \supset G_1 \supset G_2 \supset \cdots$ of the sets of underlevel points $M_c, c = c_0 > c_1 > c_2 > \cdots$, where the sequence c_r is a decreasing arithmetical progression with the difference $l(s-t)$. Then the function H is not bounded from below on G_0, which contradicts the assumptions of the lemma. Hence, the shifted images of G are disjoint. □

REMARK 5. The same proof works in a more general situation when the commutative group $\mathbb{Z}^r$ acts by shifts $T_t\colon M \to M$, $t \in \mathbb{Z}^r$, on a manifold M.

DEFINITION 8. A function $H\colon M \to \mathbb{R}$ is called *pseudoperiodic* if

$$H(T_t x) \equiv H(x) + l(t),$$

where $l\colon \mathbb{Z}^r \to \mathbb{R}$ is a monomorphism.

LEMMA 6′. *If H is bounded from below on a connected component G of the set of underlevel points of a continuous pseudoperiodic function on M, then the images $T_t G$ of G under the shifts are pairwise disjoint.*

EXAMPLE 5. Consider the affine space $\mathbb{R}^N$ with a fixed discrete group of shifts T_t, $t \in \mathbb{Z}^r$.

DEFINITION 9. A submanifold M^m is called *periodic* if it is invariant under the shifts T_t.

Let $h\colon \mathbb{R}^N \to \mathbb{R}$ be a linear pseudoperiodic function. Lemma $6'$ can be applied to the restriction H of h to a periodic submanifold M.

REMARK 6. The study of "level surfaces" (surfaces where the form vanishes) of a closed 1-form with totally incommensurable periods on a compact manifold is equivalent to the study of the restriction of a pseudoperiodic linear function in a linear space to some submanifold.

Indeed, let us choose r 1-forms forming a basis in the real 1-cohomology of the manifold V, and let us treat the integrals of these forms as functions on the covering space $\widetilde{V}$ (whose group of continuous paths is the derived subgroup of the fundamental group of V). Then we obtain a mapping $a\colon \widetilde{V} \to \mathbb{R}^r$. Suppose also that $b\colon V \to \mathbb{R}^k$ is an embedding of V in Euclidean space and $p\colon \widetilde{V} \to V$ is the covering. Then the mapping $a \oplus b \circ p\colon \widetilde{V} \to \mathbb{R}^{k+r}$ embeds $\widetilde{V}$ in $\mathbb{R}^{N=k+r}$ as a pseudoperiodic submanifold M. The 1-form we are studying can be assumed to be a linear combination of basic 1-forms. Then the potential of this form is the restriction to M of a *linear* pseudoperiodic function on $\mathbb{R}^N$.

Conversely, a pseudoperiodic function on any periodic manifold M generates a 1-form on the quotient $V = M/T$. Similar constructions can be applied to 1-forms whose number of independent periods is less than the first Betti number of V.

DEFINITION 10. Let a discrete group of shifts T_t, $t \in \mathbb{Z}^r$, act on the Euclidean space $\mathbb{R}^N$. A periodic closed submanifold of $\mathbb{R}^N$ is called *bounded* if it is a subset of a bounded neighborhood of some subspace $\mathbb{R}^r$ invariant with respect to all shifts.

In other words, the quotient manifold M/T_t must be compact.

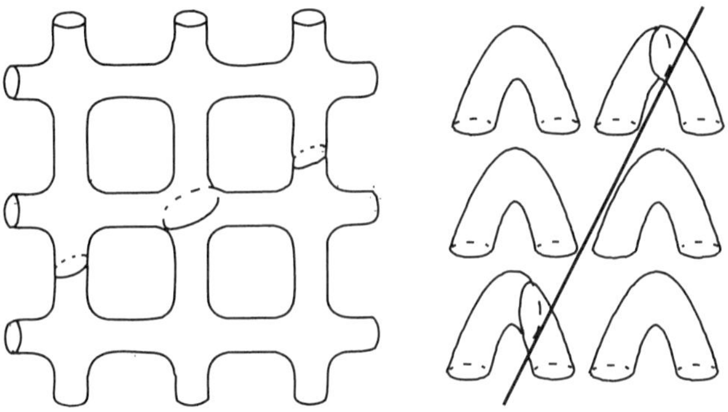

FIGURE 4

EXAMPLE 6. Figure 4 shows two bounded periodic manifolds covering a surface of genus 2 ($r = 2$, $N = 3$, $m = 2$). Intersecting these manifolds by planes, we can study the level curves of closed 1-forms with two incommensurable periods on the genus 2 surface. In the first case almost all these level curves are closed, while in the second there is an ergodic component. The difference between the two cases in Figure 4 is due to the difference in properties of two-dimensional integer planes in four-dimensional symplectic space of the one-dimensional cohomology of the genus

2 surface. A 1-form with two incommensurable periods determines such a two-dimensional plane. In the first case this plane is Lagrangian, while in the second case it is not.

DEFINITION 11. A *pseudoperiodic manifold* is the intersection of a periodic manifold with a linear subspace.

The study of closed level surfaces of a 1-form with incommensurable periods is equivalent to the study of pseudoperiodic hypersurfaces.

Consider the restriction H of a linear pseudoperiodic function $h\colon \mathbb{R}^N \to \mathbb{R}$ to a bounded periodic (with respect to some group of shifts T_t, $t \in \mathbb{Z}^r$) submanifold M^m. Fix a basis $(t_1, \dots, t_r)$ in $\mathbb{Z}^r \subset \mathbb{R}^r \subset \mathbb{R}^N$ and a complement subspace $\mathbb{R}_{N-r}$, where $h = 0$.

DEFINITION 12. A subspace $\mathbb{R}^k \subset \mathbb{R}^N$ is called *admissible* if it is spanned by some basis vectors and the subspace $\mathbb{R}^{N-r}$.

Let G be an unbounded connected component of the underlevel set

$$M_c = \{x \in M : H(x) < c\},$$

and suppose H is bounded from below on G.

LEMMA 7. *There exists an admissible subspace $\mathbb{R}^k$, $N - r < k < N$, such that there is an infinite set of shifted images of $\mathbb{R}^k$ whose intersection with G has a bounded nonempty connected component.*

PROOF. Introduce coordinates $(x_1, \dots, x_r; y_1, \dots, y_{N-r})$ in $\mathbb{R}^N$ such that the action of T_t has the form $(x, y) \mapsto (x + t, y)$, and the complementary subspace is specified by the equation $x = 0$.

One of the x-coordinates, say x_1, is unbounded on G (since G is unbounded and the coordinates y_i are bounded on G, as well as on the whole manifold M). Hence, the subset of integer values of x_1 is infinite; denote these values by s_j.

Now let us proceed by induction on r.

The case $r = 1$ is impossible, since the boundedness of H on G contradicts the unboundedness of G.

In the case $r = 2$ let us consider the intersection of G with the hyperplane $x_1 = s_j$. The restriction of the coordinate x_2 to this intersection is bounded (since the function H is bounded on G both from above and from below). The coordinates y_i are bounded on the whole manifold M. Therefore, the intersection is bounded.

Suppose $r > 2$. If at least one of the components of intersection of the domain G with at least one of the hyperplanes $x_1 = s_j$ is unbounded, then the assumptions of Lemma 7 are satisfied on this hyperplane. And so for this hyperplane, the statement is true by the induction hypothesis.

If, on the other hand, all connected components of all intersections are bounded, then $k = N - 1$, and the admissible hypersurface $x_1 = 0$ is the required one. □

Now we are going to prove a stronger version of Theorem 10 (of which Theorem 10 is an easy special case).

Let M^m be a bounded, connected, periodic with respect to shifts T_t, $t \in \mathbb{Z}^r$, analytic submanifold in the space $\mathbb{R}^N$, and let P be a hyperplane $h = c$, where $h\colon \mathbb{R}^N \to \mathbb{R}$ is a totally incommensurable pseudoperiodic linear function.

THEOREM 10′ ([**1**]). *There is precisely one unbounded component of the complement $M \setminus P$ on each side of the hyperplane P.*

REMARK 7. If the incommensurability is not total, then the statement of the theorem becomes false.

EXAMPLE 7. $r = 2$, $N = 3$, M^2: $y = x_1 + 2\sin x_1$, P^2: $x_1 = \pi$.

PROOF OF THEOREM 10′. The quotient manifold M/T_t is compact and connected. Therefore, any two points on this manifold are connected by a path such that the (absolute value of the) variation of the function h along the corresponding covering path is less than some constant C_1 independent of the chosen points.

It follows that all points of M_c whose distance from P is greater than C_1 belong to the same connected component of $M \setminus P$. Suppose there is another unbounded connected component G of the underlevel set of points M_c. Then this component is situated inside the bounded lawyer $c_1 < h < c$. Hence, the function H is bounded on G from below. By Lemma 6′, the shifted images of the unbounded connected component G are pairwise disjoint.

By Lemma 7, the domain G intersects infinitely many different shifted images $T_t\mathbb{R}^k$ of an admissible subspace $\mathbb{R}^k$, so that all connected components of the intersection are bounded. Without loss of generality, we can suppose that the space $\mathbb{R}^k$ (and, therefore, all its shifted images $T_t\mathbb{R}^k$) is transversal to M (this can be done by choosing an appropriate origin in $\mathbb{R}^N$ by means of the Sard lemma).

Now, each connected component of the intersection of G with $T_s\mathbb{R}^k$ is a bounded domain G_j in the smooth intersection $M(s) = M(T_s \cap (\mathbb{R}^k))$. The manifold $M(s)$ is a periodic and bounded, but not necessarily connected, submanifold in $T_s\mathbb{R}^k$. The function $h|M(s)$ equals $c + h(s)$ on the boundary of each domain G_j. Therefore, it has at least one critical point q_j inside G_j. The projections Q_j of these critical points on the compact quotient manifold $V = M/\mathbb{Z}^r$ are pairwise different, since the shifted images T_tG_j of the domains G_j are, by Lemma 6′, pairwise disjoint (the shifted images of G do not intersect).

The image of the submanifolds $M(s)$ under the projection on V is an analytic compact submanifold $W \subset V$. The restriction of dh to $M(s)$ determines an analytic closed 1-form on W. All points Q_j are zeros of this 1-form (since the q_j are critical points of the restriction of h to $M(s)$).

All points Q_j belong to different connected components of the set of zeros of the 1-form. Indeed, by the Sard lemma, the function $h|M(s)$ is constant (since all points of the component are critical). Therefore, the component K containing Q_i and Q_j must intersect the boundary of the projection of G_j to W. But this is impossible, since $h < c$ at the point q_j, $h = c$ on the boundary, and h is constant on the preimage K.

Hence we have obtained an analytic 1-form on a compact manifold whose set of zeros has infinitely many connected components. The contradiction thus obtained shows that the unbounded component of the underlevel set M_c is unique, and the required assertion is proved. □

REMARK 8. The same arguments prove the uniqueness of the unbounded component of the underlevel set of a generic smooth pseudoperiodic function on any bounded periodic manifold.

Indeed, for a generic h the restriction of h to $\mathbb{R}^k \cap M$ has only Morse critical points for any admissible space $\mathbb{R}^k$ transversal to M. Therefore, a generic h has a finite set of critical points Q_i.

REMARK 9. In the two-dimensional case (pseudoperiodic functions on $\mathbb{R}^2$) the statement remains true for all functions, not necessarily generic; but already in the three-dimensional case the answer in this more general situation is unknown.

References

[1] V. I. Arnold, *Topological and ergodic properties of closed 1-forms with incommensurable periods*, Funktsional. Anal. i Prilozhen. **25** (1991), no. 2, 1–12; English transl., Functional Anal. Appl. **25** (1991), 81–90.

[2] ———, *Poly-integrable flows*, Algebra i Analiz **4** (1992), no. 6, 54–62; English transl., St. Petersburg Math. J. **4** (1993), 1103–1110.

[3] I. A. Dynnikov, *Intersections of level surfaces of pseudoperiodic functions*, Uspekhi Mat. Nauk **49** (1994), no. 1, 213–214; English transl., Russian Math. Surveys **49** (1994), no. 1, 229–230.

[4] Ya. G. Sinai and K. M. Khanin, *Mixing for some classes of special flows over circle rotation*, Funktsional. Anal. i Prilozhen. **26** (1992), no. 3, 1–21; English transl., Functional Anal. Appl. **26** (1992), 155–169.

[5] D. A. Panov, *Polycomponent pseudoperiodic mappings*, Funktsional. Anal. i Prilozhen. **30** (1996), no. 1, 30–38; English transl., Functional Anal. Appl. **30** (1996), 23–29.

[6] L. E. Mata-Lorenzo, *Polyhedrons and pi-stable homotopies from 3-manifolds into the plane*, Bol. Soc. Brasil. Mat. **20** (1989), 61–85.

[7] O. Burlet and G. de Rham, *Sur certaines applications génériques d'une variété close à 3 dimensions dans le plan*, Enseignement Math. (2) 20 (1974), 275–292.

[8] A. G. Kronrod, *On functions of two variables*, Uspekhi Mat. Nauk **5** (1950), no. 1, 24–134. (Russian)

INDEPENDENT UNIVERSITY OF MOSCOW, RUSSIA
E-mail address: dima@panov.mccme.ru

Amer. Math. Soc. Transl.
(2) Vol. **197**, 1999

How Do the Leaves of a Closed 1-form Wind Around a Surface?

Anton Zorich

ABSTRACT. The paper begins with a mini-survey of some problems of pseudoperiodic topology.

In the main part of the paper, we consider analogs of irrational winding lines on a torus for arbitrary Riemann surfaces. These analogs are leaves of foliations defined by closed differential 1-forms. We study the asymptotic topological dynamics of the winding lines. We take long pieces of leaves of the foliation and consider the behavior of cycles obtained by joining the endpoints of each piece by short segments.

We prove that generically there is a flag of subspaces

$$V_1 \subset V_2 \subseteq \cdots \subseteq V_g \subseteq V \subset H_1(M_g^2;\mathbb{R})$$

in the first homology group with the following properties. The 1-dimensional subspace V_1 is spanned by the asymptotic cycle. Deviation of a cycle representing a long piece of leaf from the subspace V_j is of order $l^{\nu_{j+1}}$, $j = 1, \ldots, g-1$, where l is the length of corresponding piece of the leaf. The bound is uniform with respect to the choice of a leaf and of the position of the piece on it. The deviation of any leaf from the subspace V is uniformly bounded by a constant. "Universal constants" $0 \leqslant \nu_j < 1$ are represented in terms of Lyapunov exponents of the Teichmüller geodesic flow on the corresponding moduli space of Abelian differentials.

This statement is a corollary of a similar statement for interval exchange transformations.

Structure of the paper

In Section 1 we present a mini-survey of some problems of pseudoperiodic topology. It is independent of the rest of the paper. In Section 2 we consider foliations on Riemann surfaces defined by closed 1-forms. We show why the interesting topological dynamics of such foliations can be represented by a class of 1-forms obtained as real parts of Abelian differentials. In Section 3 we formulate the principal results. In Section 4 we reformulate the problem and the principal results in the language of interval exchange transformations. Then we prove the main theorem, using the properties of a discrete analog of the Teichmüller geodesic flow on the space of interval exchange transformations. In Appendix A we discuss the irreducibility of

---

1991 *Mathematics Subject Classification.* Primary 34D08, 57R30, 58F18; Secondary 58F11.
Key words and phrases. Measured foliation, asymptotic cycle, Teichmüller geodesic flow, interval exchange transformation, Rauzy induction, measurable cocycle, Lyapunov exponents.

the corresponding cocycle. In Appendix B we prove irreducibility for a certain particular case.

1. Overview of some problems of pseudoperiodic topology

It is difficult to define what exactly is "pseudoperiodic" or "quasiperiodic" topology. However, the following problem definitely belongs to the subject.

Consider an embedding of a closed compact manifold M^m into a torus T^n. Consider the embedding of the induced *periodic* manifold $\widehat{M}^m$ into the universal cover $\mathbb{R}^n$ over T^n:

$$\begin{array}{ccc} \widehat{M}^m & \longrightarrow & \mathbb{R}^n \\ \downarrow & & \downarrow P \\ M^m & \underset{i}{\longrightarrow} & T^n \end{array} \tag{1}$$

Now consider an affine subspace $A^l \in \mathbb{R}^n$.

GENERAL PROBLEM. *Describe the topology of*

a) *the intersection* $\widehat{M}^m \cap A^l$;
b) *the complement* $\widehat{M}^m \setminus A^l$.

Taking various values of the parameters m, n, l, and some specific embeddings $M^m \hookrightarrow T^n$, one gets different problems of pseudoperiodic topology. The present paper mostly deals with the case when $m = 2$, n is large enough, and $l = n - 1$. In this case the manifold M^m is just a Riemann surface; n "large enough" means $n \geqslant 2g$, or sometimes $n \geqslant 2g + 3$, where g is the genus of the surface. In other words, we study hyperplane sections of periodic surfaces in $\mathbb{R}^n$. Before going into details concerning this particular case, we present an outline of what is known (at least to the author) about other combinations of the parameters m, n, l, and what problems are hidden in the above general formulation.

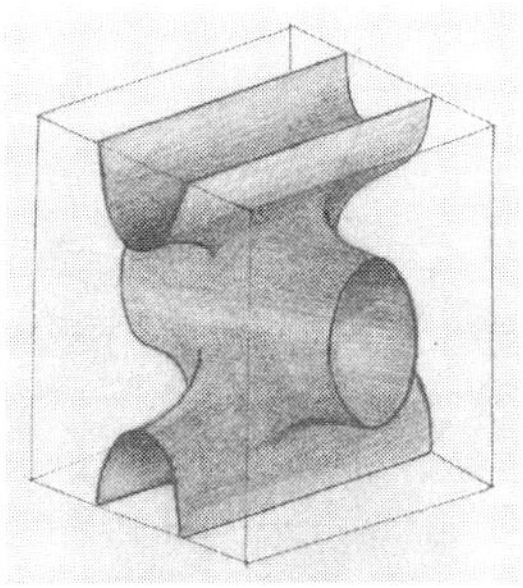

FIGURE 1. Embedding of a Riemann surface of genus 2 into a torus T^3

Topology of intersections of periodic manifolds with affine subspaces

1.1. Hyperplane sections of periodic submanifolds. The special case in which $l = n-1$, i.e., when we consider hyperplane sections of periodic manifolds, is exactly the study of *levels* of closed differential 1-forms on a closed manifold M^m. In the paragraphs below we present the construction of Arnold [**8**] identifying the two problems. Basically, the idea is in the analogy between diagram (1) and the corresponding diagram for the Abel–Jacobi map.

Locally a closed differential 1-form ω can be represented as the differential of a function. The function is defined up to a constant, so locally the level hypersurfaces of the function are well defined. Thus a closed differential 1-form defines a codimension one foliation (in general with singularities) on the manifold. Speaking about a *leaf* of the foliation, we shall usually assume that it is connected. Thus any two points of a leaf can be joined by a path γ such that the restriction $\omega|_\gamma$ vanishes, $\omega|_\gamma \equiv 0$. We shall include in a single class, which will be called a *level* of the 1-form, all the leaves which can be joined by a path in the manifold M^m along which the integral of the form ω is equal to zero.

Consider a linear function $L\colon \mathbb{R}^n \to \mathbb{R}$ such that the hyperplane under consideration is its level surface:

$$A^{n-1} = \{(x_1, \dots, x_n) \in \mathbb{R}^n \mid L(x) = a_1x_1 + \dots + a_nx_n = \text{const}\}.$$

Consider the *exact* 1-form dL on $\mathbb{R}^n$. It is easy to see that dL is induced by a *closed* differential 1-form λ on the torus, $dL = P^*\lambda$ (see (1)). Considering the torus as a unit cube with identified opposite sides, we get the following coordinate representation of the closed 1-form λ on the torus: $\lambda = a_1\,dx_1 + \dots + a_n\,dx_n$. Consider now the induced closed 1-form $\omega = i^*\lambda$ on the manifold M^m (see (1)) and the foliation on M^m by leaves of ω. By construction, any hyperplane section of $\widehat{M}^m$ is projected to a level of ω. Under certain assumptions (say, when all a_i are independent over the rationals) the projection provides a diffeomorphism of connected components.

Conversely, having an arbitrary closed differential 1-form ω on a closed manifold M^m, we can always pull back ω from a linear form on a torus T^n under some embedding $f\colon M^m \hookrightarrow T^n$. The embedding is constructed as follows. Note that we can always represent ω as a linear combination of *integer* 1-forms $\omega = a_1\alpha_1 + \dots + a_k\alpha_k$, where all periods of every closed 1-form α_i are integers, and the coefficients a_i are just real numbers linearly independent over the rationals. Every integer closed (and nonexact) 1-form α on M^m determines a mapping of M^m onto a circle S^1:

$$f_\alpha\colon M^m \to S^1, \quad \text{where } f_\alpha\colon x \mapsto \int_{x_0}^{x} \alpha \mod \mathbb{Z}.$$

Consider also an embedding $f_0\colon M^m \hookrightarrow T^{2m+1}$ such that the image of the first homology is trivial. (To construct f_0, one can just embed M^m into a "large cube" in $\mathbb{R}^{2m+1}$ using the Whitney theorem.) Taking the direct product $f_{\alpha_1} \times \dots \times f_{\alpha_k} \times f_0$ of these maps, we get an embedding $f\colon M^m \hookrightarrow T^n$, where $n = k + 2m + 1$ and $T^n = S^1 \times \dots \times S^1 \times T^{2m+1}$. The form ω is induced from the linear closed 1-form $\lambda = a_1\,dx_1 + \dots + a_k\,dx_k$ on T^n, $\omega = f^*\lambda$. By construction, every connected component of any level of ω on M^n is isomorphic to the corresponding connected component of the corresponding hyperplane section of $\widehat{M}^m \subset \mathbb{R}^n$.

The study of topology of levels of a closed 1-form on a closed manifold was initiated by Novikov in [**41, 42, 43**]. In particular, he outlined a description of the *quasiperiodic structure* of such a manifold. This structure depends on the *degree of irrationality* of the form, that is, on the number

$$d = \dim_{\mathbb{Q}}\langle p_1, \dots, p_k\rangle - 1,$$

where the p_i form a basis of periods of ω. (Sometimes the degree of irrationality is defined as $d+1$.)

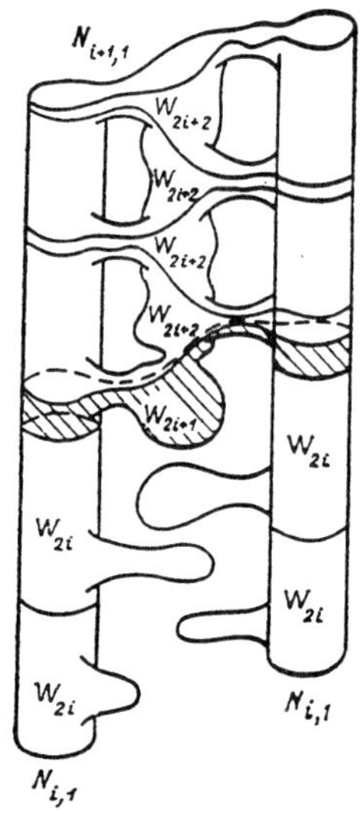

FIGURE 2. Quasiperiodic manifold of degree of irrationality 1

For $d = 1$ the quasiperiodic structure was described in the author's paper [**68**] under the additional assumption that ω is a deformation of a rational 1-form. For arbitrary d, the problem was solved by Le Tu [**34**]; see also the paper of Alaniya [**1**]. Basically, the quasiperiodic structure of a level of a closed 1-form of degree of irrationality d is similar to a $\mathbb{Z}^d$-periodic structure. But instead of having a single topological pattern with a (singular) boundary represented by d pairs of "faces", there is a finite collection of possible patterns, with some compatibility conditions for the "faces". Note that the degree of irrationality d has nothing to do with the dimension $m-1$ of the patterns (here m is dimension of the manifold M^m).

The quasiperiodic structure is basically described by the following picture. Consider a finite collection of completely irrational parallel affine hyperplanes in $\mathbb{R}^d$, where $\mathbb{R}^d$ is provided with a cubic lattice $\mathbb{Z}^d$. Take a vector of the lattice, and rescale it by an irrational factor. Consider all translations of our family of hyperplanes by the integer multiples $k\vec{t}$, $k \in \mathbb{Z}$, of the resulting vector $\vec{t}$. The lattice $\mathbb{Z}^d$ is now sliced by a periodic (with irrational period $\vec{t}$) family of parallel hyperplanes. Patterns corresponding to the cubes inside each slice bounded by two neighboring hyperplanes are the same; passing from one slice to the other, we change the pattern; patterns corresponding to slices obtained by translation by $\vec{t}$ are the same. See [**34, 68**] for a precise description.

1.2. Homology theory of periodic manifolds. Another collection of problems arising in this area is related to the homology theory of (quasi)periodic or even $\mathbb{Z}$-periodic manifolds. The corresponding homology theory of periodic manifolds and an analog of Morse theory for closed 1-forms were constructed by Novikov in [**42**]. The corresponding complexes are now defined over larger rings—for example, for a $\mathbb{Z}$-periodic manifold the complex is defined over Laurent power series in one variable. In [**23**] Farber proved the sharpness of the Morse–Novikov inequalities under certain restrictions on the topology of the manifold. Further results were obtained by Pajitnov [**48, 47**], Sikorav [**54**], and others. For more references, we refer the reader to the paper of Pajitnov in this volume. In that paper, Pajitnov studies the properties of Novikov's complex under some natural restrictions, and proves some beautiful analytic properties of the complex (for example, he proves that Laurent power series corresponding to *incidence coefficients* are represented by rational functions).

1.3. Plane sections of periodic surfaces in $\mathbb{R}^3$. It turns out that the extreme case $m = l = 2$, $n = 3$, i.e., the study of plane sections of periodic surfaces in $\mathbb{R}^3$, has immediate applications in solid state physics. In this case the hyperplane sections are of dimension one, so topologically they are trivial: every nonsingular connected component of the section is diffeomorphic either to a circle or to a line. What is important here is the way in which the open components (the lines) are embedded into $\mathbb{R}^3$.

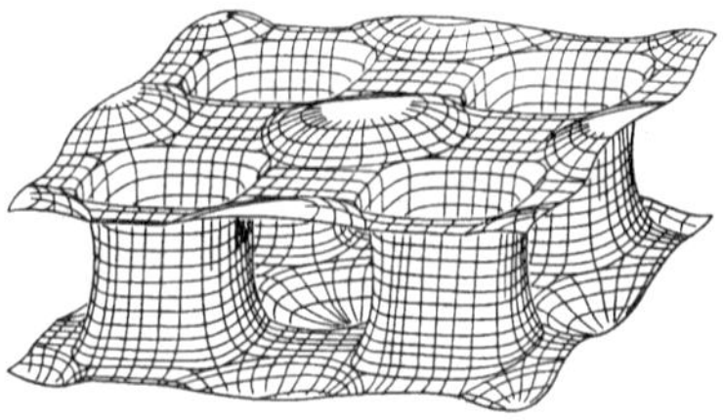

FIGURE 3. Fermi surface of tin (reproduced from [**35**], which gives [**2**] as the source). The corresponding surface $M_g^2 \hookrightarrow T^3$ is of genus 3

A periodic surface may be interpreted as a Fermi surface of some metal in the inverse lattice, and the plane can be interpreted as a plane orthogonal to a constant magnetic field. Thus the plane sections give us electron trajectories in metal in the inverse lattice in the presence of a homogeneous magnetic field. This problem was formulated by Novikov in [**42**], where he conjectured that a typical open trajectory follows a straight line (see also [**44**] and [**45**] by Novikov, and [**46**] by Novikov and Maltsev, for developments of this subject).

There are two natural approaches to this problem. We can fix the periodic surface, and consider a family of perturbations of the hyperplane, or we can fix the direction of hyperplanes and consider a family of perturbations of the periodic surface. Using the first approach, the author [**67**] established Novikov's conjecture for an open dense set of directions of hyperplanes.

Here is a more precise formulation of the result. Fix a generic periodic surface. If the direction of the hyperplane is a sufficiently small perturbation of a rational direction, then every unbounded component of any nonsingular section goes along a straight line with a bounded deviation from it. (Actually, the fact that the trajectory really *follows* a straight line from $+\infty$ to $-\infty$, i.e., that it does not "come back", was established later by Dynnikov [**17**]).

A comprehensive study of the problem was performed by Dynnikov in [**17, 18, 20, 21, 22**] (we refer the reader to [**20**] for the "state of the art" in this subject). In particular, using the second approach, Dynnikov proved in [**18**] the following statement. Let the periodic surface be a level surface of a periodic Morse function in $\mathbb{R}^3$; let a and b be the minimum and the maximum of this function. Fix a generic direction of a family of parallel hyperplanes. There is an interval $[c, d]$, $a < c \leqslant d < b$, such that for any level surface corresponding to the value outside of $[c, d]$ all connected components of the plane sections are closed. If $c < d$, then all unbounded components of plane sections of the remaining level surfaces go along straight lines with bounded deviations from them. However, Dynnikov proved that the situation in which $c = d$ is possible, and, moreover, that for this particular level surface the behavior of the plane sections might be much more complicated.

An example of a trajectory having nontypical behavior was constructed back in 1982 by Tsarev [**57**], but his example corresponds to a rather particular situation. Dynnikov elaborated a highly nontrivial construction producing numerous examples of nontypical behavior of trajectories (see [**20**] for the description of these examples, and also [**69**] for some numerical simulations of Dynnikov's examples). Thus the following problem is still open:

PROBLEM 1. *Consider a closed orientable Riemann surface embedded in T^3, and the corresponding periodic surface in $\mathbb{R}^3$. Consider the set of directions of those hyperplanes which give "nontypical" nonsingular unbounded components of intersections with the periodic surface. (Here "nontypical" means those open components which are not bounded perturbations of straight lines). Is it true that this set has measure zero in the space $\mathbb{R}\mathrm{P}^2$ of all possible directions? Describe the structure of this set. What can be said about the Hausdorff dimension of this set?*

Computer simulations performed by Dynnikov show that, even in the particular case in which a periodic surface is defined as the level surface of a trigonometric polynomial of three variables, the problem seems to be quite nontrivial. Dynnikov's recent paper [**22**] is closely related to this problem—it treats the geometry of the *stability zones* in the set of directions.

1.4. Hyperplane sections of periodic surfaces in $\mathbb{R}^n$. The case $m = 2$, $l = n - 1$ and n large enough, say $n \geqslant 2g + 3$, corresponds to the study of plane sections of periodic surfaces in $\mathbb{R}^n$. As shown in 1.1 above, this problem is in a sense equivalent to the study of behavior of the leaves of foliations defined by closed 1-forms on a Riemann surface.

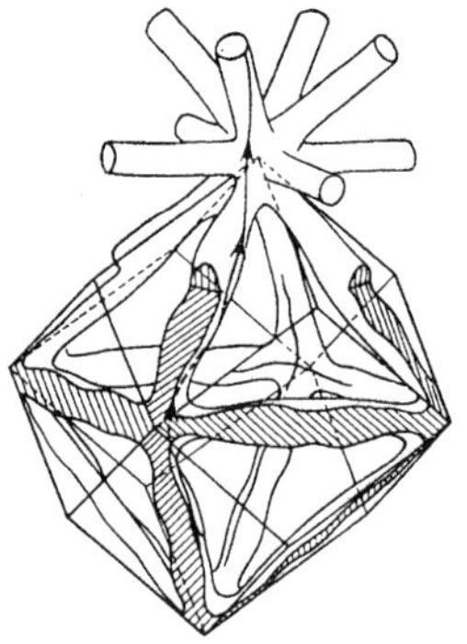

FIGURE 4. Fermi surface of iron (reproduced from [**35**], which cites [**64**] as the source)

The closed 1-forms considered in the previous section ($n = 3$) have some very specific properties. Now we consider arbitrary closed 1-forms (see 1.1). The three-dimensional situation is rather rigid, as can be shown by elementary topological arguments. A hyperplane section of a periodic surface in $\mathbb{R}^n$ has much more flexibility, and three-dimensional topological arguments are no longer applicable. However, here one can use tools from dynamics. In particular, using the results of Masur [**38**] and Veech [**59**], it is easy to prove that generically (in this paper we always use the notion "generic" in the measure-theoretical sense) the unbounded hyperplane sections of periodic surfaces follow one of several asymptotic directions. But now a deviation from this asymptotic direction is no longer bounded by a constant. This paper describes this deviation. It turns out that the description of the deviation can be obtained by means of dynamical characteristics (namely, Lyapunov exponents) of the Teichmüller geodesic flow.

Describing the hyperplane sections of a periodic surface is the same as describing the dynamics of the leaves of the corresponding orientable measured foliation on the closed orientable underlying surface. It is convenient to study the topological dynamics of leaves of a measured foliation using the first return map to a transverse interval. This first return map is an *interval exchange transformation.* We use the ergodic properties of interval exchange transformations, and then we translate them into the language of measured foliations.

* * *

In our approach we unfold the Riemann surface using the universal *abelian* cover of the surface, and we study the asymptotic behavior of leaves on the corresponding periodic surface. Actually, the study of the asymptotic behavior of leaves of a foliation on a Riemann surface was initiated by A. Weil, and then developed by D. V. Anosov (see [**3, 4**] for complete references), N. Markley [**37**], and later by S. Kh. Aranson, V. Z. Grines, E. Zhuzhoma et al. (see [**5**] for references). In this approach one lifts the leaves of a foliation on a Riemann surface to the *universal* cover, and studies their asymptotic behavior on the Lobachevskiĭ disk, comparing the unfolded leaves with the geodesics. Note that in this setting a wider class of foliations and curves is considered.

There is another problem concerning "irrational flow" on Riemann surfaces. One can consider a Morse closed 1-form ω on a Riemann surface as a multivalued

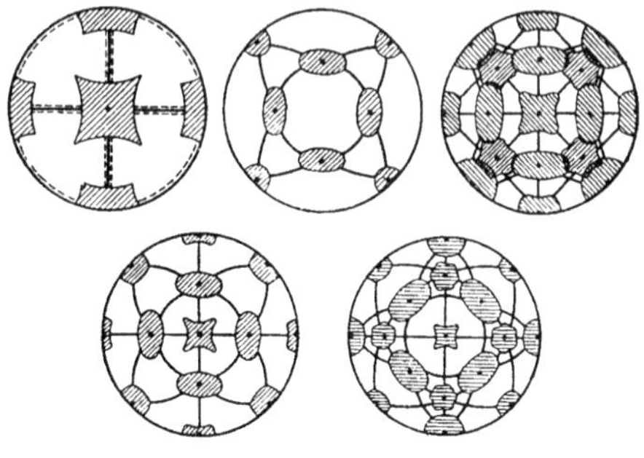

FIGURE 5. Stereographic projection of the magnetic field directions (shaded regions and continuous curves) that give rise to open trajectories for some Fermi-surfaces (experimental results in [**35**])

Hamiltonian, and consider the corresponding Hamiltonian flow along the leaves of the foliation. This flow was studied by Arnold for the torus [**8**]; Sinai and K. M. Khanin proved in [**55**] that generically such a flow on the torus is mixing. Recently the result was generalized by Khanin and Nogueira [**31**] to Riemann surfaces of arbitrary genus $g \geqslant 2$.

REMARK 1. There is a significant difference between the "Hamiltonian" parametrization of leaves and the one used in the present paper. We use a parametrization of leaves by length in some nondegenerate Riemannian metric, or equivalent parametrizations. In particular, passing close to a zero of the closed 1-form is not distinguished in this consideration, whereas in the Hamiltonian parametrization the motion along the leaves is logarithmically slow near simple zeros of the closed 1-form (see [**8**]). Moreover, in the presence of separatrix loops homologous to zero (*traps*) the flow parametrized by the natural parameter, in some sense, "does not notice" the traps: two currents of the flow which are separated by the zero merge again almost as if there were no splitting. In the Hamiltonian parametrization, the part of the flow which has to pass over the "trap" merges with the other one with considerable delay (see Remark 1 in [**8**]). Basically, this delay is exactly the source of the mixing property of the flow (see [**55**]).

1.5. Plane sections of hypersurfaces. Another extreme case of the "general problem" above is the case of $m = n - 1$, $l = 2$, i.e., the case in which we study intersections of a periodic hypersurface with a 2-dimensional affine subspace. Here the sections are again 1-dimensional, so topologically nonsingular connected components are diffeomorphic either to a circle or to a line. As an illustration, we present an example suggested by Arnold (see [**7**]). Take a level surface of the following periodic function in $\mathbb{R}^5$:

$$f(x_1, x_2, x_3, x_4, x_5) = \sum_{i=1}^{5} \cos(2\pi x_i)$$

Consider the 2-dimensional linear subspace spanned by two vectors $\vec{v}_1$ and $\vec{v}_2$ in $\mathbb{R}^5$:

$$\vec{v}_1 = \left(1, \cos\frac{2\pi}{5}, \cos\frac{4\pi}{5}, \cos\frac{6\pi}{5}, \cos\frac{8\pi}{5}\right), \quad \vec{v}_2 = \left(0, \sin\frac{2\pi}{5}, \sin\frac{4\pi}{5}, \sin\frac{6\pi}{5}, \sin\frac{8\pi}{5}\right).$$

Figure 6 presents a computer simulation of the picture of the level structure of f restricted to the plane.

The picture has an obvious rotational symmetry of order 10. It also has translational "quasisymmetry": if we choose a bounded region of the picture, then on translating it by appropriate vectors we shall see infinitely many regions where the picture is almost the same. (Since the region presented in Figure 6 is relatively small, the quasisymmetry is not quite visible at the picture.)

When a quasisymmetry of this kind (quasicrystals) was discovered in solid states physics about fifteen years ago, it was quite a sensation (see [**7**] for a popular introduction).

Although the above example has been known for more than ten years, to my best knowledge, none of the following questions of Arnold concerning this or similar examples have been given an answer.

PROBLEM 2. *Are there any nonclosed intersection lines? Is the size (length, diameter) of the closed intersection lines uniformly bounded?*

FIGURE 6. Level curves of the function $\sum_{i=1}^{5}\cos(2\pi x_i)$ restricted to the irrational 2-dimensional plane

The results of Gusein-Zade (see this volume) and of Dynnikov [**20**] seem to be relevant to this problem.

1.6. Polyintegrable flows. As a generalization of the previous problem, one can consider the particular case of our "general problem" in which the parameters l, m, n obey the following relation: $m+l=n+1$. In this case the intersection of an affine plane A^l with a periodic submanifold M^m is again one-dimensional. There is a particular case here that has been thoroughly investigated: the case in which the periodic manifold M^m is the torus T^m.

One can represent an affine subspace A^l as the intersection of $m-1$ affine hyperplanes. Playing the same game as before, we represent the affine hyperplanes as level hyperplanes of linear functions $L_1, \dots, L_{m-1}$ in $\mathbb{R}^n$. We push down the differentials dL_i to the closed 1-forms λ_i on the torus $T^n = \mathbb{R}^n/\mathbb{Z}^n$. We pull back the closed 1-forms λ_i to the forms $\omega_i = i^*\lambda_i$ on the torus T^m under the embedding $T^m \hookrightarrow T^n$. (Note that the embedding $i\colon T^m \hookrightarrow T^n$ is not linear in general!) Taking the universal Abelian cover over T^m, we get $m-1$ exact 1-forms dF_i on $\mathbb{R}^m$. Our intersection lines become the intersection lines of level hypersurfaces of $m-1$ pseudoperiodic functions $F_1, \dots, F_{m-1}$ on $\mathbb{R}^m$.

Conversely, having $m-1$ pseudoperiodic functions $F_1, \dots, F_{m-1}$ on $\mathbb{R}^m$, or, what is the same, a pseudoperiodic mapping $F\colon \mathbb{R}^m \to \mathbb{R}^{m-1}$, we can push down the differentials dF_i to closed 1-forms ω_i on the quotient $T^m = \mathbb{R}^m/\mathbb{Z}^m$. Each closed 1-form ω_i on T^m can be represented as a sum of an exact 1-form and a linear combination of standard linear forms α_i on $T^m = \mathbb{R}^m/\mathbb{Z}^m$: that is, we have $\omega_i = d\phi_i + \sum a_{ij}\alpha_j$. Consider the mapping $f_0\colon T^m \to \mathbb{R}^{m-1}$ defined by

$$f_0\colon x \mapsto (\phi_1(x), \dots, \phi_{m-1}(x)).$$

We may always choose a large cube in $\mathbb{R}^{m-1}$ such that the image of f_0 is contained in this cube. Thus we may consider the mapping f_0 as a mapping into the torus T^{m-1}. Taking the direct product of the identity map $T^m \to T^m$ and the map f_0, we get an embedding $i\colon T^m \hookrightarrow T^{2m+1}$. Each form ω_i is a pullback of a linear form on T^{2m-1} under this embedding, $\omega_i = i^*\lambda_i$.

The intersection curves of the leaves of the closed 1-forms ω_i on T^m have a natural parametrization by the leafwise 1-form $\Omega/(\omega_1 \wedge \cdots \wedge \omega_{m-1})$, where Ω is the standard volume form on T^m. Thus we get a flow along this one-dimensional foliation. This flow was introduced by Arnold in [**9**], where it was called a *polyintegrable flow*. The study of polyintegrable flows was developed by Dynnikov [**19**], who proved that all regular unbounded fibers of a pseudoperiodic map $f\colon \mathbb{R}^m \to \mathbb{R}^{m-1}$ are deformations of straight lines. Dynnikov also proved that the number of such unbounded regular fibers is constant and odd, provided f is generic. However, it was not clear whether there are any examples in which this number is different from 1, i.e., when the fiber has several unbounded components. Such examples and further results on polyintegrable flows were obtained by Panov in [**50**]. For more information on polyintegrable flows, see Panov's survey in this volume.

A one-dimensional foliation defined by $m-1$ closed 1-forms on a closed manifold M^m seems to be quite a curious object. To my best knowledge it has never been studied for any manifolds different from T^m. It seems to be a reasonable generalization of a measured foliation on a Riemann surface, where the closed 1-forms provide a sort of transverse measure. The foliation can be parametrized similarly to the case of a torus: the parametrization is given by a generic volume form Ω on M^m "divided" by the wedge product of $m-1$ 1-forms under consideration. When we fix the embedding $M^m \hookrightarrow T^n$ we can use a generic linear m-form on T^n to define m-form Ω on M^m. This parametrization of leaves enables us to consider the corresponding flow along the foliation. Similarly to the case of T^n, the flow seems to resemble a flow of an incompressible fluid (see [**9**]).

None of the natural questions, such as decomposition of the foliation into components sharing the same dynamical properties, topological dynamics of leaves, ergodic properties of the corresponding flow, etc., have been ever considered.

1.7. General case. The only result known to the author for arbitrary m, n, l is the theorem of S. M. Gusein-Zade [**25**] concerning density of topological invariants (see his paper in this volume).

Topology of the complement to an affine subspace in a periodic manifold

1.8. Unbounded components of the complement to a hyperplane in a periodic manifold. Let an embedding $M^m \hookrightarrow T^n$ be analytic. As usual, we assume that M^m is connected. Let an affine hyperplane A^{n-1} be completely irrational. Arnold proved in [**8**] that under these assumptions the complement $\widehat{M}^m \setminus A^{n-1}$ contains exactly two unbounded components, one in each of the two halfspaces defined by $A^{n-1} \subset \mathbb{R}^n$.

To my best knowledge this is the only result obtained for the complement to an affine subspace in a periodic submanifold. The question of whether the condition of analyticity of the embedding $\widehat{M}^m \to \mathbb{R}^n$ is essential is still open. Some advances in the generalization of Arnold's theorem for smooth embeddings were obtained by Chekanov [**13**].

Arnold's theorem might be thought of as a theorem on the *essential* H_0-*homology* of the complement $\widehat{M}^m \setminus A^{n-1}$ (whatever "essential homology" means).

PROBLEM 3 (V. I. Arnold). *Formulate and prove the generalization of the theorem to "essential H^{n-l-1}-homology" of the complement $\widehat{M}^m \setminus A^l$.*

Here finding an appropriate definition of "essential homology" is part of the problem.

2. Introduction

2.1. Decomposition of a measured foliation into minimal and periodic components. Consider a closed 1-form ω on a Riemann surface M_g^2. We assume that ω is a *Morse form*, i.e., that all zeros of ω are nondegenerate. This means the following. Consider a small neighborhood of a zero of ω; let $\omega = df$ in this neighborhood. A zero is called *nondegenerate* if it is a nondegenerate critical point of f. Since f is defined up to a constant, the notion is well defined. A Morse 1-form on a Riemann surface may have the same types of critical points as a function: minima, maxima, and simple saddles.

If we fix an embedding $M_g^2 \hookrightarrow T^n$, and consider closed 1-forms coming from all linear 1-forms on T^n (see 1.1), then almost all induced 1-forms will be of Morse type. (Throughout this section we use the notion of "almost all" 1-forms in the same sense without referring each time to Lebesgue measure on the linear space of linear 1-forms on $\mathbb{R}^n/\mathbb{Z}^n$.) A C^1-small perturbation of a Morse form is again a Morse form.

Consider the foliation by leaves of a closed 1-form of Morse type on a Riemann surface. It was shown by Maier [**36**] that the surface can be decomposed into several components of two types: *periodic components* and *minimal components.* Periodic components are filled with closed leaves of the foliation, while every nonsingular leaf living in a minimal component is everywhere dense in it. For example, the standard rational foliation on the torus consists of a single periodic component; the standard irrational foliation on the torus consists of a single minimal component. The boundaries of components are formed by critical leaves of the foliation: *separatrix loops* and *saddle connections.*

We distinguish separatrix loops and saddle connections for the following reason. A saddle connection, i.e., a singular leaf of the foliation joining two distinct zeros of ω, disappears under almost all small deformations of the form ω. In other words, almost all 1-forms have no saddle connections at all. The same is true for separatrix loops representing nontrivial homology cycles. On the contrary, if a separatrix loop represents a zero cycle in homology, it survives under any C^1-small deformation of ω.

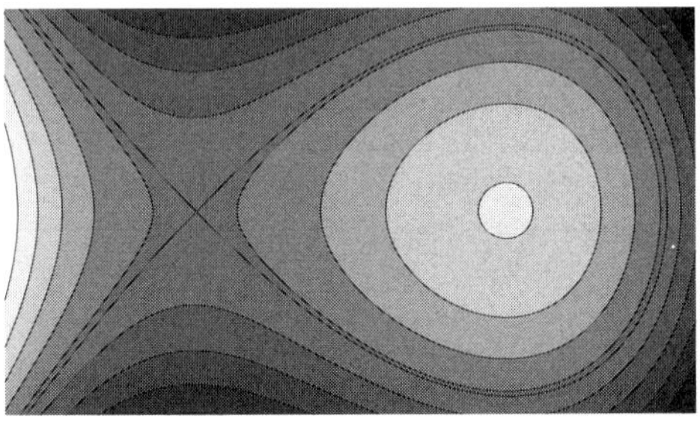

FIGURE 7. Foliation near a separatrix loop

Let us prove the latter statement. Suppose we have a separatrix loop γ_0 such that the cycle $[\gamma_0] \in H_1(M_g^2;\mathbb{R})$ is homologous to zero: $[\gamma_0] = 0$ (see Figure 7). Note that the foliation defined by a closed 1-form has a transverse measure. Thus monodromy along any closed leaf is trivial, and all leaves of the foliation passing close to γ_0 on at least one side of γ_0 are closed (see Figure 7). Thus we get a whole cylinder (or a punctured disk) filled with "parallel" closed leaves. All these closed leaves are homotopic to each other and to the separatrix loop γ_0. The separatrix loop γ_0 is one of two components of the boundary of the cylinder (punctured disk). Since $[\gamma] = 0$, the form ω is exact on the corresponding cylinder (disc). This implies that the leaves homologous to zero would survive under any C^1-small deformation of the initial closed 1-form. The boundary of the deformed disk contains a separatrix loop homologous to zero, namely the deformation of the initial separatrix loop.

The dynamics of leaves on periodic components is trivial. The interesting part of the dynamics is represented by minimal components. Thus we can cut out all cylinders filled by closed leaves homologous to zero. What we then get in the generic situation is a collection of disjoint minimal components (see, say, [**72**] for a justification). Every minimal component is represented by a Riemann surface with several holes formed by separatrix loops.

Let us shrink each hole to a point. The corresponding critical point disappears under this operation. We get several disjoint minimal components without any separatrix loops or saddle connections. It is easy to see that if we study the behavior of unfolded leaves in $\mathbb{R}^n$, or (what is almost the same) if we study the asymptotics of homology cycles obtained by joining the ends of long pieces of leaves, then the operation of "shrinking the holes" does not change the dynamics (see [**72**] for details). (Dynamics in the *Hamiltonian* parametrization (see Remark 1) of the same foliation can be drastically changed by this operation.)

2.2. Closed 1-forms versus harmonic 1-forms. We claim that the closed 1-forms obtained on minimal components after the surgery described in the previous section are harmonic in some Riemannian metric. To show this, we can use the following criterion of Calabi [**12**].

CALABI'S THEOREM. *A closed Morse 1-form ω on a closed manifold M is harmonic with respect to some Riemannian metric if and only if for every nonsingular point $x \in M$ there exists a closed path $\rho\colon [0;1] \to M$ through x such that $\omega(d\gamma/dt) > 0$ for any $t \in [0;1]$.*

The fact that a closed 1-form without separatrix loops and saddle connections on a Riemann surface is harmonic in some Riemannian metric was independently proved by Katok [**27**]. For Riemann surfaces an analog of Calabi's theorem was independently proved by Hubbard and Masur for forms having arbitrary isolated singularities (see [**26**]).

Any harmonic 1-form ω_0 on a Riemann surface can be represented as a real part of a holomorphic 1-form in an appropriate complex structure. To see this, take a 1-form $\omega_1 = *\omega_0$, where $*$ is the Hodge operator. Foliations defined by closed 1-forms ω_0 and ω_1 form a pair of transversal measured foliations. Thus they define a complex structure on the Riemann surface; the 1-form $\omega_0 + i\omega_1$ will be holomorphic in this complex structure.

The following theorem can be regarded as a dual formulation of the theorems mentioned above.

THEOREM 1. *An orientable measured foliation on a closed Riemann surface is a horizontal foliation of a holomorphic differential in some complex structure if and only if no cycle obtained as the union of closed paths following a sequence of saddle connections in the positive direction is homologous to zero.*

Theorem 1 is proved in [**33**].

The observations above show that the interesting dynamics of "generic" foliations defined by closed 1-forms on Riemann surfaces is described by closed 1-forms that are the real parts of Abelian differentials in some complex structure on the Riemann surface. The moduli space of Abelian differentials is already a finite-dimensional variety (orbifold), so what we gain by this construction is that now the dynamics is described by a certain finite-dimensional space of parameters.

2.3. Teichmüller geodesic flow. We briefly recall the basic facts concerning Teichmüller geodesic flow (see [**38, 62, 63**]). The moduli space of holomorphic quadratic differentials may be regarded as the total space of the cotangent bundle over the moduli space of complex structures on a Riemann surface of genus g. Basically, the Teichmüller geodesic flow is the geodesic flow on the moduli space of quadratic differentials with respect to the Teichmüller metric on the moduli space of complex structures. ("Basically" because the Teichmüller metric is not a Riemannian, but a Finsler metric.) More rigorously, the Teichmüller geodesic flow is defined as follows. There is a natural action of $SL(2,\mathbb{R})$ on the moduli space of quadratic differentials. The action of the diagonal subgroup generates the Teichmüller geodesic flow.

The moduli space of holomorphic 1-forms (Abelian differentials) can be regarded as a subvariety of the moduli space of quadratic differentials: to an Abelian differential $f(z)dz$ one associates a quadratic differential $f^2(z)(dz)^2$. The moduli space of Abelian differentials on a closed complex curve of genus g is naturally stratified by degrees of zeros of the Abelian differential.

The number of zeros of an Abelian differential (counting multiplicities) on a complex curve of genus g equals $2g-2$. Thus the strata are enumerated by unordered partitions $(k_1,\dots,k_s)$, where $k_1+\dots+k_s = 2g-2$ and $k_i \in \mathbb{N}$. For example, there are only two possibilities for $g=2$. Here $2g-2=2$, so either an Abelian differential has two simple zeros (partition $(1,1)$), or it has one zero of degree 2 (partition (2)). We denote the strata of Abelian differentials by $\mathcal{H}(k_1,\dots,k_s)$.

There is a natural function $A\colon \mathcal{H}(k_1,\dots,k_s) \to \mathbb{R}_+$ given by

$$A(\omega) = \frac{1}{2i}\int_{M_g^2} \omega \wedge \bar{\omega}.$$

The subvariety of Abelian differentials, all the strata, and the function A are invariant under the action of the Teichmüller geodesic flow. It was proved by Masur [**38**] and Veech [**59**] that the Teichmüller geodesic flow is ergodic on the "unit sphere" $A=1$ of each connected component of each stratum with respect to some natural finite measure.

The following proposition is widely known in the folklore (it can be extracted from a combination of [**59, 62, 63**], or from [**56**]; it can be also obtained by combining results from [**63**] and [**70**]).

PROPOSITION 1. *Consider a connected component of a stratum $\mathcal{H}(k_1,\dots,k_s)$ in the moduli space of Abelian differentials. Let $g = k_1+\dots+k_s$ be the genus of* *the surface. The collection of Lyapunov exponents of the Teichmüller geodesic flow on a connected component of* $\mathcal{H}(k_1, \dots, k_s)$ *has the following form:*

$$-2 < -(1+\nu_2) \leqslant -(1+\nu_3) \leqslant \dots \leqslant -(1+\nu_g) \leqslant \underbrace{-1 = \dots = -1}_{s-1}$$

$$\leqslant -(1-\nu_g) \leqslant \dots \leqslant -(1-\nu_2) < 0 < (1-\nu_2) \leqslant \dots \leqslant (1-\nu_g)$$

$$\leqslant \underbrace{1 = \dots = 1}_{s-1} \leqslant (1+\nu_g) \leqslant (1+\nu_{g-1}) \leqslant \dots \leqslant (1+\nu_2) < 2.$$

The numbers $0 \leqslant \nu_g \leqslant \dots \leqslant \nu_2 < 1$ *depend only on the connected component of the stratum* $\mathcal{H}(k_1, \dots, k_s)$.

3. Formulation of results

Consider a smooth closed orientable surface M_g^2 of genus g. Choose a smooth nondegenerate Riemannian metric $g_{ij}(x)$ on M_g^2. For any two points $P_0, P_1 \in M_g^2$, define a path $\rho(P_0, P_1) \subset M_g^2$ joining them. We do not assume that $\rho(P_0, P_1)$ depends continuously on the parameters P_0 and P_1, but we assume that the lengths of the paths (in terms of the metric g_{ij}) are uniformly bounded,

$$\sup_{P_0, P_1 \in M_g^2} \operatorname{length}(\rho(P_0, P_1)) = \text{const} < \infty. \tag{2}$$

For example, we can define $\rho(P_0, P_1)$ as the shortest geodesic joining P_0 and P_1. We shall make another choice of such a family of paths later on. Since M_g^2 is compact, it is easy to see that all choices satisfying (2) are equivalent for our purposes.

Take some leaf γ of ω and choose a compact connected piece of it. Since we assume that the orientation of the surface is fixed, the leaves are oriented; so let P_0 be the starting point of our piece of leaf, and let P_1 be the endpoint. Let l be the length of the piece of leaf γ bounded by the points P_0, P_1. By $c_{P_0}(l) \in H_1(M_g^2; \mathbb{R})$ we shall denote the homology class of the closed loop obtained by completion of the path from P_0 to P_1 along the leaf γ with the path $\rho(P_1, P_0)$.

THEOREM 2. *For almost all Abelian differentials* ω *in any connected component of any stratum* $\mathcal{H}(k_1, \dots, k_s)$ *and for the foliation defined by the closed* 1*-form* $\omega_0 = \operatorname{Re}(\omega)$, *there exists a flag of subspaces (depending only on* ω_0)

$$V_1 \subset V_2 \subseteq \dots \subseteq V_g \subseteq V \subset H_1(M_g^2; \mathbb{R})$$

with the following properties.

For any leaf γ *and any point* $P_0 \in \gamma$ *we have*

$$\lim_{l \to \infty} \frac{c_{P_0}(l)}{l} = c,$$

where the nonzero asymptotic cycle $c \in H_1(M_g^2; \mathbb{R})$ *is proportional to the cycle Poincaré dual to the cohomology class of* ω_0. *The one-dimensional subspace* V_1 *is spanned by* c.

For any $\phi \in \operatorname{Ann}(V_j) \subset H^1(M_g^2; \mathbb{R})$, $\phi \notin \operatorname{Ann}(V_{j+1})$, *any leaf* γ, *and any point* $P_0 \in \gamma$ *we have*

$$\limsup_{l \to \infty} \frac{\log |\langle \phi, c_{P_0}(l) \rangle|}{\log l} = \nu_{i+1} \quad \text{for } i = 1, \dots, g-1.$$

For any $\phi \in \mathrm{Ann}(V_g) \subset H^1(M_g^2;\mathbb{R})$, $\phi \notin \mathrm{Ann}(V)$, *any leaf* γ, *and any point* $P_0 \in \gamma$ *we have*

$$\limsup_{l\to\infty} \frac{\log|\langle\phi, c_{P_0}(l)\rangle|}{\log l} = 0.$$

For any $\phi \in \mathrm{Ann}(V) \subset H^1(M_g^2;\mathbb{R})$, $\|\phi\| = 1$, *any leaf* γ, *any point* $P_0 \in \gamma$, *and any length* l *we have*

$$|\langle\phi, c_{P_0}(l)\rangle| \leqslant \mathrm{const},$$

where the constant depends only on the foliation and on the choice of the norm in the cohomology.

All the above limits converge uniformly with respect to γ *and* $P_0 \in \gamma$, *i.e., their convergence depends only on* l.

The numbers $2, 1+\nu_2, \dots, 1+\nu_g$ *are the top* g *Lyapunov exponents of the Teichmüller geodesic flow on the corresponding connected component of the stratum* $\mathcal{H}(k_1,\dots,k_s)$ (*see Proposition* 1).

CONJECTURE 1. *For any connected component of any stratum of the Abelian differentials, all the Lyapunov exponents of the Teichmüller geodesic flow except the one corresponding to the tangential direction to the flow are nonzero*:

$$\nu_g > 0.$$

CONDITIONAL THEOREM 3. *Conjecture* 1 *implies that the subspaces* V_g *and* V *in Theorem* 2 *coincide. Moreover,* $V_g = V \subset H_1(M_g^2;\mathbb{R})$ *is a Lagrangian subspace in homology, where the symplectic structure is determined by the intersection form.*

CONJECTURE 2. *For any connected component of any stratum of Abelian differentials, the top* g *Lyapunov exponents of the Teichmüller geodesic flow are distinct and strictly greater than* 1:

$$1 > \nu_2 > \cdots > \nu_g > 0.$$

In other words, all Lyapunov exponents except the trivial ones occur with multiplicity one.

CONDITIONAL THEOREM 4. *Conjecture* 2 *implies that the flag*

$$V_1 \subset V_2 \subset \cdots \subset V_g = V \subset H_1(M_g^2;\mathbb{R})$$

from Theorem 2 *is a complete flag of subspaces in the Lagrangian subspace* V_g.

4. Asymptotic flags determined by interval exchange transformations

4.1. Interval exchange transformations. Recall the notion of an interval exchange transformation (see [**28**]). Consider an interval X, and cut it into m subintervals of lengths $\lambda_1,\dots,\lambda_m$. Now glue the subintervals together in another order, according to some permutation $\pi \in \mathfrak{S}_m$ and preserving the orientation. We again obtain an interval X of the same length, and hence we get a mapping $T\colon X \to X$, which is called an *interval exchange transformation.* Our mapping is piecewise linear, and it preserves the orientation and Lebesgue measure. It is singular at the points of cuts, unless two consecutive intervals separated by a point of a cut are mapped to consecutive intervals in the image.

REMARK 2. The study of interval exchanges was proposed by Arnold as an interesting problem back in the early sixties (for a particular case, see the section "Unsolved problems" in [**6**]).

An interval exchange transformation T is completely determined by a pair (λ, π), $\lambda \in \mathbb{R}_+^m$, $\pi \in \mathfrak{S}_m$. Let $\beta_0 = 0$, $\beta_i = \sum_{j=1}^{i} \lambda_j$, and $X_i = [\beta_{i-1}, \beta_i[$, so that $X = X_1 \sqcup \cdots \sqcup X_m$. Define a skew-symmetric $m \times m$ matrix by setting

$$\Omega_{ij}(\pi) = \begin{cases} 1 & \text{if } i < j \text{ and } \pi(i) > \pi(j), \\ -1 & \text{if } i > j \text{ and } \pi(i) < \pi(j), \\ 0 & \text{otherwise.} \end{cases} \tag{3}$$

Consider the translation vector $\tau = \Omega(\pi)\lambda$. Our interval exchange transformation T is defined as follows:

$$T(x) = x + \tau_i \quad \text{for } x \in X_i,\ 1 \leqslant i \leqslant m.$$

Note that if for some $k < m$ we have $\pi\{1, \dots, k\} = \{1, \dots, k\}$, then the map T decomposes into two interval exchange transformations. We consider only the class $\mathfrak{S}_m^0$ of *irreducible* permutations, those which have no invariant subsets of the form $\{1, \dots, k\}$, where $1 \leqslant k < m$.

Two interval exchange transformations sharing the same permutation π and having proportional vectors of lengths of subintervals are obviously equivalent. Thus in speaking about the *space of interval exchange transformations* it is natural to normalize the length of the interval to one: $\|\lambda\| = \lambda_1 + \cdots + \lambda_m = 1$. We can identify the space of all interval exchange transformations with the product $\Delta^{m-1} \times \mathfrak{S}_m^0$ of the standard $(m-1)$-dimensional simplex $\Delta^{m-1} = \{\lambda \in \mathbb{R}_+^m \mid \|\lambda\| = 1\}$ by the set $\mathfrak{S}_m^0$ of irreducible permutations.

4.2. Interval exchange transformations versus measured foliations. We are considering now only those measured foliations which are defined by a closed 1-form $\mathrm{Re}(\omega)$, where ω is an Abelian differential from a fixed stratum $\mathcal{H}(k_1, \dots, k_s)$. A generic foliation of this type is minimal, i.e., every leaf is dense on the Riemann surface. Taking a transverse interval X to the foliation, we get the first return map $T\colon X \to X$. The map T is an interval exchange transformation.

REMARK 3. The permutation π corresponding to the interval exchange transformation preserves all the topological information concerning the Abelian differential ω, i.e., it uniquely determines the corresponding stratum $\mathcal{H}(k_1, \dots, k_s)$, and even the connected component of the stratum.

In Section 3 we considered families of cycles obtained as follows: We chose some point P_0, then we took a piece of the leaf $\gamma_{P_0,P_1} = \gamma_{P_0}(l)$ of length l passing through P_0 in positive direction, and joined the endpoints P_0 and P_1 by the shortest geodesic to obtain a cycle $c_{P_0}(l)$. Let us consider a sequence of similar cycles related to a transversal interval X.

For every $x \in X$, consider a piece of leaf which has x as a starting point, and $T^{N-1}(x)$ as an endpoint. In other words, emit a leaf γ from the point x in the positive direction and follow it until it intersects X exactly N times (we are counting x as an intersection). Now join the endpoints of $\gamma_{x,T^{N-1}(x)}$ along X. We get a closed path; let $c_x(N)$ be the corresponding cycle in the first homology.

It seems clear that the asymptotic properties of $c_{P_0}(l)$ as $l \to \infty$ and of $c_x(N)$ as $N \to \infty$ are the same. To prove this rigorously, we must find the relationship between the discrete parameter N and the continuous parameter l. In other words, we need to know how many intersections with X a piece of leaf of length l possesses, and vice versa: we need to know what is the length of a piece of leaf obtained after

N returns to the transversal X. The answer is given by an ergodic theorem applied to the interval exchange transformation $T\colon X \to X$.

For every point P on M_g^2, define a point $x(P) \in X$ on the transversal interval X as the point of the first intersection of the leaf emitted from P in the *negative* direction with the transversal X. (The definition can be easily extended to the points P on critical leaves by some convention, see [**72**].) Having a piece of leaf $\gamma_{P_0}(l)$ of length l bounded by the points P_0, P_1, consider a piece of the same leaf having $x(P_0)$ and $x(P_1)$ as the starting point and endpoint respectively. Now the endpoints of this modified piece of leaf are already on the interval X. Let $N = N(P_0, l)$ be the number of intersections of $\gamma_{P_0}(l)$ with X, i.e., let $x(P_1) = T^{N-1}(x(P_0))$.

We formulate the following obvious result.

LEMMA 1. *The difference between the cycles $c_{P_0}(l)$ and $c_x(N)$ is uniformly bounded with respect to $P_0 \in M_g^2$, and $l \in \mathbb{R}_+$:*

$$\|c_{P_0}(l) - c_x(N)\| \leqslant \text{const},$$

where $x = x(P_0)$, $N = N(P_0, l)$.

To compare the parametrization of cycles by the length l and by the number N of returns to a transverse interval, consider the following function $l(x)$ on X. Let γ be the leaf passing through the point $x \in X$. Consider the piece $\gamma_{x,Tx}$ of γ between the points x and $T(x)$; let

$$l(x) := l(\gamma_{x,Tx}) \tag{4}$$

be its length. The function $l(x)$ is continuous on every subinterval X_i, so it is bounded:

$$0 < l_{\min} \leqslant l(x) \leqslant l_{\max} < \infty \qquad \text{for all } x \in X.$$

Recall that almost all interval exchange transformations are uniquely ergodic with respect to the Lebesgue measure on the interval (see [**38, 59**]). By $\bar{l}$ we denote the ergodic mean of $l(x)$:

$$\lim_{K\to\infty} \frac{1}{N} \sum_{K=0}^{N-1} l(T^K(x)) = \frac{1}{|X|} \int_X l(x)\, dx =: \bar{l}. \tag{5}$$

Here $|X| = \int_X \omega_0$.

REMARK 4. The statement of the ergodic theorem can be slightly strengthened for generic interval exchange transformations. It follows from [**71**] that the limits (6) and (5) converge for all points $x \in X$; moreover, the convergence is uniform.

We now see that the parametrization of cycles by the length l and by the number N of returns to a transverse interval are equivalent: $l(x, N) \sim N\bar{l}$. Taking Lemma 1 into consideration, we conclude that all asymptotic properties of the families of cycles $c_P(l)$ and of sequences of cycles $c_x(N)$ are the same. Now let us show how the asymptotic properties of $c_x(N)$ are determined by the dynamics of the interval exchange transformation $T\colon X \to X$.

Consider the function $c(x) := c_x(1)$ on X with values in the first homology group $H_1(M_g^2; \mathbb{R})$. In other words, $c(x)$ is obtained as follows: we emit a leaf γ from $x \in X$ in the positive direction, wait until it hits X for the first time (by definition it hits X at $T(x)$) and then we join x and $T(x)$ along X. Note that

$c(x)$ is constant on every subinterval X_j; we denote the corresponding values by $c_j \in H_1(M_g^2;\mathbb{R})$.

It is easy to see that

$$c_x(N) = \sum_{K=0}^{N-1} c(T^K(x)).$$

Applying the ergodic theorem to the function $c(x)$, we obtain

$$\lim_{N\to\infty} \frac{c_x(N)}{N} = \lim_{N\to\infty} \frac{1}{N} \sum_{K=0}^{N-1} c(T^K(x)) = \frac{1}{|X|}\sum_{j=1}^{m} \lambda_j c_j =: c \tag{6}$$

for almost all $x \in X$ (actually, for *all* $x \in X$; see Remark 4). Here $\lambda_j = |X_j|$ are the "lengths" of the subintervals under the exchange measured by ω_0:

$$|X_j| = \int_{X_j} \omega_0.$$

The ergodic mean c in (6) is called the *asymptotic cycle* c (see [**53**]). We formulate the following elementary lemma, which can actually be extracted from [**53**].

LEMMA 2. *The asymptotic cycle c is proportional to the cycle Poincaré dual to the cohomology class of ω_0,*

$$c = \frac{1}{|X|} \cdot D[\omega_0].$$

PROOF. The cycles c_i span the homology group $H_1(M_g^2, \mathbb{R})$ (see [**72**]). Thus it is sufficient to show that for every i, $1 \leqslant i \leqslant m$, the cycle $|X| \cdot c$ has the proper intersection number with c_i. It is easy to see that the intersection index of the cycles c_i, c_j is given by the corresponding entry of the matrix Ω (see (3)): $c_i \circ c_j = \Omega_{ij}$. Thus

$$c_i \circ (|X| c) = c_i \circ \left(\sum_{j=1}^{m} \lambda_j c_j\right) = \sum_{j=1}^{m} \lambda_j c_i \circ c_j = \sum_{j=1}^{m} \lambda_j \Omega_{ij} = \tau_i = \int_{c_i} \omega_0.$$

□

REMARK 5. The above lemma gives us the coefficient of proportionality between the asymptotic cycle in Theorem 2 and $D[\omega_0]$:

$$\lim_{l\to\infty} \frac{c_{P_0}(l)}{l} = \lim_{N\to\infty} \frac{c_{x(P_0)}(l(N))}{\bar{l}N} = \frac{1}{\bar{l}} \lim_{N\to\infty} \frac{c_{x(P_0)}(N)}{N} = \frac{c}{\bar{l}} = \frac{D[\omega_0]}{\bar{l}|X|}.$$

Let the length l be measured in the flat metric determined by the Abelian differential ω. The area of the Riemann surface measured in this flat metric equals $\bar{l}|X|$. Suppose that M_g^2 has unit area in this metric, i.e., suppose that ω is normalized by $\frac{1}{2i}\int \omega \wedge \bar{\omega} = 1$. Then we have

$$\lim_{l\to\infty} \frac{c_{P_0}(l)}{l} = D[\omega_0].$$

This calculation shows that as soon as we get any information concerning the asymptotic dynamics of the interval exchange transformation $T\colon X \to X$, we immediately get information on the asymptotic dynamics of the foliation.

4.3. The Euclid algorithm as a renormalization procedure. We need to study very long pieces of leaves of the foliation. In order to make a leaf γ wind for a very long time before the first return to X, we must choose a very short transversal X; this makes the cycles c_j corresponding to the interval exchange transformation on X very long. To implement this idea, we consider a certain procedure for shortening the transversal X, which allows us to trace modifications of c_j while passing from X to a shorter transversal. To give an idea of this procedure, we consider it for the elementary case in which the Riemann surface is a torus, the foliation is a standard irrational foliation, and the initial transversal X is closed. In this case the first return map $T\colon X \to X$ is just a rotation of a circle.

Consider the rotation of the circle $T\colon S^1 \to S^1$ by the angle α. Let the length of the circle be normalized to one. Consider the trajectory $x, Tx, T^2x, \dots$ of a point x (see Figure 8). Denote the length of the arc (x, Tx) by $\lambda = \alpha/(2\pi)$.

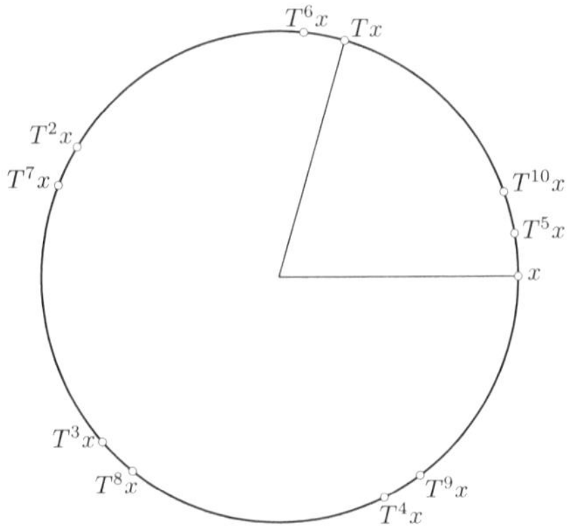

FIGURE 8

Cutting the circle at the point x, we get the interval X; the rotation of the circle generates a map of X to itself, which we denote by the same symbol $T\colon X \to X$. The map acts on X as follows: cut the unit interval X into two pieces of lengths $1-\lambda$ and λ respectively; shift the left piece up to the right endpoint of X and the right piece up to the left endpoint. The map thus obtained is the *interval exchange transformation* of the two subintervals.

Let us study how the points of the trajectory $x, Tx, T^2x, \dots$ accumulate near x, say to the left from x. Since we are not interested in the points that are far from x, we may register how the points of the trajectory visit the arc $X^{(1)} = (x, Tx)$. For the rotation presented in Figure 8, the first point of the trajectory that gets back to this arc is T^5x. Let $\lambda^{(1)} = \{1/\lambda\}$, where by $\{y\}$ we denote the fractional part of $y \in \mathbb{R}$. The length of the arc (x, T^5x) is equal to $\lambda^{(1)}\lambda$.

The next point of the trajectory $x, Tx, T^2x, \dots$ that gets to the arc $X^{(1)} = (x, Tx)$ is the point $T^{10}x$ (see Figure 8); the length of the arc $(T^5x, T^{10}x)$ is the same as the length of the arc (x, T^5x). Moreover, consider the *first return map* $T^{(1)}$ of the arc $X^{(1)}$ to itself. The image of a point $x' \in X^{(1)}$ is defined as the first point of the trajectory $Tx', T^2x', \dots$ that comes back to the arc $X^{(1)}$. If we identify the endpoints of $X^{(1)}$, this map gives us a new rotation of this new circle. If we consider the map $T_{(1)}$ as a map of $X^{(1)}$ to itself, we again obtain an interval exchange transformation of two subintervals. The length of the whole interval $X^{(1)}$ is equal to λ; the lengths of subintervals are equal to $\lambda(1-\lambda^{(1)})$ and $\lambda\lambda^{(1)}$.

Choosing the points of the trajectory $x, Tx, T^2x, \dots$ closest to x from the right, we get a sequence of points at the distances $\lambda, \lambda^{(1)}\lambda, \lambda^{(2)}\lambda^{(1)}\lambda, \dots$ from x. We can also go on with the procedure of confinement to smaller and smaller arcs of corresponding lengths. This procedure gives us not only the closest points, but a full description of how the points of the trajectory appear near the point x. The "procedure" associates to a rotation of the circle (or, what is almost the same, to an interval exchange transformation of two subintervals) a new rotation (a new interval exchange transformation). If we rescale the new interval to length one, then, starting with an exchange of two subintervals of the lengths λ and $1-\lambda$, we get an exchange of two subintervals of lengths $\{1/\lambda\}$ and $1-\{1/\lambda\}$. One can recognize the Euclid algorithm in our "confine to a smaller arc and rescale" procedure. The map

$$g\colon \lambda \mapsto \{1/\lambda\} \tag{7}$$

can be regarded as a map from "the space of rotations" to itself, or what is the same, a map from "the space of interval exchange transformations of two subintervals" to itself. The map g is ergodic with respect to the invariant probability measure

$$d\mu = \frac{1}{\log 2}\frac{d\lambda}{(\lambda+1)}, \tag{8}$$

which is called the *Gauss measure*.

4.4. Renormalization procedure for interval exchange transformations. Similarly to the case of interval exchange transformation of two subintervals, one can construct a renormalization procedure for interval exchanges of m subintervals. The particular renormalization procedure that we use is based on *Rauzy induction* [**52, 59**]. The rigorous definition of this procedure is presented in [**70**]. Below we list only those of its properties that we use in the present paper. The details and the proofs are contained in [**70**] and [**71**].

We assign to a given interval exchange transformation T corresponding to a pair (λ, π) a special subinterval $X^{(1)} \subset X$. Consider the induced map $T^{(1)} = T|_{X^{(1)}}$ of this subinterval to itself. The map $T^{(1)}$ is again an interval exchange transformation. Moreover, under a special choice of the subinterval $X^{(1)} \subset X$, we get an interval exchange transformation $T^{(1)}$ of the same number m of subintervals $X_1^{(1)}, \dots, X_m^{(1)}$.

For a point $x \in X_j^{(1)}$ in the "new" subinterval $X_j^{(1)}$, define

$$\begin{aligned} B_{ij} = {} & \text{number of visits of the trajectory } x, Tx, \dots, T^{l-1}(x)\\ & \text{to the "old" subinterval } X_i \text{ before the first return}\\ & T^l(x) \in X^{(1)} \text{ to the "new" interval } X^{(1)}. \end{aligned} \tag{9}$$

We choose the subinterval $X^{(1)} \subset X$ in such a way that for any pair $1 \leqslant i,j \leqslant m$ the number B_{ij} is the same for all $x \in X_j^{(1)}$. Moreover, the vector of lengths $\lambda_1^{(1)}, \dots, \lambda_m^{(1)}$ of subintervals $X_1^{(1)}, \dots, X_m^{(1)}$ is expressed in terms of the vector of lengths $\lambda_1, \dots, \lambda_m$ as

$$\lambda^{(1)} = B^{-1}\lambda. \tag{10}$$

The permutation $\pi^{(1)}$ corresponding to the induced interval exchange transformation $T^{(1)}\colon X^{(1)} \to X^{(1)}$ is always irreducible in our induction procedure, provided that the initial permutation π is irreducible, $\pi \in \mathfrak{S}_m^0$. Rescaling the interval $X^{(1)}$ proportionally to unit length, we obtain a map

$$\mathcal{G}\colon \Delta^{m-1} \times \mathfrak{S}_m^0 \to \Delta^{m-1} \times \mathfrak{S}_m^0$$

on the space of interval exchange transformations. Actually the set $\mathfrak{S}_m^0$ of all irreducible permutations decomposes into subsets invariant under the map $\mathcal{G}$; these subsets are called *Rauzy classes* (see [**52, 59**]).

It was proved in [**70**] that for every Rauzy class $\mathfrak{R}$ the map $\mathcal{G}$ is ergodic on $\Delta^{m-1} \times \mathfrak{R}$ with respect to an absolutely continuous invariant probability measure μ. The matrix-valued function $B(\lambda, \pi)$ defined by (9) determines two *measurable cocycles* on $\Delta^{m-1} \times \mathfrak{R}$ with respect to this measure, i.e.,

$$\int \log^+ \|B^{-1}\|\, d\mu \quad \text{and} \quad \int \log^+ \|{}^tB\|\, d\mu$$

are both finite. Here and below we denote by ${}^tA$ the transpose of the matrix A.

The map $\mathcal{G}$ is analogous to the map g defined by (7) (though they do not literally coincide for $m = 2$). Thus the map $\mathcal{G}$ represents a sort of multidimensional Euclid algorithm, while the matrices $B(\lambda, \pi), B(\mathcal{G}(\lambda, \pi)), \dots, B(\mathcal{G}^{(k-1)}(\lambda, \pi))$ play the role of the entries of the continued fraction expansion of a real number.

REMARK 6. In the matrix representation of a continued fraction expansion, one has elementary matrices of the group $SL(2, \mathbb{Z})$. Numerous multidimensional generalizations of continued fraction algorithms use matrices in $SL(n, \mathbb{Z})$. In dimension 2, the groups SL and Sp coincide, so the other way to generalize the continued fraction algorithm is to use matrices in $Sp(n, \mathbb{Z})$. One can think of the map $\mathcal{G}$ as a generalization of this type.

Continued fractions are closely related to the geodesic flow on the upper half-plane (see, say, [**11**] for a very nice exposition of this relationship). Our "*Sp*-generalization" is closely related to the Teichmüller geodesic flow (which can be considered as a generalization of the geodesic flow on the upper half-plane to genera greater than 1).

REMARK 7. The induction procedure and the renormalization map $\mathcal{G}$ constructed in [**70**] are obtained as a modification (a "speed-up") of the Rauzy induction and of the corresponding map $\mathcal{T}$ in [**59**]. In particular, the ergodicity of $\mathcal{G}$ is proved closely following the original proof of W. Veech. The relationship between the maps $\mathcal{T}$ and $\mathcal{G}$ is similar to the relation between *additive* and *multiplicative* continued fraction algorithms in [**11**]; in particular, the invariant measure corresponding to the map $\mathcal{T}$ is infinite, which was the reason for constructing the modified map $\mathcal{G}$.

4.5. Properties of the "continued fraction" cocycle. The matrix-valued function $B(\lambda, \pi)$ defined by (9) determines two *measurable cocycles* on $\Delta^{m-1} \times \mathfrak{R}$ with respect to the measure μ, i.e.,

$$\int \log^+ \|B^{-1}\|\, d\mu \quad \text{and} \quad \int \log^+ \|{}^tB\|\, d\mu$$

are both finite. Here and below we denote by ${}^tA$ the transpose of the matrix A. These cocycles are dual to each other. They play a crucial role in our study of interval exchange transformations. In this section we briefly recall their main properties. One can find the proofs in [**70, 71**].

Let

$$(\lambda^{(k)}, \pi^{(k)}) = \mathcal{G}^k(\lambda, \pi),$$
$$B^{(k)}(\lambda, \pi) = B(\lambda, \pi) B(\lambda^{(1)}, \pi^{(1)}) \cdots B(\lambda^{(k-1)}, \pi^{(k-1)}).$$

The cocycles preserve the degenerate symplectic form (3) in the following sense:

$$\Omega(\pi) = ({}^t B^{(k)}(\lambda, \pi))^{-1} \Omega(\pi^{(k)}) (B^{(k)}(\lambda, \pi))^{-1}. \tag{11}$$

As we mentioned above, the map $\mathcal{G}$ is ergodic on $\Delta^{m-1} \times \mathfrak{R}$ with respect to the absolutely continuous invariant probability measure μ similar to the Gauss measure (8). The cocycles ${}^t B(\lambda, \pi)$ and $B^{-1}(\lambda, \pi)$ have the following spectrum of Lyapunov exponents, which is the same for both cocycles:

$$\theta_1 > \theta_2 \geqslant \theta_3 \geqslant \cdots \geqslant \theta_g \geqslant \underbrace{0 = \cdots = 0}_{m-2g} \geqslant -\theta_g \geqslant \cdots \geqslant -\theta_3 \geqslant -\theta_2 > -\theta_1.$$

Here g is an integer determined by the Rauzy class $\mathfrak{R}$; topologically g is the genus of the corresponding surface.

By

$$\mathcal{H}_1(\lambda, \pi) \subset \mathcal{H}_2(\lambda, \pi) \subseteq \cdots \subseteq \mathcal{H}_g(\lambda, \pi) \subseteq \mathcal{H}(\lambda, \pi) \subset \mathbb{R}^m \tag{12}$$

we denote the corresponding flag of subspaces in $\mathbb{R}^m$ determined by the cocycle $B^{-1}(\lambda, \pi)$. This flag is defined for μ-almost all (λ, π). One has

$$\lim_{k\to\infty} \frac{\log \|(B^{(k)}(\lambda, \pi))^{-1} v\|}{k} = -\theta_j, \qquad \forall\, v \in \mathcal{H}_j,\ v \notin \mathcal{H}_{j-1},$$
$$\lim_{k\to\infty} \frac{\log \|(B^{(k)}(\lambda, \pi))^{-1} v\|}{k} > 0, \qquad \forall\, v \notin \mathcal{H}.$$

The flag of subspaces in $\mathbb{R}^{m*}$ corresponding to the dual cocycle ${}^t B(\lambda, \pi)$ is dual to the flag (12): for any linear function f in the annihilator of $\mathcal{H}_j$, $f \in \operatorname{Ann}(\mathcal{H}_j)$, such that $f \notin \operatorname{Ann}(\mathcal{H}_{j+1})$, one has

$$\lim_{k\to\infty} \frac{\log \|{}^t B^{(k)}(\lambda, \pi) f\|}{k} = \theta_{j+1}, \qquad \forall\, v \in \operatorname{Ann}(\mathcal{H}_j),\ v \notin \operatorname{Ann}(\mathcal{H}_{j+1}).$$
$$\lim_{k\to\infty} \frac{\log \|{}^t B^{(k)}(\lambda, \pi) f\|}{k} < 0, \qquad \forall\, v \in \operatorname{Ann}(\mathcal{H}).$$

The relationship between the map $\mathcal{G}$ and the Teichmüller geodesic flow is as follows (see [**59, 62, 63**]). One can consider a suspension over the space of (nonnormalized) interval exchange transformations, the space of *zipped rectangles*. One can think of a "zipped rectangle" as some way to cut the initial Riemann surface, provided with a flat structure determined by the initial Abelian differential, into rectangular pieces. Using some restrictions on the decomposition of the Riemann surface into rectangles, it is possible to define a fundamental domain in the space of zipped rectangles. This fundamental domain may be regarded as a finite (ramified) covering over a connected component of the stratum $\mathcal{H}(k_1, \ldots, k_s)$ in the moduli space of Abelian differentials. The Teichmüller geodesic flow has a very simple coordinate representation in this fundamental domain. The preimage of the subspace of interval exchange transformations of an interval of *unit* length determines a hypersurface in the fundamental domain. The Teichmüller geodesic flow determines the first return map on the hypersurface. The map $\mathcal{G}$ is the projection of this first return map. In this sense the map $\mathcal{G}$ is a discrete version of the Teichmüller geodesic flow.

Basically, this construction is quite visible even in the simplest case of genus 1. See [**11**] for a very clear presentation.

The Lyapunov exponents of the Teichmüller geodesic flow are expressed in terms of the Lyapunov exponents of the cocycles B^{-1} and ${}^tB$ as follows:

$$\nu_i = \theta_i/\theta_1 \quad \text{for } i = 1, \dots, g.$$

The connected components of the strata $\mathcal{H}(k_1, \dots, k_s)$ are in one-to-one correspondence with the *extended Rauzy classes* ([**63**]).

4.6. The function counting visits as an additive cocycle. Having an interval exchange transformation $T\colon X \to X$, one can consider two different sorts of "time" related to it. The usual "additive time" enumerates the iterations of the interval exchange transformation T, or, in the other words, it enumerates points of the trajectory $x, Tx, \dots, T^{N-1}x$.

There is also a "multiplicative" time, which enumerates the iterates of the induction procedure $\mathcal{G}$ on the space of interval exchange transformations.

In the case when there are only two intervals, any interval exchange transformation is equivalent to a rotation of the circle. In this case the "additive time" corresponds to rotations, while the "multiplicative time" corresponds to the Euclid algorithm.

Let us study the interaction between these two "times".

Let $T\colon X \to X$ be an interval exchange transformation. Let $X = X_1 \sqcup \dots \sqcup X_m$ be the corresponding partition of the interval X into subintervals under the exchange. For every index i, $1 \leqslant i \leqslant m$, define the counting function

$$S_i(X, T, x, N) := \sum_{L=0}^{N-1} \chi_{X_i}(T^L(x)),$$

where

$$\chi_Y(x) = \begin{cases} 1 & \text{if } x \in Y, \\ 0 & \text{if } x \notin Y. \end{cases}$$

Thus $S_i(X, T, x, N)$ is the number of visits of the trajectory $x, Tx, \dots, T^{N-1}x$ to the subinterval X_i. Usually we shall deal with the vector

$$S(X, T, x, N) := (S_1(X, T, x, N), \dots, S_m(X, T, x, N)).$$

Note that $S(X, T, x, N)$ is an *additive cocycle* with respect to discrete "additive time":

$$S(X, T, x, N_1 + N_2) = S(X, T, x, N_1) + S(X, T, T^{N_1}x, N_2). \tag{13}$$

4.7. Vector of visits as a multiplicative cocycle. Assume that the interval exchange transformation $T\colon X \to X$ is uniquely ergodic. According to the results in [**38**] and [**59**], this is the generic situation. Consider the subinterval $X^{(k)}$ obtained from X after k iterations of the induction procedure $\mathcal{G}$. The *induced map* $T_{(k)}\colon X^{(k)} \to X^{(k)}$ is defined for $x \in X^{(k)}$ as the first return of the trajectory $x, Tx, \dots$ to $X^{(k)}$. By construction, the map $T_{(k)}\colon X^{(k)} \to X^{(k)}$ induced by T on

$X^{(k)} \subset X$ is again an interval exchange transformation of the same number m of subintervals: $X^{(k)} = X_1^{(k)} \sqcup \cdots \sqcup X_m^{(k)}$.

Consider a finite piece of trajectory $x, Tx, \ldots, T^{N-1}x$ which starts and finishes at the interval $X^{(k)}$, i.e., $x \in X^{(k)}$ and $T^N x \in X^{(k)}$. Consider the consecutive visits of this piece of trajectory to $X^{(k)}$. By construction, they coincide with the trajectory $x, T_{(k)}x, T_{(k)}^2 x, \ldots, T_{(k)}^{N_{(k)}-1} x$ of the induced map $T_{(k)}$, where $T_{(k)}^{N_{(k)}} x = T^N x$. The counting functions $S(X,T,x,N)$ and $S(X^{(k)}, T_{(k)}, x, N_{(k)})$ corresponding to the pair of trajectories as above are related by means of the matrix $B^{(k)}$ representing the induction procedure:

$$S(X,T,x,N) = B^{(k)} \cdot S(X^{(k)}, T_{(k)}, x, N_{(k)}). \tag{14}$$

In this sense the counting function $S(X,T,x,N)$ has *multiplicative* properties with respect to the "time" k. Here the "multiplicative time" stands for the iterations of the induction procedure $\mathcal{G}$.

Of course the above relation is valid only for a very special choice of the point $x \in X$ (we assumed that $x \in X_{(k)}$), and for a very special choice of the "time" N (we assumed that $T^N x \in X_{(k)}$). Let us improve the formula to fit an arbitrary choice of x and of N. To do that, let us extend the trajectory until the extension hits the subinterval $X_{(k)}$ for the first time. Then we shall consider three parts of the extended trajectory. The first part is the part before the first visit to the subinterval $X^{(k)}$. The third part is the extension part. The second part can possibly overlap with the third one: it starts at the first visit to $X_{(k)}$ and continues until the end of the extended trajectory. By construction, the second part starts and finishes at $X_{(k)}$.

More formally, assume that the interval exchange transformation $T\colon X \to X$ is minimal, and that the first k steps of the induction procedure $\mathcal{G}$ are well defined for $T\colon X \to X$. Let

$$\begin{aligned} N^+(X,x,T,N,k) &:= \min_{0 \leqslant L} L \mid T^L(x) \in X^{(k)}, \\ N^-(X,x,T,N,k) &:= \min_{0 \leqslant L} L \mid T^{N+L}(x) \in X^{(k)}. \end{aligned} \tag{15}$$

We denote

$$x^{(k)} := T^{N^+}(x).$$

Using the additive properties (13) of $S(X,T,x,N)$ with respect to the "time" N, we get

$$\begin{aligned} S(X,T,x,N) = S(X,T,x,N^+) \\ + S(X,T,T^{N^+}x,(N - N^+ + N^-)) - S(X,T,T^N x, N^-). \end{aligned}$$

Using (14) for the middle term, we modify this relation to the following form:

$$\begin{aligned} S(X,T,x,N) = S(X,T,x,N^+) \\ - S(X,T,T^N x,N^-) + B^{(k)} S(X^{(k)}, T_{(k)}, x_{(k)}, N_{(k)}), \end{aligned} \tag{16}$$

where $x_{(k)} = T^{N^+}x$ and $T_{(k)}^{N_{(k)}} x_{(k)} = T^{N+N^-} x$.

4.8. Formulation of the main theorem in terms of interval exchanges. Now we are ready to reformulate the main result, Theorem 2, in terms of interval exchange transformations.

THEOREM 5. *For almost all interval exchange transformations* (λ, π) *corresponding to any extended Rauzy class* $\mathfrak{R}$, *the function* $S(X,T,x,N)$ *counting visits to the subintervals has the following properties.*

For any $x \in X$ *we have*

$$\lim_{N\to\infty} \frac{S(X,T,x,N)}{N} = \lambda.$$

The one-dimensional subspace $\mathcal{H}_1$ *is spanned by* λ.

For any $f \in \operatorname{Ann}(\mathcal{H}_i)$, $f \notin \operatorname{Ann}(\mathcal{H}_{i+1})$, $i = 1, \dots, g-1$, *and any* $x \in X$ *we have*

$$\limsup_{N\to+\infty} \frac{\log |\langle f, S(X,T,x,N)\rangle|}{\log N} = \frac{\theta_{i+1}}{\theta_1}.$$

For any $f \in \operatorname{Ann}(\mathcal{H}(\lambda,\pi))$, $\|f\| = 1$, *and any* $x \in X$, *we have*

$$|\langle f, S(X,T,x,N)\rangle| \leqslant \text{const},$$

where the constant does not depend on f, x *or* N.

All the above limits converge uniformly with respect to $x \in X$.

The numbers $2, 1+\theta_2/\theta_1, \dots, 1+\theta_g/\theta_1$ *are the top* g *Lyapunov exponents of the Teichmüller geodesic flow on the connected component of the stratum* $\mathcal{H}(k_1,\dots,k_s)$, *corresponding to the extended Rauzy class* $\mathfrak{R}$.

Taking into consideration the correspondence between the cycles $c_x(N)$ and $c_P(l)$, and the correspondence between two parametrizations (see 4.2), we see that Theorem 2 is an immediate corollary of Theorem 5.

4.9. Upper bound. In this subsection we prove the following result.

PROPOSITION 2. *For almost all interval exchange transformations the following upper bound is valid:*

$$\limsup_{N\to+\infty} \frac{\log |\langle f, S(X,T,x,N)\rangle|}{\log N} \leqslant \frac{\theta_{i+1}}{\theta_1}$$

for any $f \in \operatorname{Ann}(\mathcal{H}_i)$ *and any* $x \in X$.

To prove Proposition 2, we need the following lemma, which is part of Proposition 8 in [**71**]. Denote by e_j the vectors of the standard basis in $\mathbb{R}^m$.

LEMMA 3. *For almost all* $(\lambda,\pi) \in \Delta \times \mathfrak{R}$ *and for any* $\varepsilon > 0$, $\delta > 0$, $r \in \mathbb{N}$ *there exists an* $\tilde{n} = \tilde{n}((\lambda,\pi),\varepsilon,\delta,r)$ *such that for any* $n \geqslant \tilde{n}$ *one can choose a sequence of integers* $0 = n_0 < n_1 < \cdots < n_r = n$ *with the following properties: for any* $1 \leqslant l \leqslant r$

$$\max_{1\leqslant j\leqslant m} \left| \frac{\log \|B^{(n_l - n_{l-1})}(\lambda^{(n_{l-1})}, \pi^{(n_{l-1})}) \cdot e_j\|}{n_l - n_{l-1}} - \theta_1 \right| \leqslant \varepsilon.$$

Moreover, for any $0 \leqslant l \leqslant r$ *the number* n_l *is close to the corresponding entry of the arithmetic progression:*

$$\left| \frac{n_l}{n} - \frac{l}{r} \right| \leqslant \delta.$$

PROOF. One can find the detailed proof of this lemma in [**71**] (see Proposition 8 in [**71**]). Here we present only the idea of the proof, which is quite elementary.

According to [**71**], for almost all interval exchange transformations (λ,π) the columns of the matrix $B^{(k)}(\lambda,\pi)$ have the following asymptotic growth: for any $1\leqslant j\leqslant m$ we have

$$\limsup_{k\to+\infty}\frac{\log\|B^{(k)}(\lambda,\pi)\cdot e_j\|}{k}=\theta_1.$$

Of course the convergence is not uniform. Still, for any $\varepsilon>0$ we can choose an $\tilde n(\varepsilon)$ large enough so that the subset W of those interval exchange transformations for which

$$\left|\frac{\log\|B^{(k)}(\lambda,\pi)\cdot e_j\|}{k}-\theta_1\right|<\varepsilon\quad\text{as } k>\tilde n(\varepsilon)$$

is of measure arbitrary close to the complete measure. In particular, we can make the difference $1-\mu(W)$ between two measures much less than δ.

Now consider a generic interval exchange transformation (λ,π). Since the map $\mathcal{G}$ is ergodic, the trajectory $(\lambda,\pi),\mathcal{G}(\lambda,\pi),\dots,\mathcal{G}^{(k)}(\lambda,\pi)$ has only about $(1-\mu(W))k$ points outside the set W for k large enough, which by the choice of W is much less than δk. Hence, any interval of length δk placed inside the interval $[0,k]\subset\mathbb{R}$ (where n is assumed to be large enough) contains some integers $q_1,\dots$ such that $\mathcal{G}^{q_1}(\lambda,\pi),\dots\in W$. This enables us to choose n_l as desired. Lemma 3 is proved. □

Now let us prove Proposition 2.

PROOF. We start with the following definition, which is quite useful for us:

$$n(N):=\min_{k\geqslant 0}\left\{k\quad\text{such that}\quad\sum_{J=0}^{N-1}\chi_{X_{(k)}}(T^Jx)\leqslant 1\right\}. \tag{17}$$

In other words, the number $n(N)-1$ is the largest possible integer n such that the trajectory $x,Tx,\dots,T^{N-1}x$ visits the subinterval $X_{(n)}$ at least twice. The number $n(N)$ depends on the point $x\in X$; still, the following lemma shows that for large N the asymptotic behavior of $n(N)$ is the same for all $x\in X$.

LEMMA 4. *For almost all interval exchange transformations the number $n(N)$ obeys the following asymptotic relation:*

$$\lim_{N\to+\infty}\frac{\log N}{n(N)}=\theta_1.$$

The convergence is uniform with respect to the starting point $x\in X$.

PROOF. To prove the lemma, it is sufficient to prove that for almost all interval exchange transformations and for an arbitrary choice of $\varepsilon>0$ the number $n(N)$ possesses the following bounds:

$$\frac{\log N-\log 2}{\theta_1+\varepsilon}<n(N)\leqslant\frac{\log N}{\theta_1-\varepsilon}+1,$$

provided N is large enough.

According to [**71**], for almost all interval exchange transformations the following relation is valid:

$$\lim_{k\to+\infty}\frac{\log\|B^{(k)}e_j\|}{k}=\theta_1$$

for all $1 \leqslant j \leqslant m$. By $k(\varepsilon)$ denote an integer such that for all $k \geqslant k(\varepsilon)$ the above ratio differs from θ_1 at most by ε. Chose any $N \geqslant 2\exp((\theta_1+\varepsilon)k(\varepsilon))$.

Let us first prove the lower bound. Let

$$k = \left[\frac{\log(N/2)}{\theta_1+\varepsilon}\right].$$

By the choice of N, we have $k \geqslant k(\varepsilon)$. Hence

$$\max_{1\leqslant j\leqslant m} \|B^{(k)}(\lambda,\pi)\cdot e_j\| \leqslant \exp((\theta_1+\varepsilon)k).$$

Therefore,

$$\max_{1\leqslant j\leqslant m} \|B^{(k)}(\lambda,\pi)\cdot e_j\| \leqslant N/2.$$

We recall that the piece of trajectory $y, Ty, \dots$ of any point $y \in X_j^{(k)}$ before the first return to $X_j^{(k)}$ is of length $\|B^{(k)}(\lambda,\pi)\cdot e_j\|$, and any point $z \in X$ belongs to the corresponding piece of trajectory of some point $y(z) \in X^{(k)}$. Hence the above inequality implies that the trajectory of *any* point $x \in X$ of length N will visit the subinterval $X^{(k)}$ at least twice. Hence $k < n(N)$, which implies that

$$\frac{\log N - \log 2}{\theta_1+\varepsilon} < n(N).$$

The lower bound in Lemma 4 is proved.

Now let us prove the upper bound in Lemma 4. Denote

$$l = \left[\frac{\log N}{\theta_1-\varepsilon}\right] + 1.$$

By the choice of N we have $l \geqslant k(\varepsilon)$. Hence

$$\min_{1\leqslant j\leqslant m} \|B^{(l)}(\lambda,\pi)\cdot e_j\| \geqslant \exp((\theta_1-\varepsilon)l).$$

Therefore

$$\min_{1\leqslant j\leqslant m} \|B^{(l)}(\lambda,\pi)\cdot e_j\| \geqslant N.$$

Thus the above inequality implies that the trajectory of *any* point $x \in X$ of length N will visit the subinterval $X^{(l)}$ at most once. Hence $n(N) \leqslant l$, which implies that

$$n(N) \leqslant \frac{\log N}{\theta_1-\varepsilon} + 1.$$

The upper bound in Lemma 4 is proved. □

Now we are ready to prove Proposition 2. Let (λ,π) satisfy the conditions of Lemma 3. Let it be generic in the following sense as well: assume that for any $f \in \mathrm{Ann}(\mathcal{H}_i)$, $\|f\| = 1$, and for any n large enough, we have

$$\frac{\log \|{}^tB^{(n)}(\lambda,\pi)\cdot f\|}{n} \leqslant \theta_{i+1}+\varepsilon.$$

Finally, assume that Lemma 4 is also applicable to our (λ,π). Let $n = n(N)$. Choose $n_0 < n_1 < \cdots < n_r = n$ as in Lemma 3.

Let us apply (16):

$$S(X,T,x,N) = S(X,T,x,N^+) - S(X,T,T^Nx,N^-) \\ + B^{(n_1)}(\lambda,\pi)\cdot S(X^{(n_1)},T_{(n_1)},x_{(n_1)},N_{(n_1)}),$$

where $x_{(n_1)} = T^{N^+}x$ is the point of the first visit of the trajectory $x, Tx, T^2x, \dots$ to the subinterval $X_{(n_1)}$ and $T_{(n_1)}^{N_{(n_1)}}x_{(n_1)} = T^{N+N^-}x$ is the point of the first visit of the trajectory $T^N x, T^{N+1}x, T^{N+2}x, \dots$ to the subinterval $X_{(n_1)}$.

Now let us apply the same trick to $S(X^{(n_1)}, T_{(n_1)}, x_{(n_1)}, N_{(n_1)})$ by making $n_2 - n_1$ steps of the induction procedure $\mathcal{G}$. Note that the interval $(X^{(n_1)})^{(n_2-n_1)}$ obtained from $X^{(n_1)}$ by $n_2 - n_1$ steps of induction coincides with the interval $X^{(n_2)}$ obtained directly from X by $n_1 + (n_2 - n_1)$ steps of induction. The same is true for the interval exchange transformation: the interval exchange transformation $T_{(n_2)}$ obtained directly from T by n_2 steps of induction coincides with the interval exchange transformation $(T_{(n_1)})_{(n_2-n_1)}$ obtained by $(n_2 - n_1)$ steps of induction from $T_{(n_1)}$. The same is true for the point $x_{(n_2)} = (x_{(n_1)})_{(n_2-n_1)}$ and the time $N_{(n_2)} = (N_{(n_1)})_{(n_2-n_1)}$.

Hence

$$\begin{aligned} S(X^{(n_1)}, T_{(n_1)}, x_{(n_1)}, N_{(n_1)}) &= S(X^{(n_1)}, T_{(n_1)}, x_{(n_1)}, N^+_{(n_1)}) - S(X^{(n_1)}, T_{(n_1)}, T_{(n_1)}^{N_{(n_1)}} x_{(n_1)}, N^-_{(n_1)}) \\ &\quad + B^{(n_2-n_1)}(\lambda^{(n_1)}, \pi^{(n_1)}) \cdot S(X^{(n_2)}, T_{(n_2)}, x_{(n_2)}, N_{(n_2)}). \end{aligned}$$

Let us continue this procedure recursively by unfolding our expression up to the level r using the "times" $n_1, \dots, n_r$. To avoid complicated subscripts and long lists of arguments, let us denote the terms obtained at step l by $S^+_{(n_{l-1})}$, $S^-_{(n_{l-1})}$, and $S_{(n_l)}$ correspondingly. In this notation the recursion step is represented by

$$S_{(n_{l-1})} = S^+_{(n_{l-1})} - S^-_{(n_{l-1})} + B^{(n_l - n_{l-1})}(\lambda^{(n_l)}, \pi^{(n_l)}) \cdot S_{(n_l)}. \tag{18}$$

Consider the pairing of f with $S(X, T, x, N)$. Using the unfolded expression for $S(X, T, x, N)$, and taking the absolute value of the result, we obtain the following bound:

$$\begin{aligned} |\langle f, S(X,T,x,N)\rangle| &= |\langle f, S_{(n_0)}\rangle| \\ &\leqslant |\langle f, S^+_{(n_0)} - S^-_{(n_0)}\rangle| + |\langle f, B^{(n_1)}(\lambda, \pi) \cdot (S^+_{(n_1)} - S^-_{(n_1)})\rangle| \\ &\quad + \dots + |\langle f, B^{(n_{r-1})}(\lambda, \pi) \cdot (S^+_{(n_{r-1})} - S^-_{(n_{r-1})})\rangle| \\ &\quad + |\langle f, B^{(n_r)}(\lambda, \pi) \cdot S_{(n_r)}\rangle|, \end{aligned}$$

which we may continue as follows:

$$\begin{aligned} |\langle f, S(X,T,x,N)\rangle| &\leqslant \|f\| \cdot \|S^+_{(n_0)} - S^-_{(n_0)}\| + \|{}^tB^{(n_1)}(\lambda, \pi) f\| \cdot \|S^+_{(n_1)} - S^-_{(n_1)}\| \\ &\quad + \dots + \|{}^tB^{(n_{r-1})}(\lambda, \pi) f\| \cdot \|S^+_{(n_{r-1})} - S^-_{(n_{r-1})}\| \\ &\quad + \|{}^tB^{(n_r)}(\lambda, \pi) f\| \cdot \|S_{(n_r)}\|. \end{aligned}$$

Now note that for an arbitrary choice of X, T, x, and N, the norm $\|S(X,T,x,N)\|$ satisfies the following identity:

$$\|S(X,T,x,N)\| = N,$$

provided we use the norm

$$\|(v_1, \dots, v_m)\| = |v_1| + \dots + |v_m|.$$

In other words, the norm $\|S(X,T,x,N)\|$ equals the length of the corresponding piece of the trajectory.

Now note that $S^+_{(n_l)}$ corresponds to a piece of the trajectory of the map $T_{(l)}$ on $X^{(l)}$ which does not visit the subinterval $X^{(l+1)}$. Hence the length of this piece of trajectory is bounded from above by

$$\max_{1\leqslant j\leqslant m} \|B^{(n_{l+1}-n_l)}(\lambda^{(l)},\pi^{(l)})\cdot e_j\|.$$

Using our particular choice of n_l and the bound in Lemma 3, we obtain the following bounds:

$$\|S^+_{(n_l)}\| \leqslant \exp((\theta_1+\varepsilon)(n_{l+1}-n_l)) \leqslant \exp((\theta_1+\varepsilon)(1+\delta)n/r).$$

Similarly,

$$\|S^-_{(n_l)}\| \leqslant \exp((\theta_1+\varepsilon)(1+\delta)n/r).$$

Since the vectors $S^+_{(n_l)}$ and $S^-_{(n_l)}$ are positive, we finally get

$$\|S^+_{(n_l)} - S^-_{(n_l)}\| \leqslant \exp((\theta_1+\varepsilon)(1+\delta)n/r)$$

for any $0 \leqslant l \leqslant r-1$.

On the other hand, assuming n is large enough, we get the following bounds:

$$\|{}^tB^{(n_l)}(\lambda,\pi)f\| \leqslant \|f\|\exp((\theta_{i+1}+\varepsilon)n_l).$$

Finally, note that the choice of n as $n(N)$ implies the following bound:

$$\|S_{(n_r)}\| = \|S_{(n)}\| \leqslant 1.$$

Collecting these bounds together, we can develop the bound for $|\langle f, S(X,T,x,N)\rangle|$ as follows:

$$\begin{aligned}|\langle f, S(X,T,x,N)\rangle| \leqslant{}& \|f\|\cdot\exp((\theta_1+\varepsilon)(1+\delta)n/r)\\ &+\|f\|\exp((\theta_{i+1}+\varepsilon)n_1)\cdot\exp((\theta_1+\varepsilon)(1+\delta)n/r)\\ &+\cdots+\|f\|\exp((\theta_{i+1}+\varepsilon)n_{r-1})\cdot\exp((\theta_1+\varepsilon)(1+\delta)n/r)\\ &+\|f\|\exp((\theta_{i+1}+\varepsilon)n_r).\end{aligned}$$

Note that $0 \leqslant \theta_{i+1} < \theta_1$. Multiply the last term by $\exp((\theta_1+\varepsilon)(1+\delta)n/r)$; the product thus obtained dominates all the other terms. Using Lemma 4, passing to logarithms and taking the upper limit, we obtain the following bound:

$$\begin{aligned}&\limsup_{N\to+\infty}\frac{\log|\langle f, S(X,T,x,N)\rangle|}{\log N}\\ &\quad=\limsup_{N\to+\infty}\frac{\log|\langle f, S(X,T,x,N)\rangle|}{n(N)}\cdot\lim_{N\to+\infty}\frac{n(N)}{\log N}\\ &\quad\leqslant\limsup_{n\to+\infty}\frac{\log\left(\|f\|\cdot\exp((\theta_1+\varepsilon)(1+\delta)n/r)\cdot\exp((\theta_{i+1}+\varepsilon)n)\right)}{n}\cdot\frac{1}{\theta_1}\\ &\quad\leqslant\limsup_{n\to+\infty}\left(\frac{\log\|f\|}{n}+\frac{(\theta_1+\varepsilon)(1+\delta)}{r}+(\theta_{i+1}+\varepsilon)\right)\cdot\frac{1}{\theta_1}\\ &\quad=\frac{(\theta_1+\varepsilon)(1+\delta)}{r\theta_1}+\frac{\theta_{i+1}+\varepsilon}{\theta_1}.\end{aligned}$$

Taking into account the fact that we can choose ε and δ arbitrarily small, and r arbitrarily large, we arrive at the desired upper bound:

$$\limsup_{N\to+\infty}\frac{\log|\langle f, S(X,T,x,N)\rangle|}{\log N} \leqslant \frac{\theta_{i+1}}{\theta_1}.$$

Proposition 2 is proved. □

4.10. Lower bound. In this subsection we shall prove

PROPOSITION 3. *For generic interval exchange transformation* (λ, π) *the following lower bound is valid:*

$$\limsup_{N\to+\infty} \frac{\log|\langle f, S(X,T,x,N)\rangle|}{\log N} \geqslant \frac{\theta_{p+1}}{\theta_1}$$

for any $x \in X$ *and any* $f \in \operatorname{Ann}(\mathcal{H}_p(\lambda,\pi))$, $f \notin \operatorname{Ann}(\mathcal{H}_{p+1}(\lambda,\pi))$.

To prove Proposition 3, we shall use the following lemma (see Lemma 6.1 in [**71**]). Denote by $e_1 \in \mathbb{R}^m$ the vector $(1, 0, \dots, 0)$.

LEMMA 5. *For almost all interval exchange transformations* (λ_0, π_0) *one can find a collection of integers* $0 = l_1 < \dots < l_m$ *(depending on the interval exchange transformation) such that the vectors* $e_1, B^{(l_2)}(\lambda_0,\pi_0)e_1, \dots, B^{(l_m)}(\lambda_0,\pi_0)e_1$ *are linearly independent.*

We need this lemma to prove the following:

LEMMA 6. *For almost all interval exchange transformations* (λ, π) *and for any* $f \in \operatorname{Ann}(\mathcal{H}_p(\lambda,\pi))$, $f \notin \operatorname{Ann}(\mathcal{H}_{p+1}(\lambda,\pi))$, *one can find an infinite sequence of integers* $t_1 < t_2 < \cdots$ *with the following properties:*

1) $\displaystyle\lim_{i\to+\infty} \frac{1}{t_i} \log|\langle f, B^{(t_i)}(\lambda,\pi)e_1\rangle| = \theta_{p+1}$;

2) *there exists a positive integer* p *(depending on the initial interval exchange transformation* $T(\lambda,\pi)\colon X \to X$*) such that for any* $i = 1, 2, \dots$ *we have*

$$X^{(t_i+p)} \subseteq X_1^{(t_i)}.$$

PROOF. Consider a generic point (λ_0, π_0) and apply Lemma 5 to it. By construction, the vectors

$$w_1 := e_1, \quad w_2 := B^{(l_2)}(\lambda_0,\pi_0)e_1, \quad \dots, \quad w_m := B^{(l_m)}(\lambda_0,\pi_0)e_1$$

are linearly independent. Let $\Delta_0 \times \pi_0$ be a simplex containing the point (λ_0, π_0), and sharing the same matrices $B^{(l_i)}(\lambda', \pi_0) = B^{(l_i)}(\lambda_0, \pi_0)$ for all $1 \leqslant i \leqslant m$ and all $(\lambda', \pi_0) \in \Delta_0 \times \pi_0$.

Taking a smaller subsimplex if necessary, we can find an integer $p > 0$ such that for any $(\lambda', \pi_0) \in \Delta_0 \times \pi_0$ the following relation is valid:

$$X^{(l_i+p)} \subseteq X_1^{(l_i)} \quad \text{for any } 1 \leqslant i \leqslant m. \tag{19}$$

Since $\mu(\Delta_0 \times \pi_0) > 0$, and since the map $\mathcal{G}$ is ergodic, the trajectory of almost any point (λ, π) under the action of $\mathcal{G}$ will visit our subsimplex infinitely many times. By $k_1, k_2, \dots$ we denote these times of visits:

$$\mathcal{G}^{(k_i)}(\lambda,\pi) \in \Delta_0 \times \pi_0.$$

For each k_i we can choose l_{j_i} so that

$$|\langle f, B^{(k_i+l_{j_i})}(\lambda,\pi)e_1\rangle| \geqslant c\,\|{}^tB^{(k_i)}(\lambda,\pi)f\|, \tag{20}$$

since

$$\begin{aligned} |\langle f, B^{(k_i+l_{j_i})}(\lambda,\pi)e_1\rangle| &= |\langle {}^tB^{(k_i)}(\lambda,\pi)f, B^{(l_{j_i})}(\lambda^{(k_i)},\pi^{(k_i)})e_1\rangle| \\ &= |\langle {}^tB^{(k_i)}(\lambda,\pi)f, w_{j_i}\rangle|. \end{aligned}$$

Here the constant $c > 0$ does not depend on k_i.

Let $t_i := k_i + l_{j_i}$. Note that l_{j_i} is chosen from the fixed finite collection of indices $l_1, \dots, l_m$. Thus (20) implies that

$$\lim_{i\to+\infty} \frac{\log |\langle f, B^{(t_i)}(\lambda,\pi)e_1\rangle|}{t_i} = \lim_{i\to+\infty} \frac{\log c + \log \|{}^tB^{(k_i)}(\lambda,\pi)f\|}{k_i} \cdot \lim_{i\to+\infty} \frac{k_i}{t_i} = \theta_{p+1}$$

by the choice of $f \in \operatorname{Ann}(\mathcal{H}_p)$, $f \notin \operatorname{Ann}(\mathcal{H}_{p+1})$. Thus condition 1) holds for the sequence $t_1 < t_2 < \cdots$. Condition 2) follows from (19). Lemma 6 is proved. □

Now we are ready to prove Proposition 3. Assuming that the interval exchange transformation (λ, π) under consideration is generic, choose the sequence of integers $t_1 < t_2 < \cdots$ as in Lemma 6. We choose an arbitrary $x \in X$ and fix it for the duration of the proof.

Let $N^+_{(k)} = N^+(X, x, T, N, k)$. (Note that actually it does not depend on N.) Let

$$\limsup_{i\to+\infty} \frac{\log |\langle f, S(X,T,x,N^+_{(t_i+p)})\rangle|}{\log N^+_{(t_i+p)}} = \frac{\theta_{p+1}}{\theta_1} + \varepsilon. \tag{21}$$

If $\varepsilon \geqslant 0$, then Proposition 3 is proved. Suppose this inequality does not hold.

Consider the following sequence of integers:

$$N_i := \min_{J > N^+_{(t_i+p)}} \{J \quad \text{such that } T^J \in X^{(t_i)}\}.$$

In other words, we wait until the first visit of the trajectory $x, Tx, T^2x, \dots$ to the subinterval $X^{(t_i+p)}$, which happens at the time $N^+_{(t_i+p)}$, and then wait until the first visit to the subinterval $X^{(t_i)}$, which happens at the time N_i.

We are going to prove that

$$\limsup_{i\to+\infty} \frac{\log |\langle f, S(X,T,x,N_i)\rangle|}{\log N_i} = \frac{\theta_{p+1}}{\theta_1},$$

which will prove Proposition 3.

PROOF. Note that $T^{N^+_{(t_i+p)}}x \in X^{(t_i+p)}$ by Definition 15 of N^+. By the choice of t_i, we have $T^{N^+_{(t_i+p)}}x \in X_1^{(t_i)}$ in accordance with condition 2) of Lemma 6. Hence by the definition of N_i we get

$$S(X,T,x,N_i) = S(X,T,x,N^+_{(t_i+p)}) + B^{(t_i)}(\lambda,\pi)\cdot e_1.$$

Since

$$\limsup_{i\to+\infty} \frac{\log N^+_{(t_i+p)}}{t_i} \leqslant \theta_1$$

and since by assumption $\varepsilon < 0$ in (21), the second term in the sum

$$\langle f, S(X,T,x,N^+_{(t_i+p)})\rangle + \langle f, B^{(t_i)}(\lambda,\pi)\cdot e_1\rangle$$

strongly dominates the first one as $i \to +\infty$ by condition 1) of Lemma 6.

Moreover, by Proposition 2 in [**71**] we may assume that for any $1 \leqslant j \leqslant m$ we have

$$\lim_{k\to\infty} \left| \frac{\log \|B^{(k)}e_j\|}{k} \right| = \theta_1,$$

which implies that

$$\lim_{i\to+\infty}\frac{\log N_i}{t_i}=\theta_1. \tag{22}$$

Hence

$$\begin{aligned}\limsup_{i\to+\infty}&\frac{\log|\langle f, S(X,T,x,N_i)\rangle|}{\log N_i}\\ &=\limsup_{i\to+\infty}\frac{\log|\langle f, B^{(t_i)}(\lambda,\pi)\cdot e_1\rangle|}{t_i}\cdot\lim_{i\to+\infty}\frac{t_i}{\log N_i}=\frac{\theta_{p+1}}{\theta_1},\end{aligned}$$

where in the second equality we combined condition 1) of Lemma 6 and (22). Proposition 3 is proved. □

4.11. Uniform bound. Finally, we are going to prove

PROPOSITION 4. *For a generic interval exchange transformation (λ,π), there is a uniform bound. Namely, for any $x\in X$ and any $f\in\mathrm{Ann}(\mathcal{H}_g(\lambda,\pi))$, $\|f\|=1$,*

$$|\langle f, S(X,T,x,N)\rangle|\leqslant\text{const},$$

where the constant does not depend on f, x, or N.

We recall that

$$\mathrm{Ann}(\mathcal{H}(\lambda,\pi))=\{f\mid\theta(f)<0\}.$$

We start with the following elementary fact:

LEMMA 7. *For almost all interval exchange transformations (λ,π) the following property is valid. For any $\varepsilon>0$ and for all sufficiently large $n\in\mathbb{N}$ we have*

$$\|B(\lambda^{(n)},\pi^{(n)})\|\leqslant\exp(\varepsilon n).$$

PROOF. According to Proposition 1 in [**70**], the function $\log B(\lambda,\pi)$ is integrable. Hence for almost all (λ,π)

$$\lim_{n\to+\infty}\frac{1}{n}\sum_{k=0}^{n-1}\log\|B(\lambda^{(k)},\pi^{(k)})\|=\int\log\|B\|\,d\mu.$$

The convergence of this series implies that

$$\frac{1}{n}\log\|B(\lambda^{(k)},\pi^{(k)})\|\to 0,$$

which completes the proof of Lemma 7. □

Now we can prove Proposition 4.

PROOF. To unfold $S(X,T,x,N)$ we again use formula (16), as we did in 4.9. But this time we use not only the special "multiplicative times" n_i as before, but all points of the trajectory (λ,π), $\mathcal{G}(\lambda,\pi)$, $\mathcal{G}^{(2)}(\lambda,\pi),\dots$. We keep the notation from 4.9 (see (18)). We have

$$\begin{aligned}S(X,T,x,N)=(S^+_{(0)}-S^-_{(0)})&+B(\lambda,\pi)\cdot(S^+_{(1)}-S^-_{(1)})\\ &+B^{(2)}(\lambda,\pi)\cdot(S^+_{(2)}-S^-_{(2)})+\cdots.\end{aligned}$$

We prefer to use the notation as if we had an infinite sum. One should understand that actually the sum is finite: starting with some term, all subsequent terms in the tail are equal to zero.

Consider the pairing of f with $S(X,T,x,N)$. Using the expression above, we can rewrite it as follows:

$$\begin{aligned}|\langle f, S(X,T,x,N)\rangle| &\leqslant |\langle f, S^+_{(0)} - S^-_{(0)}\rangle| + |\langle f, B^{(1)}(\lambda,\pi)\cdot(S^+_{(1)} - S^-_{(1)})\rangle| + \cdots \\ &\leqslant \|f\|\cdot\|S^+_{(0)} - S^-_{(0)}\| + \|{}^tB^{(1)}(\lambda,\pi)f\|\cdot\|S^+_{(1)} - S^-_{(1)}\| + \cdots \\ &\leqslant \sum_{k=0}^{+\infty}\Big(\|{}^tB^{(k)}(\lambda,\pi)f\|\cdot 2\max_{1\leqslant j\leqslant m}\|B(\lambda^{(k)},\pi^{(k)})\cdot e_j\|\Big).\end{aligned}$$

Choose some ε such that $0 < \varepsilon < |\theta(f)|/3$. According to Lemma 7, for k large enough we have

$$\log\max_{1\leqslant j\leqslant m}\|B(\lambda^{(k)},\pi^{(k)})\cdot e_j\| \leqslant \varepsilon k.$$

On the other hand, by the definition of Lyapunov exponents, for sufficiently large k we have

$$\log\|{}^tB^{(k)}(\lambda,\pi)f\| \leqslant (\theta(f)+\varepsilon)k.$$

Thus for sufficiently large k the positive series above is majorized by the converging series $\sum\exp(-\varepsilon k)$. We have obtained a uniform bound that depends neither on x nor on N, but still depends on f. It remains to note that, having proved the statement for some basis in $\mathrm{Ann}(\mathcal{H}_g(\lambda,\pi))$, we have completed the proof of Proposition 4. □

Appendix A. Irreducibility of the cocycle ${}^tB(\lambda,\pi)$

Let $e_0 = (1,1,\ldots,1) \in \mathbb{R}^m$.

LEMMA 8. *If $\theta(e_j - \lambda_j e_0) < \theta_2$ on some subset of nonzero measure in $\Delta^{m-1}\times\pi$, then the same is true for almost all points of this simplex.*

PROOF. Consider the subspace

$$\mathrm{Ann}(H_{(2)}(\lambda,\pi)) = \{e \mid k^{-1}\log\|{}^tB^{(k)}e\| < \theta_2\}$$

and the two-dimensional plane spanned by e_0 and e_j. The above condition is equivalent to the condition that these two subspaces have nontrivial intersection on some subset of nonzero measure in $\Delta^{m-1}\times\pi$. Let

$$\dim\mathrm{Ann}(H_{(2)}(\lambda,\pi)) = d.$$

We know that $d \leqslant m-2$ (presumably $d = m-2$).

Consider a pair of vectors: one in $\Lambda^d(\mathbb{R}^m)^*$ and one in $\Lambda^2(\mathbb{R}^m)^*$, representing the subspaces $\mathrm{Ann}(H_{(2)}(\lambda,\pi))$ and $\langle e_0; e_j\rangle_{\mathbb{R}}$. The condition that the two subspaces have nontrivial intersection is equivalent to the condition that the wedge product of the corresponding vectors vanishes on some subset of nonzero measure. In other words, the one-dimensional subspace

$$L(\lambda,\pi) := \Lambda^d(\mathrm{Ann}(H_{(2)}(\lambda,\pi))) \subset \Lambda^d(\mathbb{R}^m)^* \tag{23}$$

satisfies some constant linear relation on some subset U of nonzero measure.

Let r be the maximal possible dimension of a linear subspace $R \subset (\Lambda^d(\mathbb{R}^m)^*)^*$ such that

$$L(\lambda,\pi) \subseteq \mathrm{Ann}(R) \tag{24}$$

Recall that an interval exchange transformation (λ, π) determines a closed orientable surface M_g^2 and an orientable measured foliation on it. The genus of the surface, the number and the types of the saddles are completely determined by the combinatorics of the permutation. We shall always consider only those permutations that correspond to surfaces of genera $g \geqslant 2$.

There is natural local identification between the space $\mathbb{R}_+^m$ of interval exchange transformations with fixed permutation $\pi \in \mathfrak{S}_m^0$ and the first relative cohomology $H^1(M_g^2, \{\text{saddles}\}; \mathbb{R})$ of the corresponding surface M_g^2 with respect to the set of saddles of the corresponding foliation.

Under the local identification of the space of interval exchange transformations with the relative cohomology $H^1(M_g^2, \text{saddles}; \mathbb{R})$, the quotient over the subspace $\operatorname{Ker}(\Omega)$ coincides with the absolute cohomology $H^1(M_g^2; \mathbb{R})$. The symplectic structure induced by $\Omega(\pi)$ on the quotient space coincides with the intersection form on cohomology.

Now consider the quotient cocycle. Further, consider the cocycle induced from the quotient cocycle on $\Lambda^2(\mathbb{R}^m / \operatorname{Ker}(\Omega))^*$. According to the previous subsection, the induced cocycle will have a pair of invariant subbundles (see (25)). Each subbundle is constant on every simplex $\Delta^{\pm} \times \pi$.

We may apply a parallel construction to the dual cocycle as well. Now we are ready to state several conjectures concerning the cocycles $B^{-1}(\lambda, \pi)$ and ${}^tB(\lambda, \pi)$.

A.3. Conjectures on the irreducibility of the cocycles $B^{-1}(\lambda, \pi)$ and ${}^tB(\lambda, \pi)$. We shall formulate all the statements for the cocycle ${}^tB(\lambda, \pi)$; analogous statements for $B^{-1}(\lambda, \pi)$ are implied.

CONJECTURE 3. *For any Rauzy class $\mathfrak{R}$ corresponding to genus $g \geqslant 2$, the induced cocycle on $\Lambda^2(\mathbb{R}^m / \operatorname{Ker}(\Omega))^*$ does not have any invariant subbundles constant on every $\Delta^{\pm} \times \pi$ other than those determined by the decomposition* (25) *of the representation of the symplectic group into irreducible ones.*

In fact we have a stronger conjecture.

CONJECTURE 4. *For any Rauzy class $\mathfrak{R}$, the induced cocycle on the exterior power $\Lambda^d(\mathbb{R}^m / \operatorname{Ker}(\Omega))^*$ for any $d = 1, \dots, 2g$ does not have any invariant subbundles that are constant on every $\Delta^{\pm} \times \pi$ other than those determined by the corresponding decomposition of the representation of the symplectic group $Sp(2g, \mathbb{R})$ on $\Lambda^d \mathbb{R}^{2g}$ into irreducible ones.*

The strongest conjecture, which implies Conjecture 4, is as follows.

Consider any permutation $\pi_0 \in \mathfrak{S}_m^0$. Consider the map $\Delta^{\pm} \times \pi_0 \to \Delta^{\pm} \times \pi_0$ induced from the map $\mathcal{G}$ on $\Delta^{\pm} \times \mathfrak{R}(\pi_0)$. Denote the corresponding cocycle induced from the cocycle $B^{-1}(\lambda, \pi)$ by $A(\lambda)$. Let G be the closure of the group generated by all matrices of the form $A^{(k)}(\lambda)$, where $k \in \mathbb{N}$ and $\lambda \in \Delta^{\pm} \times \pi_0$.

CONJECTURE 5. *The closure G of the group generated by all matrices of the form $A^{(k)}(\lambda)$ coincides with the group $Sp(2g, \mathbb{R})$.*

We have no ideas about how to prove Conjecture 5. On the other hand, it is not that difficult to prove Conjecture 4 for any particular Rauzy class $\mathfrak{R}$. In Appendix B, we present the proof of Conjecture 4 for the Rauzy class $\mathfrak{R}(4, 3, 2, 1)$.

Corollary 1 immediately implies the following fact.

CONDITIONAL THEOREM 6. *For any permutation* π *for which Conjecture 4 is valid and for almost all* λ *from the simplex* $\Delta^{\pm} \times \pi$, *one has*

$$\theta(e_j - \lambda_j e_0) = \theta_2 \quad \textit{for any } 1 \leqslant j \leqslant m$$

Appendix B. Proof of the irreducibility of the cocycle for the Rauzy class $\mathfrak{R}(4,3,2,1)$

B.1. Explicit form of geometric structures related to $\mathfrak{R}(4,3,2,1)$. In this section, we shall prove the validity of Conjecture 4 for a specific extended Rauzy class, namely $\mathfrak{R}(4,3,2,1)$. This Rauzy class corresponds to an orientable measured foliation with a single 6-pronged saddle on a surface of genus 2.

In this particular case, the skew-symmetric bilinear form $\Omega(\pi)$ is nondegenerate, so we do not even need to consider the quotient cocycle. For $\pi = (4,3,2,1)$, the matrix $\Omega(\pi)$ has the form

$$\Omega(4,3,2,1) = \begin{pmatrix} 0 & 1 & 1 & 1 \\ -1 & 0 & 1 & 1 \\ -1 & -1 & 0 & 1 \\ -1 & -1 & -1 & 0 \end{pmatrix}.$$

In the basis $e_1 \wedge e_2$, $e_1 \wedge e_3$, $e_1 \wedge e_4$, $e_2 \wedge e_3$, $e_2 \wedge e_4$, $e_3 \wedge e_4$ in $\Lambda^2(\mathbb{R}^4)^*$, the vector $\omega(4,3,2,1)$ representing the skew-symmetric bilinear form $\Omega(4,3,2,1)$ has the following components:

$$\omega(4,3,2,1) = (1,1,1,1,1,1) \in \Lambda^2(\mathbb{R}^4)^*.$$

The natural coupling between $\Lambda^d(\mathbb{R}^{2n})^*$ and $\Lambda^{2n-d}(\mathbb{R}^{2n})^*$ in our case determines the symmetric bilinear form $\Psi(\pi)$ on $\Lambda^2(\mathbb{R}^{2n})^*$; its value on a pair of bivectors $e_i \wedge e_j$ and $e_k \wedge e_l$ equals

$$\Psi(e_i \wedge e_j, e_k \wedge e_l) = \det e_i \wedge e_j \wedge e_k \wedge e_l. \tag{26}$$

Thus in the standard coordinates in $\Lambda^2(\mathbb{R}^4)^*$ the matrix of this bilinear form equals

$$\Psi(\pi) = \begin{pmatrix} 0 & 0 & 0 & 0 & 0 & 1 \\ 0 & 0 & 0 & 0 & -1 & 0 \\ 0 & 0 & 0 & 1 & 0 & 0 \\ 0 & 0 & 1 & 0 & 0 & 0 \\ 0 & -1 & 0 & 0 & 0 & 0 \\ 1 & 0 & 0 & 0 & 0 & 0 \end{pmatrix}.$$

The invariant codimension one subspace

$$\operatorname{Ann}(\omega(4,3,2,1)) = \{\phi \mid \Psi(\phi, \omega) = 0\} \subset \Lambda^2(\mathbb{R}^4)^*$$

is determined in components as

$$\operatorname{Ann}(\omega(4,3,2,1)) = \{\phi \mid \phi_{12} - \phi_{13} + \phi_{14} + \phi_{23} - \phi_{24} + \phi_{34} = 0\}.$$

We shall use the basis

$$\begin{gathered} a_1 = e_1 \wedge e_2 - e_3 \wedge e_4, \quad a_2 = e_1 \wedge e_3 + e_3 \wedge e_4, \quad a_3 = e_1 \wedge e_4 - e_3 \wedge e_4, \\ a_4 = e_2 \wedge e_3 - e_3 \wedge e_4, \quad a_5 = e_2 \wedge e_4 + e_3 \wedge e_4 \end{gathered} \tag{27}$$

in $\operatorname{Ann}(\omega(4,3,2,1)) \subset \Lambda^2(\mathbb{R}^4)^*$.

B.2. Choice of particular trajectories. We choose three trajectories of the map $\mathcal{G}$ more or less by chance.

The first trajectory is determined by the following path on the Rauzy graph of the Rauzy class $\mathfrak{R}(4,3,2,1)$:

$$
\begin{aligned}
&\{4,3,2,1\} \overset{b}{\to}\\
&\{4,1,3,2\} \overset{a}{\to}\\
&\{3,1,4,2\} \overset{b}{\to}\\
&\{3,1,4,2\} \overset{a}{\to} \{4,1,3,2\} \overset{a}{\to} \{3,1,4,2\} \overset{a}{\to}\\
&\{4,1,3,2\} \overset{b}{\to} \{4,2,1,3\} \overset{b}{\to}\\
&\{4,3,2,1\} \overset{a}{\to} \{2,4,3,1\} \overset{a}{\to}\\
&\{3,2,4,1\} \overset{b}{\to}\\
&\{3,2,4,1\} \overset{a}{\to}\\
&\{4,3,2,1\}
\end{aligned}
$$

The corresponding matrix ${}^tB$ is

$$
{}^tB_1 = \begin{pmatrix} 1 & 3 & 2 & 2 \\ 1 & 6 & 4 & 3 \\ 1 & 4 & 4 & 2 \\ 1 & 2 & 1 & 2 \end{pmatrix}.
$$

In the basis $e_1 \wedge e_2$, $e_1 \wedge e_3$, $e_1 \wedge e_4$, $e_2 \wedge e_3$, $e_2 \wedge e_4$, $e_3 \wedge e_4$ in $\Lambda^2(\mathbb{R}^4)^*$, the corresponding matrix ${}^tB_1^{\wedge 2}$ of the induced cocycle ${}^tB^{\wedge 2}(\lambda, \pi)$ is

$$
{}^tB_1^{\wedge 2} = \begin{pmatrix} 3 & 2 & 1 & 0 & -3 & -2 \\ 1 & 2 & 0 & 4 & -2 & -4 \\ -1 & -1 & 0 & -1 & 2 & 2 \\ -2 & 0 & -1 & 8 & 0 & -4 \\ -4 & -3 & -1 & -2 & 6 & 5 \\ -2 & -3 & 0 & -4 & 4 & 6 \end{pmatrix}.
$$

The restriction of this operator to the invariant subspace $\mathrm{Ann}(\omega(4,3,2,1))$ has the following matrix in the coordinates (27):

$$
C_1 = \begin{pmatrix} 5 & 0 & 3 & 2 & -5 \\ 5 & -2 & 4 & 8 & -6 \\ -3 & 1 & -2 & -3 & 4 \\ 2 & -4 & 3 & 12 & -4 \\ -9 & 2 & -6 & -7 & 11 \end{pmatrix}.
$$

Consider the other path on the Rauzy graph of the Rauzy class $\mathfrak{R}(4,3,2,1)$:

$$\{4,3,2,1\} \xrightarrow{b}$$
$$\{4,1,3,2\} \xrightarrow{a}$$
$$\{3,1,4,2\} \xrightarrow{b} \{4,1,3,2\} \xrightarrow{b} \{3,1,4,2\} \xrightarrow{b} \{4,1,3,2\} \xrightarrow{b} \{3,1,4,2\} \xrightarrow{b}$$
$$\{3,1,4,2\} \xrightarrow{a}$$
$$\{4,1,3,2\} \xrightarrow{b}$$
$$\{4,2,1,3\} \xrightarrow{a} \{4,3,2,1\} \xrightarrow{a}$$
$$\{2,4,3,1\} \xrightarrow{b} \{3,2,4,1\} \xrightarrow{b}$$
$$\{3,2,4,1\} \xrightarrow{a}$$
$$\{4,3,2,1\}$$

The corresponding matrix ${}^tB$ is

$$ {}^tB_2 \begin{pmatrix} 1 & 6 & 1 & 2 \\ 1 & 12 & 2 & 3 \\ 1 & 8 & 2 & 2 \\ 1 & 3 & 0 & 2 \end{pmatrix}. $$

In the basis $e_1 \wedge e_2,\ e_1 \wedge e_3,\ e_1 \wedge e_4,\ e_2 \wedge e_3,\ e_2 \wedge e_4,\ e_3 \wedge e_4$ in $\Lambda^2(\mathbb{R}^4)^*$, the corresponding matrix ${}^tB_2^{\wedge 2}$ of the induced cocycle ${}^tB^{\wedge 2}(\lambda,\pi)$ is

$$ {}^tB_2^{\wedge 2} = \begin{pmatrix} 6 & 1 & 1 & 0 & -6 & -1 \\ 2 & 1 & 0 & 4 & -4 & -2 \\ -3 & -1 & 0 & -3 & 6 & 2 \\ -4 & 0 & -1 & 8 & 0 & -2 \\ -9 & -2 & -1 & -6 & 15 & 4 \\ -5 & -2 & 0 & -6 & 10 & 4 \end{pmatrix}. $$

The restriction of this operator to the invariant subspace $\operatorname{Ann}(\omega(4,3,2,1))$ has the following matrix in the coordinates (27):

$$ C_2 = \begin{pmatrix} 7 & 0 & 2 & 1 & -7 \\ 4 & -1 & 2 & 6 & -6 \\ -5 & 1 & -2 & -5 & 8 \\ -2 & -2 & 1 & 10 & -2 \\ -13 & 2 & -5 & -10 & 19 \end{pmatrix}. $$

Consider the third path on the Rauzy graph of the Rauzy class $\mathfrak{R}(4,3,2,1)$:

$$ \{4,3,2,1\} \xrightarrow{b} \{4,1,3,2\} \xrightarrow{b} \{4,2,1,3\} \xrightarrow{b} \{4,3,2,1\} $$
$$ \xrightarrow{b} \{4,1,3,2\} \xrightarrow{b} \{4,2,1,3\} \xrightarrow{b} \{4,3,2,1\} \xrightarrow{b} . $$

The corresponding matrix ${}^tB$ is

$$ {}^tB_2 \begin{pmatrix} 1 & 0 & 0 & 2 \\ 0 & 1 & 0 & 2 \\ 0 & 0 & 1 & 2 \\ 0 & 0 & 0 & 1 \end{pmatrix}. $$

In the basis $e_1 \wedge e_2$, $e_1 \wedge e_3$, $e_1 \wedge e_4$, $e_2 \wedge e_3$, $e_2 \wedge e_4$, $e_3 \wedge e_4$ in $\Lambda^2(\mathbb{R}^4)^*$, the corresponding matrix ${}^tB_3^{\wedge 2}$ of the induced cocycle ${}^tB^{\wedge 2}(\lambda, \pi)$ is

$$
{}^tB_3^{\wedge 2} = \begin{pmatrix}
1 & 0 & 2 & 0 & -2 & 0 \\
0 & 1 & 2 & 0 & 0 & -2 \\
0 & 0 & 1 & 0 & 0 & 0 \\
0 & 0 & 0 & 1 & 2 & -2 \\
0 & 0 & 0 & 0 & 1 & 0 \\
0 & 0 & 0 & 0 & 0 & 1
\end{pmatrix}.
$$

The restriction of this operator to the invariant subspace $\mathrm{Ann}(\omega(4,3,2,1))$ has the following matrix in the coordinates (27):

$$
C_3 = \begin{pmatrix}
1 & 0 & 2 & 0 & -2 \\
2 & -1 & 4 & 2 & -2 \\
0 & 0 & 1 & 0 & 0 \\
2 & -2 & 2 & 3 & 0 \\
0 & 0 & 0 & 0 & 1
\end{pmatrix}.
$$

B.3. Proof of the irreducibility. All the matrices under consideration are integer. In our proof of irreducibility, we avoid any usage of eigenvectors and eigenvalues, restricting ourselves to integer computations only. Mostly we shall use the following obvious fact.

LEMMA 10. *If the operators A_1 and A_2 possess a common one-dimensional invariant subspace K^1, then $K^1 \subset \mathrm{Ker}(A_1A_2 - A_2A_1)$.*

PROPOSITION 5. *The cocycles $B^{-1}(\lambda, \pi)$ and ${}^tB(\lambda, \pi)$ corresponding to the Rauzy class $\mathfrak{R}(4,3,2,1)$ are irreducible on $\Delta^3 \times \mathfrak{R}(4,3,2,1)$.*

PROOF. First note that the cocycles under consideration are conjugate. So the irreducibility of one of them implies the irreducibility of the other.

Since

$$
\det({}^tB_1{}^tB_2 - {}^tB_2{}^tB_1) = 4,
$$

Lemma 10 implies that these two operators do not have any common invariant one-dimensional subspaces. Hence the cocycle ${}^tB(\lambda, \pi)$ does not have any invariant one-dimensional subspaces.

Since the cocycle ${}^tB(\lambda, \pi)$ is symplectic, the presence of an invariant subspace of codimension one implies the presence of a one-dimensional invariant subspace, that of the annihilator of the invariant subspace of codimension one with respect to the symplectic form. Thus in our case the absence of invariant one-dimensional subspaces implies the absence of three-dimensional invariant subspaces.

The absence of a two-dimensional invariant subspace of the cocycle ${}^tB(\lambda, \pi)$ is equivalent to the absence of an invariant one-dimensional subspace of the induced cocycle on the second exterior power $\Lambda^2(\mathbb{R}^4)^*$ with the additional condition that the one-dimensional subspace is generated by a two-vector. Let us proceed with our study of the induced cocycle ${}^tB^{\wedge 2}(\lambda, \pi)$.

The one-dimensional invariant subspace generated by the vector $\omega(4,3,2,1) = (1,1,1,1,1,1)$ does not obey the Plücker relations

$$
p_{12}p_{34} - p_{13}p_{24} + p_{14}p_{23}.
$$

Hence this one-dimensional subspace is not generated by a two-vector. We next prove that the cocycle ${}^tB^{\wedge 2}(\lambda, \pi)$ does not possess any other one-dimensional invariant subspaces. To do so, it is sufficient to show that the induced cocycle on $\mathrm{Ann}(\omega(4,3,2,1))$ does not possess any one-dimensional invariant subspaces.

The kernel $\mathrm{Ker}(C_1C_3 - C_3C_1)$ is spanned by the vector

$$\mathrm{Ker}(C_1C_3 - C_3C_1) = \langle(1/2, 2, 1, 1/2, 1)\rangle_{\mathbb{R}},$$

while the kernel $\mathrm{Ker}(C_2C_3 - C_3C_2)$ is spanned by the vector

$$\mathrm{Ker}(C_2C_3 - C_3C_2) = \langle(1, 5/2, 1, 1/2, 1)\rangle_{\mathbb{R}}.$$

Since these two vectors are not collinear, the operators C_1, C_2, C_3 do not possess any common one-dimensional subspaces. In other words, the restriction of the induced cocycle ${}^tB^{\wedge 2}(\lambda, \pi)$ to the subspace $\mathrm{Ann}(\omega(4,3,2,1))$ does not have any invariant one-dimensional subspaces. This in turn implies that the cocycle ${}^tB(\lambda, \pi)$ does not have any invariant two-dimensional subspaces.

The induced cocycle on $\mathrm{Ann}(\omega(4,3,2,1))$ preserves the nondegenerate symmetric bilinear form (26). Hence invariant subspaces of dimension d and of codimension d appear only in pairs. Since we have proved that this cocycle does not have any invariant one-dimensional subspaces, it does not possess any four-dimensional invariant subspaces either.

It remains to prove that the restriction of the induced cocycle ${}^tB^{\wedge 2}(\lambda, \pi)$ to the subspace $\mathrm{Ann}(\omega(4,3,2,1))$ does not have any invariant two-dimensional subspaces. This would imply the absence of any invariant three-dimensional subspaces, which would mean that there are no nontrivial invariant subspaces in $\mathrm{Ann}(\omega(4,3,2,1))$ at all, which would mean in turn that there are no nontrivial invariant subspaces in $\Lambda^2(\mathbb{R}^4)^*$ other than the one spanned by $\omega(4,3,2,1)$ and its annihilator.

To prove that the restriction of the induced cocycle ${}^tB^{\wedge 2}(\lambda, \pi)$ to the subspace $\mathrm{Ann}(\omega(4,3,2,1))$ does not have any invariant two-dimensional subspaces, it is sufficient to prove that the induced cocycle on the second exterior power $\Lambda^2\mathrm{Ann}(\omega(4,3,2,1))$ does not possess any invariant one-dimensional subspaces. Consider the matrices D_1, D_2, D_3 corresponding to our three paths. It is not difficult to verify that the two-dimensional kernel $\mathrm{Ker}(D_1D_2 - D_2D_1)$ is transversal to the four-dimensional kernel $\mathrm{Ker}(D_1D_3 - D_3D_1)$. Hence, according to Lemma 10, the induced cocycle on the second exterior power $\Lambda^2\mathrm{Ann}(\omega(4,3,2,1))$ does not possess any invariant one-dimensional subspaces.

It remains to prove that the induced cocycle ${}^tB^{\wedge 3}(\lambda, \pi)$ on the third exterior power $\Lambda^3(\mathbb{R}^4)^*$ does not possess any invariant subspaces. This is true due to the duality between $\Lambda^3(\mathbb{R}^4)^*$ and $(\mathbb{R}^4)^*$. Proposition 5 is proved. □

Acknowledgments

This paper was mainly written during my stay at the IHES in 1995, and I am very much indebted to the members and staff of the IHES for their hospitality. In particular, a shattering criticism by M. Gromov and valuable discussions with M. Kontsevich were of crucial importance for me.

I would like to thank K. Khanin, A. Pajitnov, and E. Zhuzhoma for helpful consultations.

I am grateful to V. I. Arnold for inviting me to participate in this volume.

References

[1] L. A. Alaniya, *Topological structure of level surfaces of Morse* 1*-forms*, Uspekhi Mat. Nauk **46** (1991), no. 3, 179–180; English transl., Russian Math. Surveys **46** (1991), no. 3, 211–212.

[2] N. E. Alekseevskiĭ, Yu. P. Gaidukov, I. M. Lifshits, and V. G. Peschanskiĭ, *Fermi surface of tin*, Zh. Èksper. Teoret. Fiz. **39** (1960), 1201–1214; English transl., Soviet Phys. JETP **12** (1961), 837–846.

[3] D. V. Anosov, *Flows on closed surfaces and behavior of trajectories lifted to the universal covering plane*, J. Dynam. Control Systems **1** (1995), 125–138.

[4] ———, *On the behavior of trajectories in the Euclidean or Lobachevskiĭ plane, covering the trajectory of flows on closed surfaces.* III, Izv. Ross. Akad. Nauk Ser. Mat. **59** (1995), no. 2, 63–96; English transl., Izv. Math. **59** (1995), 287–320.

[5] S. Kh. Aranson, V. Z. Grines, and E. V. Zhuzhoma, *On the geometry and topology of flows and foliations on surfaces, and the Anosov problem*, Mat. Sb. **186** (1995), no. 8, 25–66; English transl., Sb. Math. **186** (1995), 1107–1146.

[6] V. I. Arnold, *Small denominators and problems of stability of motion in classical and celestial mechanics*, Uspekhi Mat. Nauk **18** (1963), no. 6, 91–192; English transl., Russian Math. Surveys **18** (1963), no. 6, 85–191.

[7] ———, *Huygens and Barrow, Newton and Hooke. Pioneers in mathematical analysis and catastrophe theory from evolvents to quasicrystals*, Birkhäuser, Basel, 1990.

[8] ———, *Topological and ergodic properties of closed* 1*-forms with incommensurable periods*, Funktsional. Anal. i Prilozhen. **25** (1991), no. 2, 1–12; English transl., Functional Anal. Appl. **25** (1991), 81–90.

[9] ———, *Poly-integrable flows*, Algebra i Analiz **4** (1992), no. 6, 54–62; English transl., St. Petersburg Math. J. **4** (1993), 1103–1110.

[10] ———, *Remarks on the Morse theory of a divergence-free vector field, the averaging method, and the motion of a charged particle in a magnetic field*, Trudy Mat. Inst. Steklov. **216** (1997), 9–19; English transl., Proc. Steklov Inst. Math. **216** (1997), 3–13.

[11] P. Arnoux and A. Nogueira, *Mesures de Gauss pour des algorithmes de fractions continues multidimensionnelles*, Ann. Sci. École Norm. Sup. (4) **26** (1993), 645–664.

[12] E. Calabi, *An intrinsic characterization of harmonic* 1*-forms*, Global Analysis: Papers in Honor of K. Kodaira (D. C. Spencer and S. Iyanaga, eds.), Princeton Univ. Press, Princeton, NJ, and Univ. of Tokyo Press, Tokyo, 1969, pp. 101–117.

[13] Yu. Chekanov, Private communication.

[14] A. P. Cracknell, *The Fermi surface.* I, II, Adv. Phys. **18** (1969), 681–818; ibid. **20** (1971), 1–141; reprint, *The Fermi surfaces of metals*, Taylor & Francis Ltd., London, 1971.

[15] A. P. Cracknell and K. C. Wong, *The Fermi surface*, Clarendon Press, Oxford, 1973.

[16] A. Denjoy, *Le phénomène ergodique et les trajectoires sur le tore*, C. R. Acad. Sci. Paris **247** (1958), 1072–1078.

[17] I. A. Dynnikov, *A proof of S. P. Novikov's conjecture for the case of small perturbations of rational magnetic fields*, Uspekhi Mat. Nauk **47** (1992), no. 3, 161–162; English transl., Russian Math. Surveys **47** (1992), no. 3, 172–173.

[18] ———, *A proof of the conjecture of S. P. Novikov on the semiclassical motion of an electron*, Mat. Zametki **53** (1993), no. 5, 57–68; English transl., Math. Notes **53** (1993), 495–501.

[19] ———, *Intersections of level surfaces of pseudoperiodic functions*, Uspekhi Mat. Nauk **49** (1994), no. 1, 213–214; English transl., Russian Math. Surveys **49** (1994), no. 1, 229–230.

[20] ———, *Semi-classical motion of the electron. Proof of Novikov's conjecture in general position and counterexamples*, Solitons, Geometry, and Topology: On the Crossroad (V. M. Buchstaber and S. P. Novikov, eds.), Amer. Math. Soc. Transl. (2) **179** (1997), 45–73.

[21] ———, *Surfaces in* 3*-torus: geometry of plane sections*, European Congr. Math. (Budapest, 1996), Vol. 1, Progr. Math., vol. 168, Birkhäuser, Basel, 1998, pp. 162–177.

[22] ———, *Geometry of stability zones in S. P. Novikov's problem on semi-classical motion of an electron*, Uspekhi Mat. Nauk **54** (1999), no. 1, 21–60; English transl., Russian Math. Surveys **54** (1999), no. 1, 21–59.

[23] M. Sh. Farber, *Exactness of the Novikov inequalities*, Funktsional. Anal. i Prilozhen. **19** (1985), no. 1, 49–59; English transl., Functional Anal. Appl. **19** (1985), 40–48.

[24] A. Fathi, F. Laudenbach, and V. Poénaru, *Travaux de Thurston sur les surfaces*, Astérisque, No. 66–67, Soc. Math. France, Paris, 1979.

[25] S. M. Gusein-Zade, *The number of critical points of a quasiperiodic potential*, Funktsional. Anal. i Prilozhen. **23** (1989), no. 2, 55–56; English transl., Functional Anal. Appl. **23** (1989), 129–130.
[26] J. Hubbard and H. Masur, *Quadratic differentials and measured foliations*, Acta Math. **142** (1979), 221–274.
[27] A. B. Katok, *Invariant measures of flows on oriented surfaces*, Dokl. Akad. Nauk SSSR **211** (1973), 775–778; English transl., Soviet Math. Dokl. **14** (1973), 1104–1108.
[28] M. Keane, *Interval exchange transformations*, Math. Z. **141** (1975), 25–31.
[29] S. P. Kerckhoff, *Simplicial systems for interval exchange maps and measured foliations*, Ergodic Theory Dynam. Systems **5** (1985), 257–271.
[30] S. Kerckhoff, H. Masur, and J. Smillie, *Ergodicity of billiard flows and quadratic differentials*, Ann. of Math. (2) **124** (1986), 293–311.
[31] K. M. Khanin and A. Nogueira, *Mixing for suspension flows over interval exchange transformations*, to appear.
[32] M. Kontsevich and A. Zorich, *Lyapunov exponents and Hodge theory*, IHES Preprint M/97/13, 1–16.
[33] ———, *Connected components of the moduli spaces of holomorphic differentials with prescribed singularities*, to appear.
[34] Le Tu Kuok Tkhang, *Structure of level surfaces of the Morse form*, Mat. Zametki **44** (1988), 105–111; English transl., Math. Notes **44** (1988), 556–562.
[35] I. M. Lifshits, M. Ya. Azbel, and M. I. Kaganov, *Electron theory of metals*, "Nauka", Moscow, 1971; English transl., Consultants Bureau, New York, 1973.
[36] A. G. Maier, *Trajectories on closed orientable surfaces*, Mat. Sb. **12** (54) (1943), 71–84. (Russian)
[37] N. Markley, *The Poincaré–Bendixon theorem for the Klein bottle*, Trans. Amer. Math. Soc. **135** (1969), 159–165.
[38] H. Masur. *Interval exchange transformations and measured foliations*, Ann. of Math. (2) **115** (1982), 169–200.
[39] ———, *Closed trajectories for quadratic differentials with an application to billiards*, Duke Math. J. **53** (1986), 307–314.
[40] ———, *The growth rate of trajectories of a quadratic differential*, Ergodic Theory Dynam. Systems **10** (1990), 151–176.
[41] S. P. Novikov, *Multivalued functions and functionals. An analogue of the Morse theory*, Dokl. Akad. Nauk SSSR **260** (1981), 31–35; English transl., Soviet Math. Dokl. **24** (1981), 222–226.
[42] ———, *The Hamiltonian formalism and a multivalued analogue of Morse theory*, Uspekhi Mat. Nauk **37** (1982), no. 5, 3–49; English transl., Russian Math. Surveys **37** (1982), no. 5, 1–56.
[43] ———, *Critical points and level surfaces of multivalued functions*, Trudy Mat. Inst. Steklov. **166** (1984), 201–209; English. transl., Proc. Steklov Inst. Math. **1986**, no. 1 (166), 223–232.
[44] ———, *Quasiperiodic structures in topology*, Topological Methods in Modern Mathematics (J. Milnor 60th Birthday Sympos., Stony Brook, NY, 1991; A. V. Goldberg and L. R. Phillips, eds.), Publish or Perish, Inc., Houston, TX, 1993, pp. 223–233.
[45] ———, *The semiclassical electron in a magnetic field and lattice. Some problems of low dimensional "periodic" topology*, Geom. Funct. Anal. **5** (1995), 434–444.
[46] S. P. Novikov and A. Ya. Maltsev, *Topological phenomena in normal metals*, Pis′ma Zh. Èksper. Teoret. Fiz.; English transl., JETP Letters **63** (1996), 809–813; `http://xxx.lpthe.jussieu.fr cond-mat/9709007`.
[47] A. Pazhitnov (Pajitnov), *An analytic proof of the real part of Novikov's inequalities*, Dokl. Akad. Nauk SSSR **293** (1987), 1305–1307; English transl., Soviet Math. Dokl. **35** (1987), 456–457.
[48] ———, *On the sharpness of inequalities of Novikov type for manifolds with a free abelian fundamental group*, Mat. Sb. **180** (1989), 1486–1523; English transl., Math. USSR–Sb. **68** (1991), 351–389.
[49] ———, *C^0-generic properties of boundary operators in the Novikov complex*, in this volume.
[50] D. A. Panov, *Multicomponent pseudoperiodic mappings*, Funktsional. Anal. i Prilozhen. **30** (1996), no. 1, 30–38; English transl., Functional Anal. Appl. **30** (1996), 23–29.
[51] ———, *Pseudoperiodic mappings*, in this volume.

[52] G. Rauzy, *Échanges d'intervalles et transformations induites*, Acta Arith. **34** (1979), 315–328.

[53] S. Schwartzman, *Asymptotic cycles*, Ann. of Math. (2) **66** (1957), 270–284.

[54] J.-C. Sikorav, *Un problème de disjonction par isotopie symplectique dans un fibré cotangent*, Ann. Sci. École Norm. Sup. (4) **19** (1986), 543–552.

[55] Ya. G. Sinaĭ and K. M. Khanin, *Mixing of some classes of special flows over rotations of the circle*, Funktsional. Anal. i Prilozhen. **26** (1992), no. 3, 1–21; English transl., Functional Anal. Appl. **26** (1992), 155–169.

[56] J. Smillie, Private communication.

[57] S. P. Tsarev, Private communication.

[58] W. A. Veech, *Projective swiss cheeses and uniquely ergodic interval exchange transformations*, Ergodic Theory Dynam. Systems (College Park, MD, 1979–80; A. Katok, ed.), Vol. I, Progr. Math., vol. 10, Birkhäuser, Boston, 1981, pp. 113–193.

[59] ———, *Gauss measures for transformations on the space of interval exchange maps*, Ann. of Math. (2) **115** (1982), 201–242.

[60] ———, *The metric theory of interval exchange transformations.* I: *Generic spectral properties*, Amer. J. Math. **106** (1984), 1331–1359.

[61] ———, *The metric theory of interval exchange transformations.* II: *Approximation by primitive interval exchanges*, Amer. J. Math. **106** (1984), 1361–1387.

[62] ———, *The Teichmüller geodesic flow*, Ann. of Math. (2) **124** (1986), 441–530.

[63] ———, *Moduli spaces of quadratic differentials*, J. Analyse Math. **55** (1990), 117–171.

[64] S. Wakon and J. Yamashita, *Band structure of ferromgnetic iron by a self-consistent procedure*, J. Phys. Soc. Japan **21** (1966), 1712–1726.

[65] A. Weil, *On systems of curves on a ring-shaped surface*, J. Indian Math. Soc. **19** (1931), 109–114; reprinted in his *Oeuvres scientifiques*, Vol. 1 (1926–1951), Springer-Verlag, Berlin, 1980, pp. 57–62.

[66] ———, *Les familles de courbes sur le tore*, Mat. Sb. **1** (1936), 779–781; reprinted in his *Oeuvres scientifiques*, Vol. 1 (1926–1951), Springer-Verlag, Berlin, 1980, pp. 113–115.

[67] A. Zorich, *The S. P. Novikov problem on the semiclassical motion of an electron in a homogeneous magnetic field*, Uspekhi Mat. Nauk **39** (1984), no. 5, 235–236; English transl., Russian Math. Surveys **39** (1984), no. 5, 287–288.

[68] ———, *The quasiperiodic structure of level surfaces of a Morse* 1*-form that is close to a rational form—the problem of S. P. Novikov*, Izv. Akad. Nauk SSSR Ser. Mat. **51** (1987), 1322–1344; English transl., Math. USSR–Izv. **31** (1988), 635–655.

[69] ———, *Asymptotic flag of an orientable measured foliation on a surface*, Geometric Study of Foliations (Proc. Internat. Sympos./Workshop, Tokyo, 1993; T. Mizutani et al., eds.), World Sci. Publ., Singapore, 1994, pp. 479–498.

[70] ———, *Finite Gauss measure on the space of interval exchange transformations. Lyapunov exponents*, Ann. Inst. Fourier (Grenoble) **46** (1996), fasc. 2, 325–370.

[71] ———, *Deviation for interval exchange transformations*, Ergodic Theory Dynam. Systems **17** (1997), 1477–1499.

[72] ———, *On hyperplane sections of periodic surfaces*, Solitons, Geometry, and Topology: On the Crossroad (V. M. Buchstaber and S. P. Novikov, eds.), Amer. Math. Soc. Transl. (2) **179** (1997), 173–189.

Mathematical College, Independent University of Moscow, Russia and Institut Mathématique de Rennes, Université Rennes-1, Campus de Beaulieu, 35042 Rennes, France

E-mail address: `zorich@maths.univ-rennes1.fr`

Selected Titles in This Subseries

(Continued from the front of this publication)

Published Earlier as Advances in Soviet Mathematics

21 **V. I. Arnold, Editor,** Singularities and bifurcations, 1994
20 **R. L. Dobrushin, Editor,** Probability contributions to statistical mechanics, 1994
19 **V. A. Marchenko, Editor,** Spectral operator theory and related topics, 1994
18 **Oleg Viro, Editor,** Topology of manifolds and varieties, 1994
17 **Dmitry Fuchs, Editor,** Unconventional Lie algebras, 1993
16 **Sergei Gelfand and Simon Gindikin, Editors,** I. M. Gelfand seminar, Parts 1 and 2, 1993
15 **A. T. Fomenko, Editor,** Minimal surfaces, 1993
14 **Yu. S. Il′yashenko, Editor,** Nonlinear Stokes phenomena, 1992
13 **V. P. Maslov and S. N. Samborskiĭ, Editors,** Idempotent analysis, 1992
12 **R. Z. Khasminskiĭ, Editor,** Topics in nonparametric estimation, 1992
11 **B. Ya. Levin, Editor,** Entire and subharmonic functions, 1992
10 **A. V. Babin and M. I. Vishik, Editors,** Properties of global attractors of partial differential equations, 1992
9 **A. M. Vershik, Editor,** Representation theory and dynamical systems, 1992
8 **E. B. Vinberg, Editor,** Lie groups, their discrete subgroups, and invariant theory, 1992
7 **M. Sh. Birman, Editor,** Estimates and asymptotics for discrete spectra of integral and differential equations, 1991
6 **A. T. Fomenko, Editor,** Topological classification of integrable systems, 1991
5 **R. A. Minlos, Editor,** Many-particle Hamiltonians: spectra and scattering, 1991
4 **A. A. Suslin, Editor,** Algebraic K-theory, 1991
3 **Ya. G. Sinaĭ, Editor,** Dynamical systems and statistical mechanics, 1991
2 **A. A. Kirillov, Editor,** Topics in representation theory, 1991
1 **V. I. Arnold, Editor,** Theory of singularities and its applications, 1990